Developments in Agricultural and Industrial Ergonomics

VOLUME–I: General Studies

Developments in Agricultural and Industrial Ergonomics

Volume–I: General Studies

Editors

L.P. Gite
C.R. Mehta
Nachiket Kotwaliwale
Joydeep Majumder

2007
HWWE
Central Institute of Agricultural Engineering (CIAE)
Nabibagh, Bhopal 462 038, India

ALLIED PUBLISHERS PVT. LTD.

New Delhi • Mumbai • Kolkata • Lucknow • Chennai
Nagpur • Bangalore • Hyderabad • Ahmedabad

ALLIED PUBLISHERS PRIVATE LIMITED

Regd. Off. : 15 J.N. Heredia Marg, Ballard Estate, Mumbai–400001, Ph.: 022-22626476
E-mail: mumbai.books@alliedpublishers.com

12 Prem Nagar, Ashok Marg, Opp. Indira Bhawan, Lucknow–226001, Ph.: 0522-2614253
E-mail: appltdlko@sify.com

Prarthna Flats (2nd Floor), Navrangpura, Ahmedabad–380009, Ph.: 079-26465916
E-mail: ahmbd.books@alliedpublishers.com

3-2-844/6 & 7 Kachiguda Station Road, Hyderabad–500027, Ph.: 040-24619079
E-mail: hyd.books@alliedpublishers.com

5th Main Road, Gandhinagar, Bangalore–560009, Ph.: 080-22262081
E-mail: bngl.books@alliedpublishers.com

1/13-14 Asaf Ali Road, New Delhi–110002, Ph.: 011-23239001
E-mail: delhi.books@alliedpublishers.com

17 Chittaranjan Avenue, Kolkata–700072, Ph.: 033-22129618
E-mail: cal.books@alliedpublishers.com

81 Hill Road, Ramnagar, Nagpur–440033, Ph.: 0712-2521122
E-mail: ngp.books@alliedpublishers.com

751 Anna Salai, Chennai–600002, Ph.: 044-28523938
E-mail: chennai.books@alliedpublishers.com

Website: www.alliedpublishers.com

ISBN: 978-81-8424-448-9

Preface

Ergonomics, the science and technology of a person at work, addresses the issues relevant to human-machine-environment system. It aims at improving the match between a person and his/her working environment through design of work methods, equipment and physical environment. The capacities and limitations of human beings are studied and given due consideration so that the comfort, health, efficiency and safety are maximized. Thus, the main goal of ergonomics is to heighten the quality of life in work places in all walks of life.

The Indian Society of Ergonomics has started a series of conferences termed as Humanizing Work and Work Environment. The first (HWWE–2001) was held at IIT Mumbai in December 2001; the second (HWWE–2004) was held in April 2004 at NITIE Mumbai; and the third (HWWE–2005) at IIT Guwahati in December 2005. The fourth in the series (HWWE–2007) was held at CIAE Bhopal during December 10–12, 2007. The objective of this conference was to provide a platform for scientists, research workers, practitioners and decision makers from business, industry, defence, public utilities, agriculture, home management and government for discussion on various ergonomical issues. A total of 152 abstracts were selected for presentation in the conference. Of these, 90 technical papers were presented by delegates. The proceedings of HWWE–2007 titled "Development in Agricultural and Industrial Ergonomics has been brought out in two volumes, Vol. I—General Studies and Vol. II—Women at Work". The Vol. I contains sections on Anthropometry and Work Place Design, Work and Sport Physiology, Physical Environment, Cognitive/Design Ergonomics, Ergonomics in Agriculture, Ergonomics in Industry and Occupational Health and Safety. The Vol. II contains two sections namely Women in Agriculture and Women at Work and Home.

We would like to express our thanks to Dr. Mangala Rai, Secy. DARE & DG (ICAR), Dr. Nawab Ali, DDG (Engg.), ICAR, Dr. S.K. Tandon, ADG (ICAR) and Dr. M.M. Pandey, Director, CIAE and Dr. S.D. Kulkarni, Project Director, SPU Centre, CIAE for their guidance and encouragement during organizing this conference at CIAE, Bhopal.

We are also thankful to Dr. David Caple, President, International Ergonomics Association for participating in the conference and providing full support of IEA in organizing the HWWE–2007.

We are grateful to all the international and national delegates of HWWE–2007, the members of Advising Committee, Organizing Committees and the Sponsors who made it possible for us to organize this conference in a successful manner and in bringing out the proceedings.

Special thanks to Abhijit Khadatkar, Neha Jain, Mahipal Deshbratar and Ravi Raina for helping in preparation of this proceedings.

L.P. Gite
C.R. Mehta
Nachiket Kotwaliwale
Joydeep Majumder

December, 2008
Bhopal

Preface

Major Recommendations of the International Ergonomics Conference HWWE 2007 (Humanizing Work and Work Environment) Held at CIAE, Bhopal during December 10–12, 2007

1. Efforts to collect reliable anthropometric data should be intensified for different parts of India and for different types of work groups.
2. Well-established and common methodologies and tools should be used for procurement of data so as to make it easier for proper pooling of data.
3. Use of proper anthropometric and strength data be encouraged in design of machinery including agricultural tools and due consideration be given for making the tools/machines gender friendly.
4. There is a need to ascertain suitability of existing tools and machinery being used by Indian workers from ergonomics point of view.
5. Safety of workers/operators, especially in the unorganized sector is of great concern. Besides development of safe equipment and safety gadgets, involvement of society for their acceptance and legal enforcement, needs to be explored seriously.
6. All the workplaces should be designed keeping in consideration the reach dimensions, force limits and postural requirements. It will help to reduce the health problems as well as injuries to the users. Besides above, work profile and special requirements of defence, sports and IT industry should also be given due consideration.
7. The "IEA/ILO Checkpoints in Agriculture" needs to be revised keeping in view the agro-ecological requirements and socio-cultural situations in India. This will help in wider acceptability and more benefits from use of these Checkpoints.
8. There should be proper reporting on occupational diseases and injuries as it would be helpful to lay proper focus and identify the priority areas, for taking action.
9. There is a need to have National Policy on Occupational Health—which would be helpful in reducing drudgery and increasing safety, productivity, efficiency and well-being of workers.
10. Training and awareness programs for load carrying techniques are required so as to reduce occupational health hazards. ISE can step forward to publish bulletin on load carrying techniques and start a training programme in collaboration with government departments/NGOs.
11. Role of women workers in industry and agriculture is very important. The gender issue therefore needs more attention.

12. Empowerment of farmwomen through farm mechanization needs to be given priority. Ergonomically designed tools and implements can result in saving of time, money, and efforts. It can also help in reducing the drudgery and increasing the efficiency without jeopardizing their health.
13. There is a need of skill upgradation of women engaged in enterprises/cottage industries to uplift their economic status. Also, there is an urgent need to apprise women about decisions taken for their welfare. Self help groups can play a pivotal role in addressing these issues.
14. Clothing plays a pivotal role in the efficiency & output of the worker. As a woman undergoes various changes during pregnancy and lactation period, she needs special type of functional clothing. Hence, there is a dire need to design special type of clothing for women so that their efficiency could be increased.
15. Media should be involved extensively to disseminate information on ergonomically developed equipment, gender sensitization in the society and women empowerment through Ergonomics.
16. The technologies already developed by the research organizations may be licensed to the manufacturing units. This will help in faster dissemination of technologies to the users.
17. The do's and don'ts of traditional practices/improved technologies needs to be prepared from ergonomics point of view. These can be popularized through awareness camps, practical training, exhibitions, demonstrations, and mass media.
18. Number of technology transfer programmes run by government organizations are not getting success. Building of linkages between government agencies, NGOs and stakeholders would help in making such extension programmes successful.
19. Quality Research need to be encouraged and multidisciplinary approach will have to be followed to get fruitful results.
20. A need is felt for development of trained human resource for application of ergonomics in different walks of life.
21. The findings, new strategies, data presented by various researchers in this conference should be used by the designers/engineers/industry. It will help to make workers more productive, efficient, comfortable and safe.
22. All endeavours should be directed to help all the people leading a 'Quality Life'.

Contents

SECTION–I
Anthropometry and Workplace Design

SECTION–II
Work and Sports Physiology

SECTION–III

Physical Environment

SECTION–IV

Cognitive and Design Ergonomics

SECTION–V
Ergonomics in Agriculture

SECTION–VII

Occupational Health and Safety

Abstracts of Other Presented Papers

SECTION–I

Anthropometry and Workplace Design

Age and Sex Related Variation of Pinch Strength among Adult Bengalee Population

Sujaya De, Dipak Maiti, Madhumita Harh and Prakash C. Dhara

Ergonomics and Sports Physiology Division
Department of Human Physiology with Community Health
Vidyasagar University, Midnapore–721 102, West Bengal, India

ABSTRACT

Pinch strength is the ability of the thumb to apply force inward towards the palm. There are many factors that affect the pinch strength in the population. The present investigation was aimed to assess the age related changes and gender difference of pinch strength of adult male and female subjects. The study was conducted on 482 male and 232 female subjects to measure their pinch strength. The subjects were divided into five age groups: 18–21 years, 22–30 years, 31–40 years, 41–50 years and 51–60 years. Pinch strength was measured by JAMAR Hydraulic Pinch Gauge. The results showed that the men were significantly stronger in pinch strength ($p < 0.001$) than the women in both hands. It was noted that the age had a significant effect on the pinch strength. The result showed that there was a significant change in pinch strength in male ($p < 0.05$) and female ($p < 0.001$) subjects when they were divided into five age groups. The percentage difference of the pinch strength was the highest in the oldest age group (51–60 years) among all the groups. In all age groups the pinch strength of the male subject was significantly higher than that of the female subjects in both hands. Musculature development seemed to be more prominent in case of male than that of females. It was also observed that the right hand was stronger than the left hand. This might be due to the fact that the most of the selected subjects were right handed. The results of the correlation coefficient between pinch strength and other anthropometric variables *viz.*; palm length, length of digits, fist girth represented poor correlation. However, negative correlation, although non-significantly, was observed in most of the cases. The recorded database for pinch strength may be a useful guide for designing finger control devices.

Keywords: Pinch Strength, Age, Gender.

INTRODUCTION

The pinch strength is an important human characteristic. It is the back bone to the pinch grip. In the strongest of pinch grips the first knuckle of the thumb is completely extended or slightly bent. The majority of the inward force is generated by the large muscle of the lower palm just beneath the second knuckle. This force is balanced by one or more fingers of the same hand applying a force back towards the thumb. Sufficient pinch strength is necessary to successfully complete diurnal as well as industrial activities. Pinch strength is necessary for writing purpose, on-off the switch, designing finger operated control panel, machine, equipment etc. Kellor *et al.* (1971) compiled the most common norms for grip and pinch strength used in therapy clinics. Regression analysis was performed to predict the amount of right or left hand strength of an individual of a given sex and particular age. Strength of the fingers is greatly influenced by the flexion or extension angles of the wrist (Rodgers, 1983). Wells and Miller (1988) measured maximal tip pinch forces over a range of wrist angle. The American Society of Hand Therapists suggested a standardized arm positioning for hand strength tests. They concluded that the position of the upper extremity might influence measurements and recommended that the patient should be seated with his shoulder adducted and neutrally related elbow flexed at 90° and the fore arm and wrist in neutral position (Fess and Moran, 1981). Mathiowetz *et al.* (1985) documented a high correlation between grip strength and age but a low to moderate correlation between pinch strength and age. Comparison of the average hand strength of right handed and left handed subjects showed only minimal difference. In 1999, Armstrong and Oldham suggested that dominant hand (right hand) is about 10% stronger than non-dominant hand (left hand). The present investigation was aimed to assess the age related changes and gender difference of pinch strength of adult male and female subjects.

METHODS

Selection of Site

The study was conducted in different districts of West Bengal especially in West Medinipur, East Medinipur, and Purulia.

Selection of Subjects

The present study was conducted on 331 male and 132 female subjects having the age range of 18–60 years. The selected subjects were further divided into five age groups, *viz.*, 18–21 years, 22–30 years, 31–40 years, 41–50 years and 51–60 years. Each age group was divided into male and female subgroups.

Measurement of Pinch Strength

The apparatus required for the strength measurement was a pinch gauge (JAMAR Hydraulic Pinch Gauge). Red indicator remained at maximum reading until reset. Pinch strength was measured up to 20 kg. The range of the JAMAR hydraulic pinch gauge was 1–20 kg. The instrument was calibrated periodically during the study.

Before taking measurements, the subjects were requested to stand in a comfortable position. The pinch gauge was normally holded by left hand. The pointer of the pinch gauge remained in zero mark of the scale. With the thumb and 2nd digit it was pressed at the convex position of the pinch gauge. The subject was asked to press the pinch gauge as hard as possible without moving the rest of his/her body and without help of any other finger. The final pinch strength measure was taken from the pinch gauge scale, when the pointer no longer moved. The pinch gauge was the rest to zero and the test was performed in standing posture. The strength of the left hand was also measured in the same way.

Measurements of Anthropometric Dimensions

Some anthropometric dimensions were also taken along with the pinch strength measurement. These were as follows:

Weight

Weight was taken by a portable weighing machine. The machine was placed on the plane surface and the subject was asked to stand erect on the machine and the reading was taken with an accuracy of 0.5 kg.

Height

Height was taken with the help of anthropometric rod. The vertical distance from the ground to the vertex of head was measured. During the measurement the subject stood in erect position.

Length of the Digits

Lengths of the digits were taken with the help of slide calipers from the joint of metacarpal and phalanges to tip of the phalanges.

Palm Length

Palm length was taken with the help of slide calipers from the wrist joint to starting of phalanges.

Fist Girth

A Steel tape was taken to measure the fist girth by wrapping the tape around the fist and the reading was taken.

Hand Grip Diameter

The subject was asked to grip a cone so that the thumbs and middle finger could touch together. The circumference of the grip was measured from the cone and the hand grip diameter was calculated from it.

Calculation of BMI

The Body Mass Index (BMI) was calculated by dividing weight (kg) by squared value of height (m).

Statistical Analysis

Data were summarized into mean and standard deviation values using Microsoft Excel. The effect of sex on pinch strength was determined by studying the level of significance between two-group mean using t-test. The effect of hand and age on pinch strength was determined by using one way ANOVA. The correlation coefficient between pinch strength and other physical parameters *viz.*, BMI, length of the 1st digit, 2nd digit, fist girth and hand grip diameter was determined by employing product moment correlation coefficient.

RESULTS AND DISCUSSION

Pinch strength is the ability of the thumb to apply force towards the palm. To measure the pinch strength, the 2nd digit also helps. In general pinch strength gradually increases with the advancement of age. In addition there is a gender difference of pinch strength also. Table 1 shows that there is a significant difference between height and weight of all male and female subjects but there is no significant difference in BMI of the male and female subjects in all age groups excepting the subjects of 51–60 years age group. It is found that in most of the age groups there are significant differences in height and weight between male and female subjects. From Table 2 it has been noted that the percentage difference of pinch strength between all male and female subjects was 28.18% in left hand and 25.33% in right hand and pinch strength of male subject was significantly higher than that of the female subjects. From this table it has been also noted that, in 18–21 years age group the percentage difference of pinch strength between male and female was 39.06% in left hand and 37.36% in right hand. It was the highest percentage difference of pinch strength between the subjects of two sexes among the age groups. The result shows that men were stronger in their pinch strength than women because men were involved in heavier type of work than the women. So, the musculature development in the male might be more prominent than that in the females. In case of male subject there is a significant difference ($p < 0.05$) in pinch strength among five age groups (Table 3) in both hands. There is also a significant difference ($p < 0.001$) in pinch strength among five age groups of female subjects in both hands. In 1985, Mathiowetz *et al.* had established that men were stronger than the women on all hand strength tests. The present results are in conformity with results of the previous study. Data of the palm length, fist girth and hand grip diameter for both sexes (Table 4) demonstrate that there were differences between male and female but there was a little difference between left and right hand in all age groups. From the study of the length of the digits, it is noted that the length of the digit was significantly different between male and female subjects in the age group of 22–30 years and 51–60 years. The results show that physical characteristics of the subjects exert some influence on the pinch strength in their growing ages and in older ages. But in case of all male and female subjects there were significant differences (Table 5) in palm length, fist girth, hand grip diameter and length of the digit excepting length of the first digit. In case of all age groups the results of the correlation coefficient (Table 6) between pinch strength and other anthropometric variables, *viz.*, palm length; length of digits, fist girth and BMI represent negative correlation. But it has been also noted that in case of all male and female subjects there is significant correlation between pinch strength and some anthropometric variables, *viz.*, palm length, length of digits, fist girth, BMI. There is a positive correlation between pinch strength and age of all male and female subjects.

Table 1: Mean ± SD of Height, Weight and BMI of both Male and Female Subjects ($n = 463$)

Age Group (y)	*Sex*	*Height (cm)*	*Weight (kg)*	*BMI (kg/m²)*
18–21 years	Male ($n = 37$)	161.35 ± 6.77	51.27 ± 8.16	19.72 ± 3.09
	Female ($n = 14$)	150.39 ± 5.09*	45.43 ± 6.45#	20.09 ± 2.63
22–30 years	Male ($n = 132$)	159.33 ± 17.60	54.86 ± 8.47	21.04 ± 3.36
	Female ($n = 65$)	151.77 ± 4.78*	49.25 ± 6.76*	21.34 ± 2.43
31–40 years	Male ($n = 121$)	156.34 ± 22.65	54.73 ± 7.89	21.29 ± 3.16
	Female ($n = 31$)	151.39 ± 5.26	45.48 ± 7.24*	19.77 ± 2.44
41–50 years	Male ($n = 19$)	161.45 ± 6.42	54.68 ± 11.02	20.89 ± 3.54
	Female ($n = 16$)	150.54 ± 5.17*	48.13 ± 11	21.21 ± 4.98
51–60 years	Male ($n = 22$)	163.09 ± 5.21	63.05 ± 12.07	23.65 ± 4.03
	Female ($n = 6$)	143.00 ± 5.46*	38.67 ± 7.81*	18.81 ± 3.22#
All subjects	Male ($n = 331$)	158.83 ± 17.98	54.94 ± 8.95	21.15 ± 3.40
	Female ($n = 132$)	150.99 ± 5.25*	47.34 ± 7.94*	20.71 ± 2.97

w.r.t. male *$p < 0.001$ #$p < 0.05$.

Table 2: Mean ± SD and Percentage Difference of Pinch Strength between Left Hand and Right Hand of both Male and Female Subjects ($n = 463$)

Age Group (y)	*Sex*	*Pinch Strength*		*t-value*
		Left Hand	Right Hand	
18–21 years	Male ($n = 37$)	9.32 ± 2.38	9.69 ± 2.03	0.71
	Female ($n = 14$)	5.68 ± 1.42	6.07 ± 1.19	0.79
	% difference	39.06%	37.36%	–
22–30 years	Male ($n = 132$)	9.69 ± 2.13	10.01 ± 1.81	1.34
	Female ($n = 65$)	7.30 ± 1.77	8.06 ± 1.86	2.39#
	% difference	24.66%	19.48%	–
31–40 years	Male ($n = 121$)	9.55 ± 2.05	9.87 ± 1.88	1.29
	Female ($n = 31$)	6.26 ± 1.42	6.60 ± 1.61	0.88
	% difference	34.45%	33.13%	–
41–50 years	Male ($n = 19$)	8.34 ± 2.35	8.79 ± 2.51	0.57
	Female ($n = 16$)	7.09 ± 1.67	7.03 ± 1.58	0.11
	% difference	14.99%	20.02%	–
51–60 years	Male ($n = 22$)	8.52 ± 1.84	9.07 ± 1.56	1.06
	Female ($n = 6$)	5.67 ± 1.17	6.42 ± 0.74	1.33
	% difference	33.45%	29.22%	–
All subjects	Male ($n = 331$)	9.44 ± 2.15	9.79 ± 1.91	2.23#
	Female ($n = 132$)	6.78 ± 1.72	7.31 ± 1.83	2.39#
	% difference	28.18%	25.33%	–

w.r.t. left hand # $p < 0.05$.

Table 3: Comparison of Pinch Strength between 18–21 years, 22–30 years, 31–40 years, 41–50 years and 51–60 years of both Male and Female Subjects

Sex	*Hand*	*18–21 y*	*22–30 y*	*31–40 y*	*41–50 y*	*51–60 y*	*F-ratio*
Male	Variables	(*n* = 37)	(*n* = 132)	(*n* = 121)	(*n* = 19)	(*n* = 22)	–
	Left	9.32 ± 2.38	9.69 ± 2.13	9.55 ± 2.05	8.34 ± 2.35	8.52 ± 1.84	2.84#
	Right	9.69 ± 2.03	10.01 ± 1.81	9.87 ± 1.88	8.79 ± 2.51	9.07 ± 1.56	2.67#
Female	Variables	(*n* = 14)	(*n* = 65)	(*n* = 31)	(*n* = 16)	(*n* = 6)	–
	Left	5.68 ± 1.42	7.30 ± 1.77	6.26 ± 1.42	7.09 ± 1.67	5.67 ± 1.17	4.93*
	Right	6.07 ± 1.19	8.06 ± 1.86	6.60 ± 1.61	7.03 ± 1.58	6.42 ± 0.74	7.10*

*$p < 0.001$, #$p < 0.05$.

Table 4: Mean ± SD of Length of Palm, Fist Girth, Hand Grip Diameter and Length of the Digit of both Male and Female Subjects of Five Age Groups

Parameter			*18–21 years*		*22–30 years*		*31–40 years*		*41–50 years*		*51–60 years*	
			Male (n = 37)	*Female (n = 14)*	*Male (n = 132)*	*Female (n = 65)*	*Male (n = 121)*	*Female (n = 31)*	*Male (n = 19)*	*Female (n = 16)*	*Male (n = 22)*	*Female (n = 6)*
Palm Length		LH	10.00 ±0.58	9.15 ±0.62*	10.65 ±8.38	9.36 ±0.48	9.88 ±0.60	9.15 ±0.59*	9.88 ±0.73	9.29 ±0.62#	10.20 ±0.64	8.93 ±0.58*
		RH	10.01 ±0.52	9.14 ±0.45*	9.95 ±0.58	9.35 ±0.49*	9.91 ±0.51	9.08 ±0.56*	9.80 ±0.66	9.25 ±0.55#	10.21 ±0.51	9.23 ±0.28*
Fist girth		LH	24.06 ±2.08	22.55 ±2.37#	24.18 ±2.50	21.81 ±1.82*	23.81 ±2.87	22.73 ±1.28#	25.79 ±2.30	23.40 ±1.68§	26.61 ±1.75	21.98 ±3.56*
		RH	24.28 ±2.22	23.17 ±2.71	24.32 ±2.58	22.15 ±2.04*	23.89 ±2.98	23.31 ±1.29	25.81 ±2.60	24.01 ±1.78#	27.08 ±1.77	22.47 ±3.79*
Handgrip Diameter		LH	3.32 ±1.93	5.40 ±0.39*	3.10 ±1.85	3.02 ±1.79	2.51 ±1.48	4.50 ±1.29*	4.81 ±1.18	4.66 ±1.23	5.16 ±0.87	5.32 ±1.15
		RH	3.25 ±1.85	5.37 ±0.43*	3.07 ±1.82	3.03 ±1.73	2.49 ±1.45	4.42 ±1.24*	4.83 ±1.19	4.55 ±1.18	5.11 ±0.82	5.21 ±1.16
Length of the digit	1st digit	LH	5.29 ±0.65	6.20 ±0.92*	5.23 ±0.72	4.92 ±0.80§	5.20 ±0.64	5.69 ±0.58*	5.86 ±0.82	5.94 ±1.03	6.29 ±0.55	5.63 ±0.44#
		RH	5.35 ±0.67	6.04 ±0.56§	5.28 ±0.78	4.96 ±0.78§	5.19 ±0.63	5.71 ±0.58*	5.97 ±0.87	5.79 ±0.72	6.35 ±0.63	5.58 ±0.28*
	2nd digit	LH	6.90 ±0.41	6.74 ±0.61	6.88 ±0.40	6.54 ±0.50*	6.81 ±0.45	6.48 ±0.35*	6.80 ±0.56	6.63 ±0.52	7.11 ±0.40	6.38 ±0.66§
		RH	6.92 ±0.36	6.69 ±0.39#	6.87 ±0.41	6.57 ±0.47*	6.79 ±0.45	6.47 ±0.31*	6.72 ±0.53	6.72 ±0.53	7.00 ±0.47	6.27 ±0.48§
	3rd digit	LH	7.77 ±0.46	7.39 ±0.47#	7.77 ±0.43	7.29 ±0.50*	7.74 ±0.41	7.20 ±0.39*	7.61 ±0.70	7.61 ±0.70	7.92 ±0.38	7.17 ±0.55*
		RH	7.69 ±0.46	7.36 ±0.49#	7.68 ±0.43	7.30 ±0.50*	7.70 ±0.42	7.14 ±0.40*	7.54 ±0.70	7.54 ±0.70	7.78 ±0.44	7.17 ±0.61§
	4th digit	LH	7.04 ±0.42	6.80 ±0.36	7.02 ±0.39	6.56 ±0.51*	6.98 ±0.38	6.62 ±0.34*	7.08 ±0.61	7.08 ±0.61	7.25 ±0.36	6.73 ±0.30§
		RH	7.10 ±0.40	6.88 ±0.42	7.04 ±0.39	6.58 ±0.50*	7.03 ±0.35	6.65 ±0.36*	7.04 ±0.65	7.04 ±0.65	7.23 ±0.31	6.67 ±0.44§
	5th digit	LH	5.45 ±0.66	5.59 ±0.80	5.21 ±0.70	4.73 ±0.62*	5.19 ±0.68	5.27 ±0.43	5.63 ±0.67	5.63 ±0.67	5.91 ±0.51	5.35 ±0.34#
		RH	5.32 ±0.64	5.56 ±0.52	5.17 ±0.70	4.73 ±0.62*	5.17 ±0.68	5.22 ±0.40	5.51 ±0.71	5.41 ±0.58	5.87 ±0.45	5.32 ±0.22§

w.r.t. male *$p < 0.001$, §$p < 0.01$, #$p < 0.05$.

Table 5: Mean ± SD of Length of Palm, Fist Girth, Hand Grip Diameter and Length of the Digit of all Male and Female Subjects

Parameter			*All Subjects*	
			Male (n = 331)	*Female (n = 132)*
Palm Length		Left Hand	10.22 ± 5.31	9.26 ± 0.55#
		Right Hand	10.24 ± 5.25	9.25 ± 0.51#
Fist girth		Left Hand	24.29 ± 2.65	22.31 ± 1.92*
		Right Hand	24.43 ± 2.77	22.77 ± 2.12*
Hand Grip Diameter		Left Hand	3.13 ± 1.81	3.93 ± 1.75*
		Right Hand	3.10 ± 1.77	3.88 ± 1.69*
Length of the digit	1st digit	Left Hand	5.33 ± 0.74	5.39 ± 0.92
		Right Hand	5.37 ± 0.78	5.38 ± 0.81
	2nd digit	Left Hand	6.87 ± 0.43	6.55 ± 0.49*
		Right Hand	6.85 ± 0.43	6.55 ± 0.43*
	3rd digit	Left Hand	7.76 ± 0.44	7.29 ± 0.47*
		Right Hand	7.69 ± 0.45	7.27 ± 0.48*
	4th digit	Left Hand	7.03 ± 0.41	6.64 ± 0.45*
		Right Hand	7.06 ± 0.39	6.66 ± 0.46*
	5th digit	Left Hand	5.30 ± 0.70	5.05 ± 0.68*
		Right Hand	5.25 ± 0.69	5.04 ± 0.63#

w.r.t. male $^{*}p < 0.001$ $^{\S}p < 0.01$ $^{\#}p < 0.05$.

CONCLUSION

There is a change in pinch strength of left and right hand with the change of age. Significant difference has been observed in pinch strength between left hand and right hand as well as between male and female subjects. Significant correlation has been observed between pinch strength and the dimensions of the hand. It may be concluded that the design of finger operated controls should be considered in different ways for the females as there is an appreciable difference of pinch strength between male and female.

Table 6: Correlation Coefficient (r) between Pinch Strength and Age and Other Physical Variables, *e.g.*, Hand Grip Diameter, Palm Length, Fist Girth, BMI, Length of the Digit of both Male and Female Subjects

Parameter		*18–21 years*		*22–30 years*		*31–40 years*		*41–50 years*		*51–60 years*		*All Subject*	
		Male (n = 37)	*Female (n = 14)*	*Male (n = 132)*	*Female (n = 65)*	*Male (n = 121)*	*Female (n = 31)*	*Male (n = 19)*	*Female (n = 16)*	*Male (n = 22)*	*Female (n = 6)*	*Male (n = 331)*	*Female (n = 132)*
BMI		0.09	–0.37	–0.18#	0.30#	–0.00	0.39#	0.08	0.07	0.17	–0.23	–0.5*	0.27*
Palm length		–0.01	0.01	–0.17	–0.10	–0.06	0.33	–0.03	0.10	0.10	–0.18	–0.07	0.13§
Fist girth		–0.24	0.37	–0.08	–0.04	–0.05	–0.28	–0.30	0.04	–0.10	–0.48	–0.14§	–0.15§
Hand grip diameter		–0.21	–0.05	–0.08	–0.27#	–0.06	–0.50§	–0.70*	–0.10	–0.24	0.20	–0.16§	–0.46*
Length of the digit	1st	–0.44#	0.09	–0.16	–0.32§	–0.02	–0.36#	–0.17	0.02	–0.08	–0.35	–0.19*	–0.43*
	2nd	–0.10	0.45	–0.16	–0.01	0.09	0.30	0.26	0.20	0.24	–0.86#	0.00	0.09*
	3rd	0.00	0.53#	–0.09	–0.02	0.10	0.31	0.17	0.03	0.21	–0.63	0.04	0.10#
	4th	–0.20	0.45	–0.11	–0.04	0.07	–0.00	0.13	0.11	–0.09	–0.56	–0.04	–0.06
	5th	–0.30	0.43	–0.01	–0.16	–0.01	–0.19	–0.03	–0.17	–0.06	–0.17	–0.09	–0.31*
Age		0.09	0.04	–0.08	–0.04	–0.09	–0.28	–0.26	–0.32	–0.54*	–0.12	–0.15§	–0.18*

*$p < 0.001$ §$p < 0.01$ #$p < 0.05$.

REFERENCES

[1] Armstrong, C.A. and Oldham, J.A. (1999). A comparison of dominant and non-dominant hand strength. *J. Hand Surg*, 24B (4): 421–5.

[2] Fess, E.E. and Moran, C. (1981). Clinical Assessment Recommendations, Indianapolis, In. *American Society of Hand Therapists*.

[3] Kellor, M., Frost, J., Silberberg, N., Iversen, I. and Cummings, R. (1971). Hand strength and dexterity. *Am. J. Occup Ther*, 25: 7743.

[4] Mathiowetz, V., Kashman, N., Volland, G., Weber, K., Dowe, M. and Rogers, S. (1985). Grip and pinch strength: Normative data for adults. *Arch. Phys. Med. Rehabil.*, 66: 69–72.

[5] Rodgers, S.H. (1983). Ergonomic Design for People at Work. *Van Nostrand Reinhold*, 2.

[6] Wells, R.P. and Miller, M. (1988). The influence of Wrist Flexion and Extension on Pinch Strength; A comparison of Experimental and Computer Stimulation Results. *Proceedings of the Fifth Biennial Conference of the Canadian Society for Biomechanics, Spodym Publishers, London, Ontario*, 168–169.

Concept of Metropolitan City Students on the Design of a Safe School Bus: A Participatory Ergonomics Study

Samrat Dev, Tarannum Ara, Tirthankar Ghosh, Tamal Das, Goutam Ghoshal, Banibrata Das and Somnath Gangopadhyay

Occupational Ergonomics Laboratory
Department of Physiology, University of Calcutta
Kolkata, West Bengal, India

ABSTRACT

In urban India, school buses play a significant and important role in transportation of students. These are considered as the safest mode of transports in city areas in spite of poor ergonomic interior design and predominant unsafe condition. The present participatory ergonomic study is formulated in such a way to get a virtual concept of a safe, ergonomically designed school bus. For this study a questionnaire was developed and performed among sixty urban students of four different schools of Kolkata. The students were divided into three of equal strength major groups: junior, middle and high. Each group had ten male and ten female students. From the analysis of questionnaire it was observed that all students were facing many non-safety and poor ergonomics related problems while travelling in the school buses. Moreover, they had a very little concept of safety and ergonomics in bus design. It was further revealed that design of Indian school buses should be improved immediately. At the end of study a virtual design of school bus is sketched on the basis of suggestion rendered by students.

INTRODUCTION

In metropolitan cities of India school bus is one of the most predominant and important modes of transportation for students. They are commonly painted yellow for purposes of visibility, safety and equipped with specialized traffic warning devices. Full-size school buses can accommodate 59 to 90 passengers, but in many districts smaller vehicles are used as well. Such smaller vehicles are commonly known as "short buses", and are often associated with private schools (National Highway Traffic Safety Administration, 1993).

In the recent years, there has been a significant increase in the numbers of infants school children who are being transported in school buses.

According to annual school bus transportation statistics of USA, 450,000 public school buses are present which are used by 24 million students and travel 4 billion miles. These school buses carry out 10 billion student trips and 20 billion times a student gets on or off from this school buses (NHTSA, FARS, 1999).

In Kolkata only 95 registered school buses are present, which are used by 5700 students and travel 4750 km per day (Public Vehicle Department, 2006).

It is commonly believed that school buses are safer than other transports. But several tragedies or problems related to the school buses indicate how unsafe it is. The safety of the general people and students both within the bus as well as those in the roads are given least priority (Gangopadhyay *et al.*, 2005).

According to official school bus accident statistics, approximately 27 children die in school bus accidents every year (CASTO, 2001). As the number of school buses operating on the roadways increased, several serious tragedies occurred involving school buses. In 1939, some "standards" and recommendations for school buses were developed in USA and Canada (NHTSA, FARS, 1999).

In India, the conditions of the school buses are poor and the students suffer from a lot of problems while travelling in such buses (Dev *et al.*, 2007). However, no study regarding the internal design of the school bus has been recorded. The present investigation is made to find out the responses of the school children for the design of future school bus.

METHODS

Participatory ergonomics study seeks to maximize the involvement of the user in a particular design. This participatory approach to ergonomics relies on actively involving workers in implementing ergonomic knowledge, procedures and changes with the intention of improving safety, productivity, quality and comfort.

Selection of Subjects

In this study, the experiment was performed on three different groups of students (junior, middle and high school students). The groups were chosen in such a manner that they all availed school bus facilities on a regular basis. Twenty such students from each group, constituting 10 male and 10 female were chosen. To avoid selection bias, all the subjects were selected randomly from four different schools of Kolkata.

Questionnaire

A questionnaire was developed for the selected students and applied on them. It consisted of a series of objective type questions with multiple-choice responses and was divided into two sections. The questionnaire was formulated to find out their response regarding the school bus design and also evaluate the concept of safety among the users of the school buses.

Statistical Analysis

Student t-test was done to find out the significant difference of physical parameters between the male and female students of the groups (Das and Das, 1993).

RESULTS AND DISCUSSION

From the analysis of questionnaire, the following results have been tabulated in the subsequent tables. Table 1 represents the physical parameters of male and female students of the three groups, which were represented as Junior school students: Group A, Middle school students: Group B and High school students: Group C.

Table 1: Physical Parameters of Male and Female Students

Parameters		*Male*		*Female*		*Remarks*
		Mean	*±SD*	*Mean*	*±SD*	
Age (Years)	Group A	7.5	1.26	7.7	1.25	Not Significant
	Group B	11.2	0.78	11.2	0.78	Not Significant
	Group C	14.5	1.08	14.2	0.78	Not Significant
Height (cm)	Group A	129.4	8.06	126.2	12.13	Significant
	Group B	147.3	6.73	141.1	6.58	Significant
	Group C	164.6	6.55	151.0	4.50	Significant
Weight (kg)	Group A	34.4	5.19	27.3	2.00	Significant
	Group B	49.2	4.49	37.9	3.57	Significant
	Group C	54	3.09	47.0	3.39	Significant

It was observed that different groups of individuals—the junior, middle and high school students, both male and female, had significant values in terms of height and weight.

Table 2 shows the positive and negative response of the school bus users regarding the school bus design. From the questionnaire analysis performed among the three groups of the school students it has been found that they like school bus for their journey. They can travel in them as a group together and thus enjoy the journey thoroughly. However according to the maximum number of students, the seats need improvement as mostly these seats bear torn and tattered cushions. Students had a difficulty in sitting and when brakes are applied suddenly, they suffer from back injury while sitting on those seats. They felt that the seats should be cushioned from top to bottom along with proper armrests for their comfort. Moreover the armrests may protect them from sudden fall on the floor of the bus. According to students of junior section the seat length as well as the leg space is sufficient but for the students in the middle and high schools both these parameters are inadequate. So they want enhancement of seat length and leg space. But according to junior school students the height of the seat pan was high for them, but it was sufficient for middle and high school students. Most of them do not prefer seat belt in school buses because according to them seat belts will restrict their normal movements. It is revealed from the questionnaire that most of the students of three groups want their bus windows to be larger in size and also felt that the number of such windows should be increased. This is required because according to them the internal environment of school bus is warm and suffocating. There should be more than one emergency door in the buses to facilitate exit in times of crisis. They also felt that the passage between the two rows of seats should be enhanced so that they could move freely inside the bus. According to junior school students the height of the handles attached to the ceiling was too high for them, but this height was sufficient for middle and high school students. The junior school students met with accidents inside the school bus. They had cut injuries due to sharp edges, exposed sharp nails; swelling of head on being struck against the ceiling handles and also suffered from chest injuries as they get hit against the back portion of the preceding seat on sudden application of brakes. Presence of first aid boxes should be there so that every one could use it. Fire extinguisher also must be present in all school buses and proper training should be provided to the drivers and high school students. From the above discussion it is clearly evident that a number of changes have been suggested by the groups of students for the overall improvement of the school buses.

Table 2: Response of School Bus Users Regarding the School Bus Design

S. No.	*Questionnaire*	*Group A*		*Group B*		*Group C*	
		Yes	*No*	*Yes*	*No*	*Yes*	*No*
1.	Do you prefer existing school bus for your journey? Y/N and Why?	15	5	11	9	15	5
2.	If school bus seats are comfortable for sitting? Y/N What is your concept regarding comfort? Explain.	10	10	8	12	3	17
3.	Do you prefer the sitting arrangement of the school bus? Y/N and Why?	14	6	13	7	10	10
4.	If the seat length is sufficient to sit with comfort? Y/N	18	2	10	10	2	18
5.	If leg space is sufficient or comfortable? Y/N	20	0	8	12	0	20
6.	If armrest is required in the seats of the school bus? Y/N	13	7	18	2	20	0
7.	If the height of seat is sufficient? Y/N	18	2	8	12	5	15
8.	If seat belts are required in the school bus? Y/N and Why?	5	15	6	14	10	10
9.	If position, size and number of windows are appropriate? Y/N and Why?	10	10	8	12	14	6
10.	If more than one emergency door are required? Y/N	10	10	14	6	18	2
11.	Do you prefer more widen walking passages in the school bus? Y/N and Why?	18	2	10	10	13	7
12.	Do all the buses have handle bars inside to hold? Y/N	4	16	5	15	10	10
13.	Do you want any changes in the existing school bus-design? Y/N and Why?	10	10	13	7	18	2
14.	Do you ever have any accident? Y/N	10	10	6	14	5	15
15.	Are fire extinguishers and first aid boxes required in the school buses? Y/N	11	9	16	4	20	0

Virtual Design

Depending upon the responses of the students, a virtual design of bus has been done. The virtual design (Figure 1) has again been shown to the same students for confirming their preferences. After obtaining their feedback, Table 3 has been drawn wherein it is clearly observed that most of them are satisfied with the virtual design of the school bus.

Table 3: Commuters Preferences over the Virtual Design of the School Bus

Parameters		*Response*	
		Positive	*Negative*
Arrangement of seat	(a) Seat Design	45 (75%)	15 (25%)
	(b) Leg Clearance	60 (100%)	0 (0%)
	(c) Arm Rest	60 (100%)	0 (0%)
	(d) Passage between the two rows of seats	60 (100%)	0 (0%)
	(e) Height of the seat	60 (100%)	0 (0%)
Height of the handlebars		49 (81.7%)	11 (18.3%)
Emergency Door		60 (100%)	0 (0%)
Windows		60 (100%)	0 (0%)
First aid boxes and fire extinguisher		60 (100%)	0 (0%)

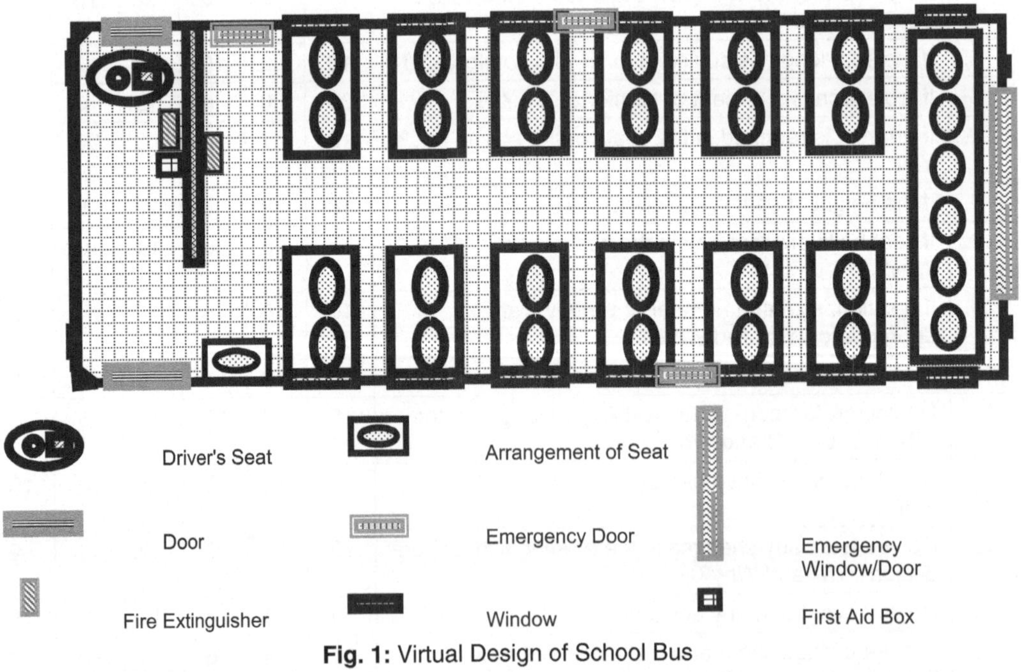

Fig. 1: Virtual Design of School Bus

Views of Commuters

Design of the School Bus	*Suggestions of Commuters Towards the Betterment of the Existing Design*	*Reasons behind Suggestions Rendered by the Commuters*
Arrangement of seat (a) *Seat Design:* Most of the seats bear cushions only on the sitting surface and the seats are uncomfortable due to insufficient seat length. All the metallic portions of the seats are highly exposed.	The seat as well as back portion of the immediate front seat should be highly cushioned. Exposed metallic portion should also be covered. The seat length should be increased according to the body measurement of the students.	Difficulty in sitting and when brakes are applied suddenly, the students suffer from back injury while sitting in those seats. An increase in seat length will protect them from sudden fall on the floor of the bus.
(b) *Leg Clearance:* Leg Clearance is not sufficient for students.	Leg Clearance must be sufficient according to the body measurement of the students.	To provide adequate space for free movement of legs while sitting.
(c) *Armrest:* In existing buses armrest is not present.	Armrest should be present in the school buses with cushion, rounded in shape without sharp edges.	Armrests may protect them from sudden fall on the floor of the bus.
(d) *Passage between the two rows of seats:* Passage between the two rows of seats is not sufficient.	Passage must be sufficient according to the body measurement of the students.	To move freely inside the bus.
(e) *Height of the seat:* Seat height is non-adjustable according to the body measurement of the students.	Seat height should be adjustable.	To provide support to the feet and avoid dangling of feet.
Height of the handlebars: Inability to hold the handlebars attached to the ceiling of the bus at a greater height above their reach.	The handlebars should be installed over the seats with rubber cushion so that the students can hold it below shoulder level while standing.	To hold the handlebars in standing position to prevent from fall due to sudden brakes.
Emergency Door: Presence of only one emergency door.	Presence of four emergency doors-one at the front part of the bus, one each on either side at the middle of the bus and the entire rear portion of the bus will bear an emergency exit. All such doors should be such that they can be pushed easily rather than being pulled.	To facilitate exit in times of crisis.
Windows: The windows are provided with ordinary glasses and the rims bear sharp edges.	The rims of the windows should also be cushioned and instead of glasses, fibreglass should be used. The windows should operate from downward to upward direction instead of sideways movement.	For airy atmosphere inside the school bus.
First aid boxes and fire extinguisher: Absence of first aid boxes and fire extinguisher in the school buses.	Presence of first aid boxes in such place so that every one can use it and two fire extinguishers—one in the driver's cabin and the other near the students.	The first aid box is highly essential in times of injury and the fire extinguishers are a must at times of emergency and crisis.

CONCLUSION

A school bus is one of the most important and biggest modes of transport for students mainly in metropolitans as well as in small cities. At the end of the present investigation it is found that due to poor design of the school buses, students face many safety related problems. They want some major changes in the existing design of the bus.

ACKNOWLEDGEMENT

Authors express their sincere gratitude to all those school bus drivers, school students and their parents who rendered immense cooperation towards the completion of this study.

REFERENCES

[1] Das, D. and Das, A. *Statistics in Biology and Psychology* (Academic Publishers, Calcutta), 1993.

[2] Dev, S., Das, B., Ghoshal, G., Das, T., Ara, T. and Gangopadhyay, S. "A study on Indian School Buses and Drivers Concept of safety in School Buses of Kolkata", Book of Abstract of the *International Conference on Frontier Researches in Integrative Physiology*, University of Calcutta, Kolkata, January 8–10, 2007.

[3] Gangopadhyay, S., Das, T., Ghoshal, G. and Biswas, P. "Questionnaire Study to Evaluate the Predominant Causes Behind City Bus Accidents in Calcutta, India", *The Indian Journal of Physiology and Allied Sciences*, 59 (2005) 33.

[4] National Highway Traffic Safety Administration (NHTSA), Fatal Analysis Reporting System (FARS) and *School Bus Safety: Safe Passage for America's Children* August 16, 1999.

[5] National Highway Traffic Safety Administration. *School Bus Safety Report*, May 1993.

[6] Press release from International Truck and Engine Corporation, February 13, 2001 and an Op/Ed piece written by Larry Laxson, President, and California Association of School Transportation Officials (CASTO), February 28, 2001.

[7] Public vehicle department, Government of India, 2006.

Anthropometric and Strength Data Bank of Indian Agricultural Workers

L.P. Gite[1] and Joydeep Majumder[2]

AICRP on Ergonomics and Safety in Agriculture
Central Institute of Agricultural Engineering, Nabi Bagh, Bhopal–462 038, India
E-mail: [1]lpgite@ciae.res.in, [2]jm78ergo@gmail.com

ABSTRACT

In India, the traditional agriculture utilized mainly manual and animal power whereas use of mechanical power has also come on a large scale. Equipment for different agricultural operations and suitable for manual, animal and motor power is commercially available in the country and many more are being developed in various research organizations. For design of this equipment, it is necessary to give due consideration to the anthropometric and strength data of agricultural workers so as to have enhanced performance along with better comfort and safety of operators.

The All India Coordinated Research Project on Ergonomics and Safety in Agriculture through its cooperating centres as well as ICAR adhoc research schemes collected anthropometric data of 12525 agricultural workers on 79 body dimensions and strength data of 5937 agricultural workers on 16 strength parameters useful in farm equipment design from 12 states of India. The mean stature and weight of agricultural workers ranged 163.3 ± 6.8 cm and 54.7 ± 8.7 kg for male workers and 151.5 ± 6.1 cm and 46.3 ± 7.8 kg for female workers.

The 16 strength parameters on which the data were collected, include hand grip strength, push and pull strength with both hands/one hand in standing/sitting postures, leg as well as foot strength and torque strength in standing and sitting postures.

In the data bank, the data have been arranged state-wise and sex-wise. All India values for mean, standard deviation, 5th percentile and 95th percentile have also been calculated and given for male as well as female workers. The paper gives the methodology adopted for data collection, compilation and analysis. An example for application of these data in farm equipment design is also presented.

Keywords: Anthropometry, Muscular Strength, Agricultural Workers, Data Bank, Farm Women.

INTRODUCTION

In agriculture, the farmers and labourers work with hand tools as well as hand operated, animal drawn and power operated equipment to carry out various field operations. Many times it has been noticed that occupational health problems and low levels of productivity are the results of inadvertent neglect of ergonomical aspects in the design of equipment or workplace. Therefore, ergonomics has a very important role to play in the design and use of agricultural equipment for better performance as well as comfort.

Anthropometric measures vary considerably with gender, race as well as age. Therefore, it is extremely important to give due consideration to these aspects while collecting the anthropometric data. In western countries a large amount of anthropometric data are available for reference and use. Some anthropometric data on Indian subjects are available at Defence Institute of Physiology and Allied Sciences, Delhi. However, these data are on armed forces people. The National Institute of Design, Ahmedabad has published a monogram on anthropometric data of Indians. They have given data on 1000 subjects all over the country from student-community as well as other occupational groups. There are very few studies available on anthropometric data on Indian agricultural workers (Gite and Singh, 1997). Again most of these are case studies and generally only male workers have been covered in these studies.

METHODS

Considering all the above points, a coordinated programme for collection of anthropometric and strength data on agricultural workers of India was worked out. It included measurement of 79 body dimensions and 16 strength parameters useful in farm equipment design.

Methodology for Data Collection

A major anthropometric concern in design and evaluation of engineering products is the statistical description of the user population. It is usually impossible or uneconomical to measure every individual in a population. Therefore, a sample is selected and measurements are carried out on the individuals in that sample. The larger the sample selected, the more likely it is to be representative of the total population. However, cost and time becomes the limiting factors. The sample must be large enough to assure the accuracy and representativeness of the results while being small enough for the survey to stay within budget and time schedule constraints (NASA, 1978). Considering all these aspects, a sample size of 12200 was taken for anthropometric survey and 4500 for strength survey.

Collection of Anthropometric Data

AICRP on ESA has seven centres, *viz.* CIAE, Bhopal (Madhya Pradesh), TNAU, Coimbatore (Tamil Nadu), OUAT, Bhubaneswar (Orissa), PAU, Ludhiana (Punjab), IIT, Kharagpur (West Bengal), NERIST, Nirjuli (Arunachal Pradesh) and MPUAT, Udaipur (Rajasthan). These centres have collected data for their respective states. Adhoc schemes in GAU, Junagarh (Gujarat), ICAR Research Complex, Barapani (Meghalaya), CAET, Etawah (Uttar Pradesh), KKV, Dapoli, MAU, Parbhani (Maharashtra) and SKUAST, Srinagar (Jammu & Kashmir) collected anthropometric and strength data of agricultural workers.

For measurement of body dimensions the equipment used was either the Harpenden Anthropometer or Siber Hegner Anthropometer or IIT Kharagpur Anthropometer. For measurement of strength parameters, a strength measurement set-up developed at CIAE Bhopal was used. The anthropometric data bank was created at CIAE Bhopal and the raw data from the centres were compiled and stored in the bank. The data collected in anthropometric data bank were analyzed and presented in the form necessary for use in agricultural equipment design. The data available here are state wise, region wise as well as on the gender basis. The data available on body dimensions and strength parameters include mean, standard deviation, range, 5th and 95th percentile values (Gite & Majumder, 2005).

RESULTS

Till date data on 79 body dimensions for 12525 agricultural workers from twelve states namely Arunachal Pradesh, Gujarat, Jammu & Kashmir, Madhya Pradesh, Maharashtra, Meghalaya, Mizoram, Orissa, Punjab, Tamil Nadu, West Bengal and Uttar Pradesh have been collected. Of these 8025 were male workers and 4500 were female workers. The strength data on 16 parameters have been collected for 5937 workers (3423 male and 2514 female) from six states namely Madhya Pradesh, Maharashtra, Tamil Nadu, Gujarat, Jammu & Kashmir and Orissa (Gite, 2007). Based on these data, the mean stature and weight of agricultural workers worked out to be 163.3 ± 6.8 cm and 54.7 ± 8.7 kg for male workers and 151.5 ± 6.1 cm and 46.3 ± 7.8 kg for female workers. The age of the workers surveyed varied from 16 to 65 years. The mean values of push strength with both hands in standing posture were 224 ± 56 N and 143 ± 39 N for male and female workers respectively. The detailed data on various body dimensions and strength parameters have been analyzed and compiled.

Application of Anthropometric and Strength Data in Wheel Hoe Design

Weeding is one of the important field operations. It consumes about 15% of the total energy spent in crop production. The traditional tools for weeding is hand hoe (Khurpi) and generally women carry out weeding operation using this tool in squatting posture. As output is low, in a typical situation, about 250 man hours are required to cover one hectare area. As the time for weeding is limited wheeled type manual weeders (wheel hoe) have been developed at various institutions and at present more than 15 designs of wheel hoe are available in the country. These types of weeder can be operated by a person in standing posture and only 50 to 100 man hours are required to cover one hectare area. Though these wheel hoes are having higher output, their adoption in field is yet to pick up. Many times, the workers complain of fatigue during operation of the wheel hoes.

Criteria for Ergonomic Design

- Design within the capabilities of human worker.
- Use of proper posture of the operator for most efficient performance of the tool at a lesser fatigue.
- Suitability of the tool for workers of varying age and body dimensions.

Force Limits for Standing Work

For better muscular efficiency, the dynamic effort of a repetitive nature should not exceed 30% of 5th percentile of force exertion. As per the data collected, the 5th percentile values of push and pull force with both hands in standing posture for female workers are 79 N and 95 N, respectively. Therefore, weeding tools should be such that the operator does not have to exert more than 24 N push or 29 N pull. If more force is needed than this, the operator will have to take frequent rest breaks in between the work schedule depending on severity of work.

Handle Height: Experiments have shown that for operation of push-pull type weeder, 0.7 to 0.8 of the shoulder height offers the least fatigue during operation. Its corresponding height ranges from 93 to 103 cm, *i.e.* 0.8 of the 5th percentile value of shoulder height for female workers and 0.7 of the 95th percentile of male workers.

Handle Length: Length of the handle tool and the angle of operation are interdependent. Angle of operation is based on functional design and geometry of the tool and generally lies between 300 and 450 degree. Once the height of handle is determined and the point of attachment is known, the length of handle can be calculated.

Handle Cross Bar Length: The wheel hoe is operated in push and pull mode. Here, the hands positioning should be such that they are close to their natural position. Thus the length of the cross bar will depend on elbow-elbow breadth. The 5th percentile and 95th percentile values of elbow-elbow breadth for female workers are 29 cm and 41 cm, respectively. Corresponding values for male workers are 30 cm and 45 cm. Therefore, it is proposed that the handle arm can be taken as 40 cm.

Handle Grips: Cylindrical shape of handle grip is recommended. The diameter of grip should be such that while an operator grips the handle, his longest finger should not touch his palm. At the same time, it should not exceed the internal grip diameter. Therefore, the grip diameter should not exceed 35 mm. The length of grip should be 90 mm.

Posture: A good working posture is one which can be sustained with a minimum of static muscular effort and in which it is possible to perform the given task more effectively and with least muscular discomfort. It has been observed that weeding either in squatting or bending posture did not cause a marked difference in energy expenditure. But, the drudgery caused due to bending is reflected in terms of postural discomfort experienced by the worker. Therefore, when work can be done in sitting/standing posture, it should not be done in a bending posture. Use of long handled tools can help to avoid the bending posture of the operator during work.

CONCLUSION

An effort is made in this paper to impart brief information about the anthropometric and strength data generated for agricultural workers through a countrywide survey. An example of application of these data in wheel hoe design has also been presented here.

REFERENCES

[1] Gite, L.P. (2007). Project Coordinator's Report of AICRP on Ergonomics and Safety in Agriculture, Technical Bulletin No. CIAE/ESA/2007/346, Central Institute of Agricultural Engineering, Bhopal, pp. 9–10.

[2] Gite, L.P. and Majumder, J. (2005). Anthropometric and Strength Survey of Indian Agricultural Workers, *Proceedings of International Conference on Humanizing Work and Working Environment*, IIT Guwahati, Dec. 10–12, 2005, pp. 755–760.

[3] Gite, L.P. and Singh, G. (1997). Ergonomics in agriculture and allied activities in India, *Technical Bulletin No. CIAE/97/70*, Central Institute of Agricultural Engineering, Bhopal, p. 51.

[4] NASA (1978). Anthropometric Source Book, Volume II: *A handbook of anthropometric data,* NASA reference publication 1024, National Aeronautics and Space Administration, Washington, USA, p. 421.

REFERENCES

SECTION–II

Work and Sports Physiology

Anaerobic Threshold and Oxygen Consumption as a Measure of Training Status in Females

Ajita D. Singh[1] and Asis Goswami[2]

[2]Department of Sports Sciences, Punjabi University, Patiala–147 002, India
[2]Sports Authority of India, Kolkatta–700 091, India
E-mail: [1]ajita_dsingh@yahoo.co.in

ABSTRACT

Exercise is a precise tool for studying functioning of human organism. The importance of physical performance has been recognized from the studies on economical and optimal work production to achieve high performance in sports. There are many factors such as body composition, physiological functions, psychological state, environment and socio-cultural context, which play a pivotal role in performance in every sphere of life. Physical work capacity of the performer depends upon his aerobic capacity, therefore maximal oxygen uptake has been used extensively as an objective measurement of physical work capacity for both trained and untrained subjects. Trained subjects accumulate less lactate than untrained subjects at a given submaximal workload. This has prompted the consideration of anaerobic threshold as a determinant of fitness. Furthermore, the concept of anaerobic threshold is more appealing because it may be more sensitive to training-induced adaptations than VO_{2max} alone. Therefore, in the present study, an endeavour has been made to study maximum aerobic capacity as well as anaerobic threshold level of different sportive events. A total of 278 subjects have been studied belonging to different sportive events. The national campers of Sports Authority of India were recruited as the subjects for the present study. It was observed that the middle distance runners possessed the highest values of VO_{2max} *i.e.*, 51.2 ml/kg/min, followed by the hockey players (45.9 ml/kg/min), then basketball players (42.2 ml/kg/min) and then football players (39.5 ml/kg/min). Actually middle distance runners undergo endurance type of training and endurance training increases the capillary and mitochondrial density in skeletal muscles and these changes contribute to a higher rate of oxygen consumption at the onset of exercise. The highest vales of oxygen consumption at maximum level as well as anaerobic threshold level of hockey players contribute to the fact that with the advancement of technology, *i.e.* with the replacement of grassy field with the Astroturf the game is now much faster and the energy demands of the game has been increased.

Keywords: Aerobic Threshold, Players.

INTRODUCTION

Physical work capacity of a performer depends upon his aerobic capacity. Therefore, maximal oxygen uptake has been used extensively as an objective measurement of physical work capacity for both trained and untrained subjects (Astrand and Rodahl, 1986).

It is a fact that the maximum oxygen consumption values of endurance athletes are about twice that of normal persons. The major contributing factors are the genetic endowment and the type of training undertaken. The maximum oxygen consumption values can give the information about the training status of athletes. It has been reported by many studies that individuals with similar VO_{2max} have variability in endurance capacity and that highly trained athletes perform at a higher percentage of their VO_{2max} with minimal lactate accumulation (Tanaka *et al.*, 1984; Withers *et al.*, 1981). Furthermore, trained subjects accumulate less lactate than untrained subjects at a given submaximal workload. This has prompted the consideration of anaerobic threshold as a determinant of fitness. During exercise of increasing intensity there is a rise of blood lactate concentration and this response was first reported half a century ago (Owles, 1930; Bang, 1936). While appearance of lactate in blood during exercise is the result of an increased glycogenolysis, it is important to recognise that its concentration, at any time, is the result of balance between the rate of production and removal (Brooks, 1986).

Nevertheless, during exercise of increasing intensity, the rise in blood lactate concentration is an indication of increase of glycogen metabolism. This increase in blood lactate has been interpreted as a reflection of the onset of hypoxia in skeletal muscles and the exercise intensity at which anaerobic metabolism complements the regeneration of ATP by aerobic metabolism has been called the anaerobic threshold. The anaerobic threshold *vis-à-vis* the oxygen consumption at anaerobic threshold level, is also an important parameter to consider. A high aerobic-anaerobic transition level (AT-Level) reflects the high fractional utilisation of oxygen for an endurance activity without building up significant amount of anaerobic metabolites. Thus, a runner who is able to run with a high percentage of VO_{2max} (closer to his maximum capacity) without building up anaerobic metabolites like lactic acid will certainly perform better. Thus, anaerobic threshold is another most important parameter to determine the training status of athletes.

With increasing involvement of young children in competitive sports there have been many studies on cardiorespiratory response of children to endurance training. Therefore, it is very important to study that how these physiological parameters vary with age. The performance of an individual depends upon his physiological parameters like aerobic capacity, fractional utilisation of VO_{2max} and anaerobic threshold. These are very important or we can say that the determining factors of the performance or training status in continuous or endurance events.

For non-continuous or intermittent events also these are very important as aerobic capacity is important for maintaining the basic endurance and anaerobic threshold is used as a training tool.

Keeping in view the above points, the present study aims to evaluate the AT of different groups of sportswomen and to assess the age related variations in AT level.

MATERIAL AND METHODS

Subjects

A total of 278 subjects have been covered in different age groups. The incumbents of various sports schemes of Sports Authority of India have been taken as the experimental group. Higher age group subjects have been recruited from state and national level athletes.

Table 1: Number of Subjects Covered in each Age Group

Age Group	*No. of Subjects Covered*
Group I (8–12)	30
Group II (13–16)	88
Group III (17–20)	117
Group IV (21–25)	43
Total	**278**

Table 2: Number of Subjects Covered in each Event

Events	*No. of Subjects*
Middle Distance Runners	60
Basketball	14
Hockey	92
Volleyball	47
Football	35
Table Tennis	11
Weight Lifting	19

Exercise Protocol

The subjects were administered a graded exercise protocol on a Bicycle Ergometer to measure the physiological changes during exercise. The initial load of the graded exercise protocol was 1 W/kg with 60 rpm and was increased at a rate of 0.5 W/kg. The subject was asked to perform exercise until his exhaustion. The criteria for arrival of maximal level were taken as:

- Volitional exhaustion by the subject.
- Arrival of plateau of ventilation, HR and oxygen consumption.
- Attainment of age related HR_{max}.

All the physiological parameters, *i.e.* Heart Rate (HR), Ventilation (VE), Oxygen uptake (VO_2), Carbondioxide production (VCO_2) and Respiratory Quotient (RQ) were recorded breath by breath on a computerised Gas Analyzer, Oxycon Champion. (Erich Jaeger, Germany).

Data Analysis

In most of the studies, anaerobic threshold has been studied by measuring blood lactate, the abrupt increase in arterial lactate signifying increasing reliance on anaerobic metabolism or the onset of metabolic acidosis and the method with which it is measured is known as lactate threshold method.

But it can also be determined by the assessment of exchange of gases during exercise and the method is known as Gas Exchange Method (Wassermann *et al.*, 1973; Beaver *et al.*, 1985; Ghosh *et al.*, 1987).

In the present study, the method used to determine Anaerobic threshold was the gas exchange method. Oxygen consumption and heart rate at maximal level and at the level of Anaerobic threshold were calculated and recorded. Maximum oxygen consumption was expressed in absolute and relative basis, *i.e.* in terms of litres per minute reflecting total body energy output or by dividing this value by body weight in kilogram.

RESULT AND DISCUSSION

The results of the study indicate that physical measurements, *i.e.* height and weight show an increasing trend with increasing age and this increase is accelerated at adolescence following the achievement of its maximum level at the age (Table 3). Group II shows the highest values of relative as well as absolute VO_{2max} followed by the subjects of Group III (Table 4). Actually, VO_{2max} depends upon the oxygen transport in the body which can be determined by studying maximum cardiac output and maximum arteriovenous oxygen difference (Bar-or, 1989; Shephard, 1971).

Table 3: Mean and SD Values of Physical Parameters

Age Group	*Age (year) Mean (SD)*	*Height (cm) Mean (SD)*	*Weight (kg) Mean (SD)*
Group I (8–12)	11.9 (0.70)	153.7 (6.45)	42.3 (6.30)
Group II (13–16)	15.4 (0.97)	159.5 (6.80)	50.3 (5.9)
Group III (17–20)	18.7 (1.40)	159.5 (7.14)	52.5 (7.74)
Group IV (21–25)	23.6 (1.62)	157.5 (4.91)	53.9 (6.60)

Table 4: Mean and SD Values of the Physiological Variables at Maximum Level

Age Group	*VO_2 (l/min) Mean (SD)*	*VO_2 (ml/kg/min) Mean (SD)*	*Heart Rate (beats/min) Mean (SD)*
Group I (8–12)	1.68 (0.33)	40.3 (5.7)	192.2 (9.02)
Group II (13–16)	2.22 (0.43)	44.7 (9.30)	189.8 (8.64)
Group III (17–20)	2.20 (0.41)	42.5 (9.32)	185.4 (8.1)
Group IV (21–25)	2.12 (0.31)	40.1 (7.40)	183.5 (10.6)

Similarly, in Table 5, no statistically significant difference has been observed in oxygen consumption and heart rate at anaerobic threshold level, although there is difference in the mean values. It has been observed that oxygen consumption as well as percentage utilisation of oxygen is maximum in case of Group-III. It predicts that Group-III is having maximum lactate tolerance. Oxygen consumption at anaerobic threshold level indicates the fractional utilisation of oxygen without building up of anaerobic metabolites leading to muscle anaerobiosis (sheer fatigue). It has been observed that AT VO_2 is maximum in Group-III and followed by Group-II, Group-I and Group-IV.

Table 5: Mean and SD Values of the Physiological Variables at Anaerobic Threshold Level

Age Group	*VO_2 (ml/kg/min) Mean (SD)*	*AT VO_2% (ml/kg/min) Mean (SD)*	*Heart rate (beats/min) Mean (SD)*
Group-I (8–12)	30.6 (5.33)	76.3 (8.34)	163 (5.0)
Group-II (13–16)	33.0 (6.94)	74.4 (8.54)	161.4 (5.54)
Group-III (17–20)	33.1 (7.41)	78.1 (9.04)	161.4 (5.80)
Group-IV (21–25)	29.4 (5.86)	74.4 (9.13)	159.5 (5.2)

Among the court games, basketball and volleyball, basketball players exhibit higher relative VO_{2max} than the volleyball players. But if we consider absolute VO_{2max}, it is higher in case of volleyball players than the basketball players. Table 6 highlights that these players exhibit lower VO_{2max} values and even it is not comparable to that of previous Indian basketball players as reported by Ghosh *et al.* (1985).

Table 6: Mean and SD Values of the Physical Parameters of the Subjects belonging to Different Sportive Events

Event	*Age (y) Mean (SD)*	*Height (cm) Mean (SD)*	*Weight (kg) Mean (SD)*
Middle distance runners	16.8 (3.13)	156.6 (5.3)	45.5 (4.7)
Basketball	12.5 (1.8)	156.6 (8.2)	43.6 (8.2)
Hockey	16.9 (1.4)	166.3 (6.1)	55.1 (5.3)
Volleyball	17.8 (3.1)	158.3 (5.4)	51.4 (6.6)
Football	21.5 (3.0)	156.3 (5.6)	53.0 (6.3)
Table tennis	17.5 (3.5)	153.0 (7.0)	57.8 (11.6)
Weight lifting	16.8 (3.5)	157.1 (5.6)	49.9 (4.4)

The physiological fitness levels required for playing match depends upon the work rate demands of the game which varies with the level of competition. The present study indicates that aerobic capacity (39.5 ml/kg/min)) as well as anaerobic threshold level (29.4 ml/kg/min) of the footballers is very low and the players with such a level of aerobic capacity cannot fulfil the physiological requirement of the game.

Table 7: Mean and SD Values of the Physiological Parameters of the Subjects belonging to Different Sportive Events

Event	*VO_{2max} (l/min)*	*VO_{2max} (ml/kg/min)*	*AT-VO_2 (ml/kg/min)*	*ATVO_2 %*	*HR_{max} (beats/min)*	*ATHR (beats/min)*
Middle distance runners	2.31 (0.42)	51.2 (7.8)	39.7 (6.2)	78.3 (7.9)	187.7 (7.9)	162.0 (4.8)
Basketball	1.84 (0.44)	42.2 (3.3)	33.7 (2.3)	80.1 (7.1)	192.4 (9.4)	162.3 (5.0)
Hockey	2.11 (0.40)	39.0 (5.0)	30.2 (4.1)	78.2 (8.7)	186.0 (9.6)	161.1 (7.4)
Volleyball	2.35 (0.35)	45.9 (6.2)	35.0 (4.8)	76.7 (7.7)	189.0 (8.0)	162.0 (5.1)
Football	2.10 (0.30)	39.5 (5.1)	29.4 (4.3)	74.7 (9.1)	180.8 (9.7)	158.1 (5.1)
Table Tennis	1.60 (0.22)	28.0 (3.6)	22.0 (2.3)	78.5 (9.2)	187.3 (9.8)	161.3 (3.5)
Weight lifting	1.83 (0.21)	36.8 (5.1)	26.4 (3.1)	72.6 (10.3)	190.3 (10.8)	162.6 (6.6)

CONCLUSIONS

- At any given level, the cardiac output of children is lesser than that of adults reflecting lesser oxygen demands of smaller viscera in case of children. Therefore, VO_{2max} of the children is lesser than that of adults as observed in the present study.
- The Anaerobic Threshold (AT) represents how efficiently the muscles use oxygen to produce energy or work. As such, AT represents the level of work the body can sustain over an extended period. When the body is called upon to perform above AT, lactic acid builds in the muscles creating fatigue (the burning sensation felt). The ability to continue at this level will be very limited (a few seconds to a few minutes, depending on the fitness level). Improving the anaerobic threshold is paramount to develop aerobic fitness. Generally, AT may be between 52 and 95 per cent of maximum heart rate. The more aerobically fit, the higher the AT.

REFERENCES

[1] Astrand, P.O. and Rodahl, K. (1986). *Textbook of Work Physiology: Physiological bases of exercise,* McGraw-Hill, New York.

[2] Bang, O. (1936). "The Lactate Content of Blood during and after Muscular Exercise in Man", *Scand. Arch. Physiol*, 74: 51–82.

[3] Bar-or, O. (1989). Trainability of the prepubescent children, *Sports Med.*, 17: 65–81.

[4] Billat, V.L., Bocqet, V., Slawinski, J., Laffite, L., Demarle, A., Chassaing, P. and Koralsztein, J.P. (2004). Effect of a prior intermittent run at $VO_{2\,max}$ on the oxygen kinetics during an all-out severe run in humans, *Sports Med. Phys. Fitness*, 40: 185–194.

[5] Brooks, G.A. (1984). "Anaerobic threshold: Review of the concept and directions for future research", *Medicine and Science in Sports and Exercise*, 17(1): 22–31.

[6] Bunc, V., Heller, J., Leso, J., Sprynarova, S. and Zdanowicz, R. (1987). "Ventilatory threshold in various groups of highly trained athletes", *Int. J. Sports Med.*, 8: 275.

[7] Ghosh, A.K., Ahuja, A., Khanna, G.L., Kelley, R. and Sam, J. (1985). "Intergroup variations in physiological characteristics of Indian basketball players", *Proc. Fisu Conf,* Kobe Japan.

[8] Ghosh, A.K., Mazumdar, P., Goswami, A. and Khanna, G.L. (1988). "Aerobic-anaerobic transition level of Indian middle & long distance runners", *Indian J. Med. Res.*, 88: 371–375.

[9] Hagberg, J.M., Hickson, R.C. and Ehsam, A.A. *et al.* (1980). "Faster adjustment to and recovery from submaximal exercise in the trained state", *J. Appl. Physiol,* 48: 218–224.

[10] Henry, F.M. (1954). "Aerobic Oxygen Consumption and Alactic Debt in Muscular Work", *J. Appl. Physiol,* 3: 427–438.

[11] Hollmann, W. (1959). "The relationship between pH, lactic acid, potassium in arterial and venous blood, the ventilation and pulse frequency during increasing spiroergometric work in endurance-trained and untrained persons", *Pan-American Congress for Sports Medicine, Chicago.*

[12] Kenney, W.L. and Hodgson, J.L. (1985). "Variables Predictive of Performance in Middle-distance Runners", *Br. J. Sports Med.,* 19: 207.

[13] Khanna, G.L., Ghosh, A.K., Sharma, J.G. and Malhotra, M.S. (1983). "Physiological Status of some Judoists in India", *SNIPES J.,* 6: 21–27.

[14] Knuttgen, H. (1962). "Oxygen Debt, Lactate, Pyruvate and Excess Lactate after Work in of Applied Physiology", 17: 639–644.

[15] Kumagai, S., Tanaka, K., Matsura, Y., Matsuzaka, A., Hirakoba, K. and Asano, K. (1982). "Relationship of the anaerobic threshold with the 5 km, 10 km and 10 mile races", *Eur. J. Appl. Physiol,* 49:13–23.

[16] Owles, W.H. (1930). "Alterations in the Lactate Content of Blood as a Result of Light Exercise and Associated Changes in the CO_2 Combining Power of the Blood and the Alveolar CO_2 Pressure", *J. Physiol.*, 69: 214–237.

[17] Tanaka, K., Matsura, Y., Matsuzaka, A., Hirakoba, K. and Kumagai, S. (1984). "A longitudinal assessment of anaerobic threshold and distance running performance", *Med. Sci. Sports and Exerc.,* 16: 278.

[18] Wasserman, K. and Mcllory, M.B. (1964). "Detecting the Threslold of Anaerobic Mechanism in Cardiac Patients during Exercise", *Am. J. Cardiol.,* 14: 844–859.

[19] Wasserman, K. and Whipp, B.J. (1975). "Exercise Physiology in Health and Disease", *Am. Rev. Res. Disease*, 112: 219–249.

[20] Wasserman, K., Hansen, J.E., Sue, D.Y. and Whipp, B.J. (1994). "Principles of Exercise Testing and Interpretation", *Philadelphia*, Lea & Febiger.

[21] Wasserman, K., Whipp, B.J., Koyal, S.N. and Beaver, W.L. (1973). "Anaerobic Threshold and Respiratory Gas Exchange during Exercise", *J. Appl. Physiol.,* 35, p. 236.

[22] Whipp, B.J. (1981). "The Control of Exercise Hyperpnea. In: The Regulation of Breathing", Editor: Hornbein, T. New York, Dekker, pp. 1069–1139.

[23] Whipp, B.J. and Ward, S.A. (1990). "Physiological Determinants of Pulmonary Gas Exchange Kinetics during Exercise", *Med. Sci. Sports Exerc.*, 22(1): 62–71.

[24] Withers, R.T. and Roberts, R.G.D. (1981). "Physiological Profiles of Representative Women Softball, Hockey and Netball Players", *Ergonomics*, 24: 583–591.

[25] Yoshida, T. and Watari, H. (1993). "Metabolic Consequences of Repeated Exercise in Long Distance Runners", *Eur. J. Appl. Physiol.*, 6(3): 261–265.

Comparison of Aerobic Fitness, Excess Post Exercise Oxygen Consumption (EPOC) and Recovery Patterns of Different Groups of Sub-Elite Sports Persons

Suman Mukhopadhyay[1], Neha Chitkara[2], Ajita[2] and Rupinder Mokha[2]

[1]Physical Ergonomics Laboratory, National Institute of Industrial Engineering, Mumbai, India
[2]Department of Physiotherapy and Sports Sciences, Punjabi University, Patiala, India

ABSTRACT

A study was carried out on 36 male sub-elite sports persons attending various state level camps at Netaji Subhas National Institute of Sports, Patiala, consisting of 12 each of middle distance runners, boxers and judokas, to ascertain among others, the possible correlation between EPOC and VO_{2max} and recovery oxygen kinetics.

Exercise tests were performed on an electronically operated bicycle ergometer till exhaustion and cardiopulmonary transients were recorded using portable computerized metabolic analyzer, CosmedK4, during the entire period of test and 15 minutes of active recovery. EPOC was calculated by the analysis software of K4 system.

The results revealed that VO_{2max} values were comparable in middle distance athletes (54.9 ± 7.712) and boxers (57.04 ± 7.038), and were significantly higher than those of judokas (47.16 ± 6.822). The recovery oxygen kinetics however, differed widely among the three groups, with the middle distance athletes displaying significantly faster decrement in oxygen consumption in to the first minute of recovery than the judokas and the boxers exhibiting significantly higher decrements in the second minute of recovery. No significant differences in the magnitude of EPOC were observed among three groups.

The study elucidated a close association between VO_{2max} and recovery oxygen consumption kinetics, although EPOC was not found to be correlated to VO_{2max}, suggesting that it is more possibly related to lactate acidosis or the degree of anaerobiosis during exercise, than to VO_{2max}.

Keywords: VO_{2max}, EPOC, Bicycle Ergometer, Cardiopulmonary, Anaerobiosis.

INTRODUCTION

Traditionally, the concepts of fitness have often been associated with athleticism. However, the American College of Sports Medicine (ACSM, 1991), describes fitness as the ability to perform moderate to vigorous level of physical activity with undue fatigue. The American Alliance for Health, Physical Education, Recreation and Dance (AAHPERD) has categorized fitness component into two different categories, health related and motor skill related. Health related includes five key components: cardiovascular efficiency, muscular strength and muscular endurance, flexibility and body composition. Motor fitness components such as speed, coordination, reaction time and agility are less important, although important for participation in athletics does not appear to emphasize health related fitness.

Maximum oxygen uptake capacity (VO_{2max}) has long been regarded as the best variable that describes aerobic fitness. It represents the ability of the cardiovascular and respiratory systems to provide for the oxygen needs of the muscular system over a sustained period of time and therefore provides a quantitative statement of an individual's capacity for aerobic energy transfer. High oxygen uptake values are required to achieve higher performance at national and international levels.

During the period of recovery following exercise, although the VO_2 declines, it remains elevated above the resting level for a duration depending on the severity of the exercise. The volume of oxygen consumed during recovery above the resting volume is known as Excess Post-exercise Oxygen Consumption (EPOC). EPOC reflects the body's recovery requirements after exercise in terms of replenishment of the body's oxygen and ATP-CP stores and oxidation of lactate produced during exercise.

A strong relationship between aerobic fitness and VO_{2max} in determining the magnitude of the oxidative response is already established (Tomlin and Wenger, 2001; Mukhopadhyay *et al.*, 2001; Sukhjinder *et al.*, 2004). It would therefore, be pertinent to hypothesize that a greater VO_{2max} should hasten the recovery response following exercise. Moreover, the possible association between VO_{2max}, EPOC and recovery VO_2 kinetics has not been studied, although it is tempting to speculate that the above transients will be intimately related. It is with this background, the present study was undertaken.

MATERIALS AND METHODS

Subjects

The study was conducted on 36 volunteers, all sub-elite sports persons, consisting of three groups, *viz.*, middle distance runners ($n = 12$), boxers ($n = 12$) and judokas ($n = 12$). The volunteers were taken from sub-elite male sports persons belonging to various camps that were held in Sports Authority of India, Netaji Subhas National Institute of Sports, Patiala. All volunteers were adults, within the age range of 18–26 years. Each volunteer was first subjected to a physical examination that included measurement and /or recording of age, stature and weight. Body surface area was computed using prediction equation of Du Bois and Du Bois (1916).

Modalities of the experiment were explained and a signed informed consent was obtained from each of the volunteer. Thereafter, the volunteers were subjected to Cardiopulmonary Exercise Testing (CPET), as described below.

Methods

Volunteers were asked to report to the laboratory at least 3 hours after last meal and 36 hours of limited exertion. Tests were carried out in an air-conditioned room with temperature 23˚C (±2˚C) and relative humidity levels of 55% (±5%).

Exercise test was performed on an electronically operated computerized bicycle ergometer (ER 900: Erich Jaeger, Germany) using a test protocol that consisted of incremental exercise. The volunteers were first asked to warm up and then cycle at 60 RPM for 2 minutes without any load. This constituted the 'reference phase'. The test phase followed next. The initial load in the test phase was fixed at 50 watts, and increased by 50 watts every 2 minutes till exhaustion (Dey, 1993), consequent to which, the volunteers were asked to cycle briskly at a rate of 60 RPM without any load during the entire period of recovery for a total period of 15 minutes (active rest: recovery phase).

All measurements were carried out using the computerized portable metabolic analyzer, Cosmed K4 (Cosmed Srl, Italy). The volunteers breathed through a low resistance mouthpiece and a volume transducer (bi-directional digital turbine). Respired gases were withdrawn automatically from the mouthpiece and passed onto the CO_2 and O_2 analyzer where the values were measured at a preset interval of 15 seconds. Data regarding relative humidity and ambient temperature were fed to the equipment, and the pressure sensors of the equipment measured the atmospheric pressure automatically so that all the gas volumes were automatically corrected for and expressed in STPD. The CO_2 and O_2 analyzers as well as the volume transducers were calibrated before each test using open air calibration (approx 20.95% O_2 and 0.03% CO_2), and a 3-litre calibration syringe (Cosmed srl, Italy), respectively.

A test was considered maximal only if a change in load did not produce a corresponding significant change in heart rate and VO_2 (±4 beats/min and ±100 ml respectively), and if the heart rate values were within 15% of the age-predicted heart rate maximum (Bruce *et al.*, 1974; Jones *et al.*, 1985). The RQ was also greater than 1.13.

The Anaerobic Threshold (AT) point was determined from deviation point of the VE-VO_2 linearity (Beaver *et al.*, 1986) and confirmed from plots of VCO_2 *vs.* VO_2 (modified V-slope technique (Sue *et al.*, 1988). The Excess Post-exercise Oxygen Consumption (EPOC) was calculated by the analysis software of the equipment (K4-rq system, Cosmed srl, Italy).

Statistical Analysis

Means (±SD) of the selected physical and physiological transients were calculated by the statistical software package of Microsoft Office (MS 7.0)/MS excel (Microsoft Corporation, USA).

Students 't-test' was then applied to compare the results and elucidate whether the groups differed significantly or otherwise, with reference to the selected physical and physiological transients.

RVO_2 1, RVO_2 2 and RVO_2 3 were computed as the oxygen consumption in the first, second and third minute of recovery, respectively. δVO_2 1, δVO_2 2 and δVO_2 3 were calculated as the decrement in oxygen consumption in the first, second and third minute of recovery, respectively. Multicorrelations were computed using statistical software package of Microsoft Office (MS 7.0)/MS excel (Microsoft Corporation, USA), among VO_{2max}, AT–VO_2, RVO_2 1, RVO_2 2, RVO_2 3, δVO_2 1, δVO_2 2 and δVO_2 3 to elucidate the possible relationship between VO_{2max} and the recovery oxygen transients.

RESULTS AND DISCUSSION

Table 1 compared the mean (±SD) of the physical/corporal data of the athletes who participated in the study, *viz.*, sports persons belonging to middle distance running (n = 12), boxing (n = 12) and judokas (n = 12). The height, weight and the BSA did not differ significantly amongst the groups.

Table 1: Comparison of Means (±SD) of Physical Profiles of Sub-Elite Athletes of Different Groups under Study

Event	*(n)*	*Age (yrs)*	*Height (cm)*	*Weight (kg)*	*BSA (sqm)*
Middle distance runners	12	22.58 (±3.80)	175.75 (±4.663)	66.33 (±7.726)	1.82 (±0.134)
Boxers	12	20.25 (±0.753)	172.17 (±8.518)	65.75 (±16.125)	1.77 (±0.227)
Judokas	12	23.92 (±2.891)	170.5 (±6.929)	70.92 (±9.331)	1.82 (±0.142)
Grand total	36	22.23 (±1.856)	172.857 (±6.927)	67.66 (±2.829)	1.80 (±0.028)

Table 2 compared the mean (±SD) of the physiological transients among the three groups of sports persons at the AT level. When students' 't-test' was conducted, the results revealed that the AT-VO_2 was found to be highest in the boxers, significantly so of middle distance runners and judokas. This indicated that the boxers were endowed with a significantly larger aerobic zone as well as a significantly higher level of fractional utilization of oxygen consumption at the level of the anaerobic threshold. A higher aerobic-anaerobic transition zone represents a large aerobic component, highly correlated with endurance performances (Kumagai, 1982). An earlier study by Mukhopadhyay *et al.* (2001) found that elite boxers had significantly higher AT-VO_2 as compared to elite judokas and kabaddi players. When Excess Post-exercise Oxygen Consumption (EPOC) was computed (both in ml and ml/kg), using analysis software of Cosmed K4 analyzer, no significant differences were observed among the three groups. Since, EPOC is primarily related to rephosphorylation of ATP-CP stores and the magnitude of oxidation of lactate produced during exercise, the result suggests that the three groups of athletes share similar or nearly similar degree of anaerobiosis. This follows that sports persons with different endurance capacities can have similar or nearly similar EPOC if they undergo a similar degree of anaerobiosis during exercise, which may be quite independent of their VO_{2max}.

Table 2: Comparison of Means (±SD) of Some Selected Physiological Transients of Sub-Elite Athletes of Different Groups under Study at the AT Level

Event	AT VO_2 (l/min)	AT VO_2 (ml/kg/min)	AT HR (beats/min)	AT VE (l/min)	EPOC (ml/min)	EPOC (ml/kg)
Middle distance runners	2.441 (±0.395)	34.60 (±7.0177)	154.25 (±10.661)	66.25 (±15.937)	5047.417 (±1904.93)	67.174 (±32.973)
Boxers	2.849 (±0.487)	4.030 (±7.222)	159 (±14.109)	73.366 (±10.309)	4665.583 (±2301.268)	72.958 (±38.368)
Judokas	2.450 (±0.4249)	34.64167 (±4.625)	157.75 (±13.071)	67.666 (±11.713)	5959.5 (±1821.363)	87.166 (±34.311)

Table 3 compared the maximum oxygen uptake capacity (VO_{2max}) recovery VO_2 in first, second and third minute of recovery (RVO_2 1), RVO_2 2 and RVO_2 3 respectively, as well as the decrement in oxygen consumption from maximum to first minute, first minute to second minute and second minute to third minute (δVO_2 1), (δVO_2 2) and (δVO_2 3) of recovery respectively.

Table 3: Comparison of Means (±SD) of Recovery O_2 Consumption and O_2 Decrement Kinetics among the Different Groups of Sub-Elite Athletes

Event	VO_{2max} (ml/kg/min)	R VO_2 1 (ml/kg/min)	R VO_2 2 (ml/kg/min)	R VO_{2max} 3 (ml/kg/min)	δ VO_2 1 (ml/kg/min)	δ VO_2 2 (ml/kg/min)	δ VO_2 3 (ml/kg/min)
Middle distance runners	54.879 (±7.712)	43.550 (±10.571)	24.775 (±9.7551)	14.615 (±3.929)	11.245 (±6.117)	18.834 (±9.602)	10.16 (±7.692)
Boxers	57.038 (±57.038)	51.236 (±6.937)	25.082 (±6.304)	15.754 (±3.654)	7.34 (±4.314)	26.451 (±4.691)	9.329 (±4.787)
Judokas	47.159 (±6.822)	41.334 (±6.937)	22.180 (±4.235)	15.406 (±3.501)	6.140 (±2.4944)	20.029 (±6.335)	6.7658 (±1.782)

When students, 't-test' was computed, the results revealed that the boxers displayed the highest maximum oxygen uptake capacity, significantly higher than both judokas ($p < 0.01$) as well as middle distance runners ($p < 0.05$). This again exhibited that the aerobic component is most well pronounced in boxers. The VO_{2max} values of the sub-elite boxers of the present study were, as a matter of fact, comparable to those demonstrated by the elite boxers in the study of Mukhopadhyay *et al.* (2001).

The recovery oxygen consumption $RVO_2 1$ in the first minute was found to be highest in the boxers significantly so from middle distance runners ($p < 0.05$) and judokas ($p < 0.01$). The δVO_2 1, or decrement in oxygen consumption in first minute recovery was found to be maximum in middle distance runners than judokas ($p < 0.05$) as well as boxers (NS) indicating, a faster recovery in the middle distance runners than the other two groups, during the first minute of recovery.

The RVO_2 2 (recovery oxygen consumption in the second minute), did not vary significantly among the groups. However, the δVO_2 2 (decrement in oxygen consumption between first and second minute of recovery) was observed to be quickest in boxers, significantly so, from middle distance runners ($p < 0.05$) and judokas ($p < 0.01$). There was however, no significant observable difference in decrement of oxygen consumption in the second minute between middle distance runners and judokas.

The RVO_2 3 or the recovery oxygen consumption in the third minute was also found to be comparable and no significant difference were observed among the groups. However, the δVO_2 3 was found to be maximum in middle distance runners followed by boxers, whereas the judokas displayed the least decrement but the difference was not found to be significant.

When multicorrelations were computed, it was found that $\delta VO_2 1$ and δVO_2 2 were well correlated with VO_{2max} indicating that a larger VO_{2max} ensure a quicker recovery. This is in agreement with the earlier studies by Tomlin and Wenger (2001), Mukhopadhyay *et al.* (2001) and Sukhjinder *et al.* (2004).

The overall trend noticed in the recovery oxygen response was that both the boxers and the middle distance runners had quicker recovery VO_2 dynamics, when compared to judokas. The boxers however outdid the middle distance runners also, possibly as a long term effect of specificity of training in augmenting a quicker recovery response.

Table 4: t-values of Recovery O_2 Consumption and O_2 Decrement Kinetics among the Different Groups of Sub-Elite Athletes

Event	*VO_{2max} (ml/kg/min)*	*R VO_2 1 (ml/kg/min)*	*R VO_2 2 (ml/kg/min)*	*R VO_2 3 (ml/kg/min)*	*δ VO_2 1 (ml/kg/min)*	*δ VO_2 2 (ml/kg/min)*	*δ VO_2 3 (ml/kg/min)*
Middle distance runners	–1.22594	–2.14561*	–0.0917	–0.73538	1.807039	–2.4691*	0.317631
Boxers	3.89483**	3.531525**	1.33316	–0.23957	0.835332	2.83733**	1.747724
Judokas	–2.47044*	–0.59984	0.92243	0.524585	–2.6789*	0.360763	–1.48964

* $p < 0.05$, ** $p < 0.01$

CONCLUSIONS

The study concluded that the boxers were endowed to the highest oxygen uptake capacity and significantly higher fractional oxygen utilization capacity at the AT level. This group also displayed significantly greater levels of decrement in the recovery oxygen consumption dynamics. A strong correlation was also exhibited between VO_{2max} and δVO_2 1 and δVO_2 2 indicating that a larger VO_2 can indeed hasten recovery VO_2 kinetics. Further, although the groups significantly differed in their VO_{2max}, they exhibited comparable EPOC suggesting that the degree of anaerobiosis tolerated or lactate produced to be the reason rather than their maximum aerobic capacity. A bigger study involving a larger sample size may be taken up to bring out the kinetics of lactate disappearance during recovery and its possible relationship with EPOC.

REFERENCES

[1] Beaver, W.L., Wassermann, K. and Whipp, B.J. (1986). "A new method for detecting anaerobic threshold by gas exchange", *J. Appl. Physiol*, 60: 2020–2027.

[2] Bruce, R.A., Gey Jr., G.O., Cooper, M.N., Fisher, L.D. and Peterson, D.A. (1974). "Seattle heart watch: Initial clinical, circulatory and electrocardiographic responses to maximal exercise", *Am. J. Cardiol*, 33: 459–565.

[3] Dey, S.K., Khanna, G.L. and Batra, M. (1993). "Morphological and physiological studies on Indian National kabaddi players", *Br. J. Sp.*, 27(4): 237–242.

[4] DuBois, D. and DuBois, E.F. (1916). "A formula to estimate the approximate surface area if height and weight be known", *Arch Int Med.*, 17: 863–71.

[5] Jones, N.L., Makrides, L., Hitchcock, C., Chypchar, T. and McCartney, N. (1985). "Normal standards for an incremental progressive cycle ergometer test", *Am. Rev. Respir. Dis.*, 131: 700–708.

[6] Kumagai, S., Tanaka, K., Matsura, Y., Matsuzaka, A., Hirakoba, K. and Asano, K. (1982). "Relationship of anaerobic threshold with the 5 km, 10 km and 10 mole races", *Eur. J. Appl. Physiol,* 49: 13–23.

[7] Mukhopadhyay, S., Verma, S.K. and Sidhu, L.S. (2001). "A comparison of selected cardiopulmonary responses of elite Indian combative sports discipline of kabaddi, judo and boxing", Chapter 10 in *The Science of man in service of man*, Ed: Dr. M.K. Bhasin and Dr. S.L. Malik, Department of Anthropology, University of Delhi, India.

[8] Sue, D.Y., Wasserman, K., Moricca, B. and Casaburi, R. (1988). "Metabolic acidosis during exercise in patients with chronic obstructive pulmonary disease", *Chest.*, 94: 1143–1150.

[9] Sukhjinder, Mukhopadhyay S., Sood, Monica and Verma, S.K. (2004). "Aerobic fitness and recovery cardiopulmonary dynamics: A comparison between sprinters and long distance athletes", *presented in the Physiology section of Indian Science Congress,* 3–8 Jan., 2004.

[10] Tomlin, D.L. and Wenger, H.A. (2001). "The relationship between aerobic fitness and recovery from high intensity intermittent exercise", *Sports Med.*, 31(1): 1–11.

A Comparison of Cardiopulmonary Fitness Levels and Recovery Responses among Sedentary Indian Males and Females

Sanchita Ghosh[1], Suman Mukhopadhyay[1], Amitabha De[1] and Tushar Kanti Ghosh[2]

[1]TIFAC-CORE, NITIE, Mumbai, India

[2]Department of Physiology with Community Health
Vidyasagar University, Midnapore, India

ABSTRACT

A nation's health and fitness level of its people are intimately related. Cardiovascular fitness, synonymous with aerobic fitness is perhaps the single most important determinant of physical fitness, representing the cardiovascular system's ability to supply the O_2 requirement of the muscular system over a sustain period of time. VO_{2max} has been considered as an excellent predictor of cardiovascular fitness. A low level of cardiorespiratory fitness has been implicated with increased morbidity and mortality risks and a lower quality of life. In India, death due to coronary heart disorder has risen from mere 4% to a staggering 33% in the last 30 years.

Cardiopulmonary dynamics during the process of recovery, following exhaustive exercise, is perhaps as important as those during exercise. The ability to recover quickly is also an indicator of cardiopulmonary fitness status. It is already known that persons predisposed to a higher level of VO_{2max} can recover faster.

A study was therefore conducted on 15 male and 15 female college going students to ascertain the cardiovascular fitness levels and to decipher their recovery cardiopulmonary kinetics, so that those can be compared with cardiorespiratory fitness levels of developed countries.

The results reveled, among others, that predictably, the VO_{2max} was significantly higher in Indian males. However, interestingly, the females displayed a significantly higher fractional utilization of O_2 consumption at the anaerobic threshold. The recovery O_2 kinetics also was significantly different between the males and females implying different fitness levels among them.

Keywords: Cardiovascular Fitness, VO_{2max}, Anaerobic Threshold.

INTRODUCTION

A nation's health and fitness levels of its people are intimately related. A nation comprising healthy or physically fit people will be able to promote and sustain growth and development, improving the production as well as contributing to the economy of the country. Fitness is a trait that may depend on personal characteristics like age, sex, family history and/or lack of physical activity. The physiological transients that may affect fitness levels include blood pressure, cholesterol level and obesity. Further, social determinants like a poor social and economic condition, poverty and unemployment, among others, can negatively influence the fitness and health of a population.

What is Fitness

Conventionally, the concept of fitness and fitness tests are associated with athleticism. However the Oxford Dictionary's definition of fitness is simply, 'Good Condition' or 'Good Health'. Likewise, the American College of Sports Medicine describes fitness as the ability to perform moderate to vigorous level of physical activity without undue fatigue. According to Brynteson and Brynteson (1993), fitness is the capacity to adjust to increased demands of the body's cells for oxygen.

Components of Fitness

The AAHPERD (American Alliance for Health, Physical Education, Recreation and Dance) has categorized fitness into two different components—Health related and Motor Ability related. Health related fitness includes five key components: cardiovascular efficiency, muscular strength, muscular endurance, flexibility and body composition. Motor ability components are speed, coordination, reaction time and agility. Additionally, health related fitness is also associated with a selectively stress free environment and, among others, an optimal level of hemoglobin, cholesterol and triglycerides in the blood.

Aerobic Fitness or Cardiopulmonary Fitness

The term 'Aerobic fitness', however, is probably the single most important component of physical fitness. It represents the cardiovascular and respiratory systems' ability to supply the oxygen needs of the muscular systems over a sustained period of time. A number of other terms are associated with aerobic fitness, like cardiopulmonary fitness, cardio-respiratory endurance, aerobic endurance and physical work capacity, and essentially have the same meaning.

Evaluation of Cardiopulmonary Fitness

The most common measure of cardiopulmonary fitness that evaluates the effectiveness of the physiological functions, particularly the cardio-respiratory fitness, is maximal oxygen consumption or VO_{2max}. Taylor *et al.* (1955), defined maximum aerobic capacity or VO_{2max} as the capacity of an individual to consume maximum amount of oxygen from outer atmosphere at a maximal effort in the last thirty seconds of an incremental exercise at sea level. It is well documented that aerobic fitness alone, whether measured directly or estimated, is an excellent predictor of cardiopulmonary fitness (Myers, 2002, Kavanagh, 2002).

Studies (ACSM, 2006) have shown that the primary cardiopulmonary and cardio-respiratory fitness parameter, or the maximum aerobic capacity (VO_{2max}), as well as others like ventilation (VE), blood levels of steroid hormones, may have relations with cardiovascular and pulmonary health. Low cardiopulmonary fitness level increases morbidity and mortality risk and lowers the quality of life (Blair *et al.*, 1989, Wasserman *et al.*, 1987, ACSM, 1993).

The importance of assessment of cardiopulmonary fitness assumes more significance because cardiovascular disease in recent years is reported to be one of the leading causes of human mortality. In 1990, 5.3 million deaths had occurred in the developed countries due to cardiovascular disease, of which 2.7 million and 1.4 million deaths were due to primary Coronary Heart Disease (CHD) and stroke respectively. Medical records compiled by the Institute of Preventive Cardiology (India) had indicated that the percentage of death due to CHD had risen from 4% to 33% in India over last 30 years (IPC, 2005). Further it has also been reported that during the last 25 years (1979–2004), the nature of jobs in India has changed substantially. The percentage of people employed in agriculture has decreased from 64% to 54%, while the employment in manufacturing sector and the service sector has increased from 15% to 19% and 20% to 27% respectively. The service sector has grown at a faster pace than the manufacturing sector. It is expected that by the year 2020, the employment pattern in India will be around 44% in agriculture, 21% in manufacturing and 35% in service sectors (Kalam, 2005). The jobs in service sector are predominantly desk jobs or assisted by mechanization and automation. Those are mostly sedentary in nature and are less physically demanding due to the change in the nature of job and in many cases human power is substituted by the machine power.

Role of Aerobic Fitness in Enhancing Recovery

The return of the muscle homeostasis to its pre-exercise state following exercise is known as recovery. A strong relationship between aerobic fitness and recovery from high intensity intermittent exercise (Tomlin and Wenger, 2001), and from continuous graded exercise till exhaustion (Mukhopadhyay *et al.*, 2005) has already been established. Some studies have also supported an association between aerobic fitness and lactate removal (Tomlin and Wenger, 2001).

Aerobic fitness has been shown to enhance recovery from high intensity intermittent exercise through increased aerobic response, improved lactate removal and enhanced phosphor creatine regeneration (Tomlin and Wenger, 2001). Individuals with high maximal aerobic power exhibit increased concentration of aerobic enzymes, increased mitochondrial number, size and surface area, contributing to an improved oxygen extraction by muscle (Holoszy and Coile, 1984).

The ability to recover quickly is also an indicator of cardiopulmonary fitness status and it is already known that persons predisposed to higher level of VO_{2max} can recover faster.

With this background therefore, a study was conducted on 15 male and 15 female college going students, with the following aims and objectives:

- To objectively quantify the aerobic (cardiopulmonary) fitness levels of the Indian adult sedentary population (male and female).
- To compare these data with those of other International population to understand the levels of fitness of the Indian population *vis-à-vis* their international counterparts.
- To ascertain the kinetics of recovery oxygen consumption dynamics and to comprehend whether the recovery O_2 consumption dynamics can be used as a predictor or pre-requisite of cardiopulmonary fitness status.

MATERIALS AND METHODS

Subjects

15 male and 15 female volunteers between 20 and 29 years of age were randomly selected from post graduate students of NITIE, Mumbai. No previous history of serious disorders was found. The volunteers never participated in sports/physical activity (except for recreational purposes). Modalities of exercise were explained and 'signed informed consent' was obtained from the volunteers.

Each volunteer was first subjected to a physical examination that included measurement and/or recording of the following: Date of birth, Stature, Weight, Body Mass Index (BMI) and Body Surface Area (BSA). Those were computed by using standard derivative equations.

Methods

Volunteers were asked to report to the laboratory at least 3 hours after last meal and 36 hours of limited exertion. Tests were carried out in an air-conditioned room with temperature 23°C (±2°C) and relative humidity levels of 55% (±5%) controlled.

Cardio-pulmonary Exercise Testing (CPET) was conducted which involved graded maximum treadmill (H/P/COSMOS, COS10198, Germany) running/walking, till exhaustion. Volunteers breathed through a low resistant facemask and expired air was passed into the computerized breath-by-breath metabolic analyzer, "Oxycon Pro System" (Viasys Healthcare GmbH, Germany). All gas volumes were automatically corrected for, and expressed in STPD, and VE was expressed as, VE BTPS. Oxycon PRO was calibrated before each and every testing using a precision gas mixture (Viasys Healthcare GmbH, Germany).

A test was considered maximal only if a change of load did not produce a corresponding significant change in heart rate (±4 beats per minute) and if the heart rate values were within 15% of the age predicted HRmax. The RQ was greater than 1.13 also. A change of load did not produce a corresponding significant change in VO_2 (±100 ml or 2.1 ml/kg/min) and the ST segment of ECG depressed by at least 0.5 mm. (Bruce *et al.*, 1974; Jones *et al.*, 1985).

The Anaerobic Threshold point was determined from the deviation point of VE-VO_2 linearity (Beaver *et al.*, 1986) and confirmed from plots of VCO_2 Vs VO_2 (modified

V-Slope technique, Sue *et al.*, 1988). Volunteers were given passive recovery comprising 5 minute of standing on the stationary treadmill.

RVO_2 1, RVO_2 2, RVO_2 3 were computed as the recovery O_2 consumption at the 1st, 2nd and 3rd minutes of recovery respectively.

$\delta\ VO_2$ was calculated as below:

$VO_{2max} - RVO_2\ 1 = \delta\ VO_2\ 1$ (ml/kg/min)
$RVO_2\ 1 - RVO_2\ 2 = \delta\ VO_2\ 2$ (ml/kg/min)
$RVO_2 2 - RVO_2\ 3 = \delta\ VO_2\ 3$ (ml/kg/min)

Statistical Analysis

Means (±SD) of the selected physical and physiological transients were compared between males and females, by students, t-test and by analysis of variance (ANOVA). Multi-correlations were also computed between VO_{2max} and $\delta\ VO_2\ 1$, $\delta\ VO_2\ 2$ and $\delta\ VO_2\ 3$.

RESULTS AND DISCUSSION

The means (±SD) of the physical transients/corporal data, is presented in Table 1. The results show that the males and females of the study are age matched, with comparable BMI.

Table 1: Computation of Mean (±SD) of the Physical Transients of the Volunteers of the Study

	Age (years)	*Height (cm)*	*Weight (kg)*	*BSA (m^2)*	*BMI (kg/m^2)*
Male	23.5 (±1.30)	172.3 (±5.02)	66.1 (±8.52)	1.8 (±0.13)	22.2 (±2.43)
Female	23.3 (±1.35)	156.9 (±4.91)	57.9 (+10.85)	1.6 (±0.15)	23.4 (±3.67)

Table 2 displays mean (±SD) of certain selected cardiopulmonary transients. The males seem to display the VO_{2max} 44.9 ml/kg/min with fractional utilization of 58.6% O_2 consumption at the AT level. Females on the other hand, displayed a VO_{2max} 30.6 ml/kg/min with a comparatively high fractional utilization of 64.9% O_2 consumption at the AT level (NS).

Table 2: Computation of Certain Selected Physiological Transients of the Volunteers of the Study

	VO_{2max} (ml/kg/min)	*AT VO_2 (ml/kg/min)*	*AT VO_{2max} %*	*HR_{max} (beats/min)*	*AT HR (beats/min)*	*VE_{max} (l/min)*
Male	44.9 (±6.10)	26.3 (±5.32)	58.6 (±10.13)	192 (±5.60)	147.5 (±16.72	91.7 (±14.68)
Female	30.6 (±4.44)	20.5 (±5.62)	64.9 (±12.18)	19 (±5.60)	152.1 (±13.70)	57.7 (±15.82)

Table 3 compared the VO_{2max} displayed by the male sedentary population of the present study with those of their international counterparts (ACSM, 1996).

Table 3: Comparison of VO_{2max} of Males of Present Study with Those of Sedentary Population of Other Nationals

Studies	*VO_{2max} (ml/kg/min)*	*Fitness level as per ACSM (1996) norms*
Present study	44.9	Average
Chinese	46.0	Average
Russian	45.7	Average
Japan	44.4	Average
American	44.0	Average

As has been hypothesized, the VO_{2max} displayed by the sedentary Indian males compared favourably with those of the other populations, and somewhat higher than the American and Japanese sedentary population (NS).

Table 4 compared the VO_{2max} of present sedentary male population with some other Indian studies on special population like male sports persons and agricultural workers. The table reveals that the VO_{2max} displayed in the table below compares favourably well with those of Defense personal (Sengupta *et al.*, 1974), non athletes (Das *et al.*, 1995) and higher than those of agricultural workers (Gite *et al.*, 1996; Nag *et al.*, 1980), miners and industrial workers. However, VO_{2max} values displayed the present sedentary falls well short of those displayed by elite Indian athletes (Mukhopadhyay *et al.*, 2005).

Table 4: Comparison of the VO_{2max} of Present Sedentary Male Population with Some Other Indian Studies

Occupation	*VO_{2max} (ml/kg/min)*	*Fitness Level as per ACSM (1996) Norms*	*References*
Agriculture	39.0	Fair	Gite *et al.*, 1996
Agriculture	413	Fair	Nag P.K *et al.*, 1980
Miners (combined)	37.3	Poor	Sen Sharma (Ph. D. thesis), 1975
Industrial Workers (moderately heavy to very heavy job)	41.3	Fair	Saha *et al.*, 1972
Load lifter and Manual material handler (heavy)	46.0	Average	Saha *et al.*, 1979
Defense personnel	44.3	Average	Sengupta *et al.*, 1974
Athletes (elite-aerobic)	74.9	Excellent	Mukhopadhyay *et al.*, 2005
Non-Athletes	42.4	Average	Das *et al.*, 1995
Young sedentary men	36.8	Poor	Verma *et al.*, 1979
Young sedentary men	36.8	Poor	Sukhjinder, 2005
Present Study	44.9	Average	Ghosh *et al.*, 2007

Table 5 compared the selective cardiopulmonary transients of sedentary Indian adult males with a similar study (Sukhjinder *et al.,* 2005). The sedentary males of present study displayed a significantly higher VO_{2max} and AT VO_2 levels than those of Sukhjinder. There was no significant differences in maximum HR, AT HR and maximum ventilation values between the two groups.

Table 5: Table Displays and Compares Means (±SD) of Selected Cardiopulmonary Transients of Sedentary Males

Parameters	*n*	VO_{2max} *(ml/kg/min)*	*AT* VO_2 *(ml/kg/min)*	HR_{max} *(beats/min)*	*AT HR (beats/min)*	VE_{max} *(l/min)*
Present study	15	44.9 (±6.10)	26.3 (±5.32)	192 (±5.60)	147.5 (±16.72)	91.7 (±14.68)
Sukhjinder *et al.* (2005)	25	36.75 (±9.47)	23.45 (±6.42)	189.3 (±7.71)	152.6 (±10.80)	91.3 (±14.80)

Table 6 compared the cardiopulmonary transients of sedentary females of the present study with those of Sukhjinder *et al.* (2005). As it is evident from the table, no significant differences were observed in any of the cardiopulmonary transients.

Table 6: Table Displays and Compares Means (±SD) of Selected Cardiopulmonary Transients of Sedentary Females

Parameters	*n*	VO_{2max} *(ml/kg/min)*	*AT* VO_2 *ml/kg/min)*	HR_{max} *(beats/min)*	*AT HR (beats/min)*	VE_{max} *(l/min)*
Present study	15	30.6 (±4.44)	20.5 (±5.62)	192.0 (±5.60)	152.1 (±13.70)	57.7 (±15.82)
Sukhjinder *et al.* (2005)	25	29.13 (±4.58)	19.6 (±4.23)	188.5 (±8.40)	156.6 (±8.52)	58.37 (±7.76)

Table 7 compared the Recovery O_2 kinetics following maximum exercise in the male and female sedentary volunteers of the study. As it is evident from the table, significant differences were observed in δVO_2 1, and δ VO_2 2 at 0.05 level between the males and the females. The decrement in VO_2 consumption was significantly larger in the males than the females in both the 1st and 2nd minutes of recovery. This would seem to suggest that gender difference existed in the kinetics of the recovery cardiopulmonary dynamics.

Table 7: Comparison of Means (±SD) of Recovery O_2 Kinetics among Sedentary Males and Females

Parameters	δ VO_2 *1 (ml/kg/min)*	δ VO_2 *2 (ml/kg/min)*	δ VO_2 *3 (ml/kg/min)*
Male	16.4 (±2.94)*	8.2 (±6.10)*	2.8 (±2.18)
Female	12.3 (±4.93)*	4.4 (±3.94)*	2.4 (±2.12)

The VO_{2max} values as well as those of δ VO_2 1, δ VO_2 2, δ and VO_2 3 of the all the volunteers of the study (male and female) were bunched together and Pearson's correlation and ANOVA were computed to elucidate the relationship between VO_{2max} and recovery O_2 kinetics. The results are tabulated in Table 8. As it can be deciphered, VO_{2max} was found to be highly correlated to both δVO_2 1 and δVO_2 2 indicating that recovery responses, at least in the initial part of the recovery were strongly related to the maximum O_2 uptake capacity. The hypothesis was therefore proven true.

Table 8: Correlation between VO_{2max} and δ VO_2 1, δ VO_2 2, δ VO_2 3 of the Combined Group of Male and Female Volunteers

	'r' value	*F ratio*	*Comments*
VO_{2max} and δ VO_2 1	0.504	9.518	Significant at 0.05 level
VO_{2max} and δ VO_2 2	0.559	12.717	Significant at 0.05 level
VO_{2max} and δ VO_2 3	0.100	0.282	Not significant

CONCLUSION

The aerobic fitness levels displayed by the volunteers of the present study (VO_{2max}: 44.9 ml/kg/min), rank "Average" by ACSM standards, but those are comparable to the US national average (44.0/kg/min, for 20–29 years: ACSM, 1996), as well as to those of Chinese (44 ml/kg/min); Russian (45.7 ml/kg/min); Japanese (44.4 ml/kg/min). However, the volunteers of the present study were college going students with a relatively active lifestyle, and perhaps with a credible socio-economic (not investigated) background. These values therefore may or may not reflect the true national average.

In the present study, the VO_{2max} and AT VO_2 data of male population were found to be significantly higher than a similar study carried out on age-matched sedentary population of Punjab by Sukhjinder (2005). However, no such differences were observed in case of female sedentary volunteers.

VO_{2max} was found to be highly correlated to both δVO_2 1 and δ VO_2 2 indicating that recovery responses, at least in the initial part of the recovery are strongly related to the maximum O_2 uptake capacity.

The study therefore concluded that the cardiopulmonary fitness levels of male sedentary volunteers of the present study were comparable to those of their age matched counterparts of USA, China and Japan. However, it is credible to note that the population of USA and other developed countries may be anything but fit. Two out of three Americans are over-weight, one out of three are obese, with significant increasing in obesity related disorders like hypertension and diabetes (WHO, 2005). The Indians therefore need to augment their cardiopulmonary fitness further in order to be able to enjoy disease-free, wholesome life.

The close association between VO_{2max} and δVO_2 1 and δ VO_2 2 supports our hypothesis that the ability to recover faster would to a large extent depend on the capacity to mobilize a greater volume of oxygen.

REFERENCES

[1] Banket India, "Banking and technology portal", Economic survey 2004–2005, Indian Budget 2005–2006: < http://www.banknetindia.com/banking/eco055.htm>

[2] Beaver, W.L., Wassermann, K. and Whipp, B.J. (1986). "A new method for detecting anaerobic threshold by gas exchange." *J. Appl. Physiol.,* 60, pp. 2020–2027.

[3] Blair, S.N., Kohl, H.W., Paffenbarger, R.S., Clark, D.G., Cooper, K.H. and Gibbons, L.W. (1989). "Physical fitness and all-cause of mortality, A prospective study of healthy men and women", *JAMA*, 262: 17.

[4] Bruce, R.A., Gey (jr), G.O., Cooper, M.N., Fisher, L.D. and Peterson, D.A. (1974). "Seattle heart watch: Initial clinical, circulatory and electrocardiographic responses to maximal exercise", *Am. J. Cardiol.,* 33, pp. 459–565.

[5] Brynteson, P. and Brytenson, D. (1993). *Essentials of Physical Activity*, Dubuque: Eddie Bowers.

[6] Dacie and Lewis, *Practical haematology*, Churchill Livingstone, 2002, p. 12.

[7] Gite, L.P. (1996). "Some investigation on a side lever operated knapsack sprayer from mechanical and ergonomical consideration", Ph.D. Thesis (Unpublished), Indian Institute of Technology, Kharagpur.

[8] Goswami, A., Ghosh, A.K. and Ganguli, S. (1984). "Aerobic Capacity of Severely Disabled Indians." *Ergonomics*, 27 (12), pp. 1267–1269.

[9] Institute of preventive Cardiology, Mumbai, International cardiovascular disease statistics (2005) from <http:www.ipc-india.com/heart-encyclopedia/details.asp?Id=23, 2005>

[10] Kalam, A.P.J. (2005). "Action Plan for Employment Generation", *Address to the nation by the President of India, Dr. A.P.J. Abdul Kalam on the eve of 56th Republic day 2005*, from <http://meaindia.nic.in/speech/2005/01/25ss02.htm>

[11] Kavanagh, T., Mertens, D.J. and Hamm, L.F. (2002). "Prediction of long-term prognosis in 12169 men referred for cardiac rehabilitation", *Circulation*, 106, pp. 666–671.

[12] Mishra, D. and Singh, H.P. (2003). Kuppuswami's Socio-economic Status Scale—A review, *Ind. J. Pediatrics*, 273–274.

[13] Mukhopadhyay, Suman Sukhjinder, Sidhu, L.S. and Verma, S.K. (2005). "The Kinetics of Cardiopulmonary Dynamics During Recovery following Maximal Exercise", *J Ex Sc and Physioth*, (1), (1 and 2).

[14] Myers, J., Prakash. M. and Froelicher, V.F. (2002). "Exercise capacity and mortality among men referred for exercise testing", *N Engl J. Med.,* 346, pp. 793–801.

[15] Sen Gupta, J., Srinivasulu, N., Sampath Kumar, T. and Malhotra, M.S. (1974). "Influence of age on Maximum oxygen uptake and Maximum heart rate of Indians during work", *The Indian Journal of Medical Research*, 62: 1228–1232.

[16] Sue, D.Y., Wasserman, K., Moricca, B. and Casaburi, R. (1988). "Metabolic acidocis during exercise in patients with chronic obstructive pulmonary disease", *Chest*, 94, pp. 1143–1150.

[17] Sukhjinder *et al.* (2005). "A Comparison of Aerobic Fitness and Cardiopulmonary Recovery Patterns in Selected Groups of Elite Indian Athletes". Punjabi University, Patiala.

[18] Taylor, H.L., Buskirk, E. and Henschel, A. (1955). "Maximum oxygen intake as an objective measure of Cardio-respiratory performance:, *J. Appl. Physol.,* 8, pp. 73–80.

[19] Tomlin, D.L. and Wenger, H.A. (2001). "The relationship between aerobic fitness and recovery from high intensity intermittent exercise", *J. Sports Med.* 31(1), pp. 01–11.

[20] Verma, S.K., Sidhu, K.S. and Kansal, D.K. (1979). "Aerobic work capacity in young sedentary men and active athletes in India (age 17–25 y)", *Sports Med.,* 13, pp. 8–102.

[21] Wasserman, K., Hansen, J.E., Sue, D.Y. and Whipp, B.J. (1987). "Principles of Exercise testing and Interpretation", *Lea and Febiger, Philadelphia,* pp. 27, 67, 157, 168.

[22] WHO (2005). "Fit and Well: Core Concepts and Labs in Physical Fitness and Wellness, Chapter 1", *McGraw-Hill Higher Education.*

An EMG Based Muscle Fatigue Analysis of Mobile SMS Users

J. Srinivasan, Priyanka Gopal and Venkatesh Balasubramanian

Rehabilitation Bioengineering Group
Department of Biotechnology, IIT Madras, Chennai, India
E-mail: chanakya@iitm.ac.in; chanakyaaa@yahoo.com

ABSTRACT

Mobile phones have become an essential part of personal and business communication. One of the cheaper and faster communicating modes using a cell phone is the Short Message Services (SMS). To send a SMS, the user inputs the message by keying-in the number keys (in alphabet mode) available on the phone. Although there are many programs and machine utilities that help in messaging, there still are issues with muscle fatigue and hence repetitive strain injury. Both of these issues are more pronounced among the younger generation who tend to use SMS more frequently. The aim of this study was to examine muscle fatigue on flexor pollicis brevis, abductor digiti minimi, abductor pollicis brevis and brachio radialis muscles during SMS typing.

Eight healthy volunteers (both male and female) participated in the experimental procedure whose mean age was between 20–27 years. Surface Electromyogram (sEMG) signals were collected bilaterally from a selected group of muscles for 15 minutes, while persons were messaging using SMS. The collected sEMG signals were filtered using Butterworth filter with a pass band of 20–400 Hz and stop band for removing the line interference noise. Fatigue was calculated using mean power frequency difference. Results indicate that persons with the habit of text messaging for more than 15 minutes could have muscle fatigue. On extrapolation, this activity over a period of time could lead to weakening of muscles and hence result in repetitive strain injury.

This study offers an insight on product development like position and size of keys for easy fingering in the case of continuous messaging. Also this could throw light upon mobile models that are not suited for the purpose of text messaging.

Keywords: Surface EMG, Stochastic Signal Processing.

INTRODUCTION

With the benefits of current technology, mobile phones are now a means of entertainment and also provide access to the internet. Mobile phones have become one of the main modes of communication in recent times in business and social circles as people are always on the travel. A mobile phone is very much functional in making information move faster, after the introduction of the Short Messaging Services (SMS).

The usefulness of mobile communication has led it to take a plunge into the field of medicine with its use in telemetry services in the case of emergencies (Kyriacou *et al.*, 2003).

Along with its usefulness come its drawback and related health issues. One of the main issues that are being dealt with in recent times is the exposure to the electromagnetic radiations due to the use of mobile phones (Kumar, 2001).

We are dealing with another possible drawback of the mobile phone in relation to the risk of developing repetitive strain injury or musculoskeletal disorder due to muscular fatigue in relation to text messaging using mobile phones. Muscle fatigue is the point of distinction where the muscles are unable to deliver the required output for the necessary work that has been assigned. Fatigue is most often referred to as incompetence or the inability to perform (Clark *et al.*, 1984). The intensity of the muscular fatigue depends on the level of physical fitness and the person's ability to text message.

It is believed that improper postures or motions when repeated continuously over a period of time may lead to injury often known as the repetitive strain injury (Voge and Dingwell, 2003). Similarly, there is a possibility that the continuous messaging using a narrow keypad of a mobile phone will lead to repetitive strain injury over a long duration. sEMG has been used to display muscle activation patterns and is often extensively employed to interpret both dysfunction and functional muscle recruitment patterns during various activity (De Luca, 1979; Oddsson and De Luca, 2003). The sustained muscular contractions externally associated with not being able to maintain a certain force, lead to physiological fatigue, tremor or pain, localized in the specific muscle. This is called muscular fatigue (Oddsson and De Luca, 2003), associated with a compression of the Power Spectral Density (PSD) of the sEMG towards lower frequency (Gregor *et al.*, 1991), measured by MF computed from the sEMG PSD. In general, fatigue is viewed as an unavoidable and negative consequence of physical activity. In this study, we compared the EMG patterns acquired from a group of subjects while typing text message using mobile phone.

METHODOLOGY

Data Acquisition

There were eight volunteers who participated in the experimental procedure consisting of four males and four females. The subjects who volunteered were in the mean age of 23.2 years (range 18–30 years) and with a mean height 162.4 cm and weight of 58.85 kg (range 52–68 kg). Customized program was written in LabVIEW$_{TM}$, National Instruments, Inc., to collect the sEMG signal. The sampling was performed at the rate of 1000 Hz. The sEMG were recorded continuously for 20 minute, during text messaging using MCC DAQ card. All experimental data were collected at the Rehabilitation Bioengineering Group (RBG) laboratory at IIT Madras, India.

Protocol

All volunteers were made aware of the experimental details prior to obtaining their consent to participate in the study and signed an informed consent. Subjects were made to type pre-prepared messages regularly for 20 minutes at the Rehabilitation Bioengineering

Lab. The volunteers were instructed to maintain comfortable postures during the full length of experiment. This text typing activity involved the exertion of several muscle groups but the most dominant muscle was alone taken for this study. The dominant muscles considered were, flexor pollicis brevis, abductor digiti minimi, abductor pollicis brevis muscle groups of the hand and brachio radialis muscle of the forearm.

Analysis

Raw EMG signals were filtered using second-order Butterworth filter with a pass band of 20–400 Hz. A second order notch filter was used to eliminate AC line interference. The line interference frequency is time dependent. Therefore, a power spectrum analysis was performed for each subject and line noise was removed. Filtered signals were normalized with respect to the mean, and then full-wave rectified. sEMG signals were analyzed in frequency domain using power spectrum in frequency domain.

Frequency Domain Analysis

Power spectrum of EMG has been observed to change under the influence of muscle fatigue (Van der Hoeven *et al.*, 1993; Srinivasan and Balasubramanian, 2005). As muscle fatigues, there is a concomitant change in power spectrum of EMG signals, where there is an increase in the amplitude of lower frequency band and a relative decrease in higher frequency band (Van der Hoeven *et al.*, 1993). Central tendency measures such as mean, peak, median and ratios of the power of high and low frequency bands extracted from the power spectrum are indicators of muscle fatigue (Gregor *et al.*, 1991). Decrease in Mean Power Frequency (MPF) of EMG profiles is a recognized method of determining fatigue during an isometric muscle action (Van der Hoeven *et al.*, 1993).

In this study, Hamming window algorithm was applied on the rectified sEMG signals and MPF was extracted from the power spectrum. Fatigue difference was calculated using MPF difference as given in equation 1 (Srinivasan and Balasubramanian, 2005). These were used to determine the level of muscle fatigue.

$$\textit{MPF difference} = (\textit{MPF after} - \textit{MPF before}) \quad \ldots (1)$$

A linear regression analysis was also performed to determine the muscle fatigue for each individual during the 15 minutes of text typing in mobile. All statistical testing was carried out using the Statistical Package for Social Sciences (SPSS v. 10.0.1).

RESULT

Sample power spectrum of EMG acquired from right flexor pollicis brevis is shown in Figure 1. Table 1 shows the regression value for all the subjects and negative value indicates the negative slope.

Hunter *et al.* (2001) reported that a decrease in the MPF can be observed as the muscle fatigue progressively increases. Although central and peripheral factors both influence the decreasing trend of the MPF the main contributing factor can be attributed to the peripheral which is, a slowing of the muscle action potential conduction velocity.

We used MPF to classify the fatigue of muscle.

In our study, MPF of various muscles decreased while using SMS in mobile phone, that has been shown using "*" in Table 1. Figure 2 shows a sample regression slope of right Flexor Pollicis brevis of one subject.

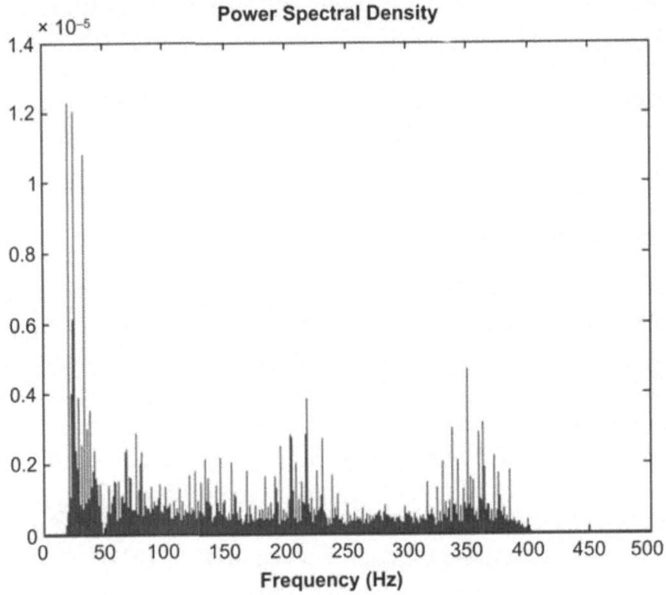

Fig. 1: Power Spectrums for Entire 1 minute of sEMG Signal of Right Flexor Pollicis Brevis

Table 1: Regression Value "*r*" for all Muscle Groups and * Indicate the Negative Slope

No. of Subject	*Right*				*Left*			
	Flexor Pollicis Brevis	*Abductor Digiti Minimi*	*Brachio Radialis*	*Abductor Pollicis Brevis*	*Flexor Pollicis Brevis*	*Abductor Digiti Minimi*	*Brachio Radialis*	*Abductor Pollicis Brevis*
1.	0.994	0.968	0.982	0.895	1	0.892	0.994	1.00
2.	0.725	0.00	–0.03*	–0.931*	–1*	0.196	–0.118*	–0.915*
3.	0.98	–0.381*	–0.004*	–0.281*	0.817	0.109	0.857	0.043
4.	0.982	–0.922*	–0.984*	0.026	0.046	0.91	–0.06*	0.049
5.	0.666	0.981	0.99	0.000	–0.493*	–0.19*	0.739	–0.095*
6.	–1*	–0.766*	–0.068*	0.062	0.022	0.099	0.012	–0.264*
7.	–0.009*	–0.065*	0.044	0.997	–0.398*	–0.014*	0.012	0.003
8.	0.867	0.615	0.046	–0.355*	0.693	0.781	0.073	0.618

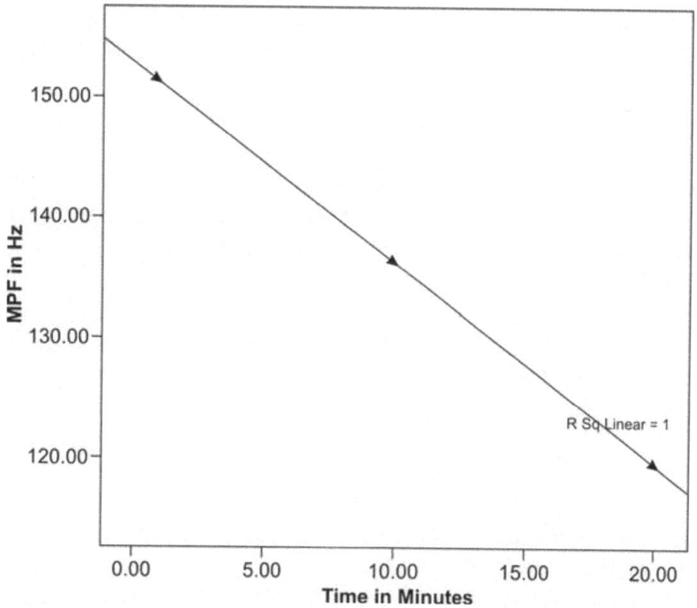

Fig. 2: The Regression Slope for 1st minute, 10th minute and 20th minute of MPF of Right Flexor Pollicis Brevis

CONCLUSION

This study helped us to understand the consequences of text typing using mobile phone. A sudden fall of MPF indicates muscle fatigue. From the results, it was evident that SMS typing for short duration could worsen significantly for subjects. The study could be extended in future to analyze shoulder elevation of the joint articulation associated with the trapezius and different postures adopted by the subjects during SMS activates.

ACKNOWLEDGEMENT

The authors wish to thank all other members of Rehabilitation Bioengineering Group (RBG) at IIT Madras and the volunteers who participated in this study.

REFERENCES

[1] Clark, G.T., Beemsterboer, P.L. and Jacobson, R. (1984). "The effect of sustained submaximal clenching on maximum bite force in myofascial pain dysfunction patients", *J. Oral Rehabilitation,* 11: 387–391.

[2] De Luca, C.J. (1979). "Physiology and mathematics of myoelectric signals", *IEEE Trans Biomed Engg.*, 26: 313–325.

[3] Gregor, R.J., Broker, J.P. and Ryan, M.M. (1991). "The Biomechanics of Cycling. Exerc. Sport", *Sci. Rev.,* 19, 1991, pp. 127–169.

[4] Hunter, M.A., Gibson, Allan St. Clair, Lambert, Michael and Noakes, Timothy D. (Sept. 2001). "EMG normalization method for cycle fatigue protocols", *Medical and Science in sports and Exercise*, 857–861.

[5] Kumar, A. (2001). "Electromagnetic Radiation and Biological Effects", *IEEE Journal* 2001, 0-7803-6569-0/01.

[6] Kyle, R. Voge and Jonathan, B. Dingwell (2003). "Relative Timing of Changes in Muscle Fatigue and Movement Coordination During a Repetitive One-Hand Lifting Task", *Proceedings of the 25th Annual International Conference of the IEEE Engineering in Medicine and Biology Society,* Cancum, Mexico, September 17–21, 2003, pp. 1807–1810.

[7] Kyriacou, E., Pavlopoulos, S., Berler, A., Neophytou, M., Bourka, A., Georgoulas, A., Anagnostaki, A., Karayiannis, D., Schizas, C., Pattichis, C., Andreou, A. and Koutsouris, D. (2003). "Multi-purpose HealthCare Telemedicine Systems with mobile communication link support", *Biomed Central:* Open Access Published: 24 March 2003 BioMedical Engineering OnLine, 2: 7.

[8] Lars, I.E. Oddsson and Carlo, J. De Luca (2003). "Activation Imbalances in Lumbar Spine Muscles in the Presence of Chronic Low Back Pain", *Journal of Applied Physics* 94: 1410–1420, 2003. First published December 13, 2002; doi:10.1152/japplphysiol.01183.2001 8750–7587, Vol. 94, Issue 4, 1410–1420.

[9] Srinivasan, J. and Venkatesh, Balasubramanian (2005). "Low Back Pain and Muscle Fatigue Due to Road Cycling—An sEMG study", *Journal of Bodywork and movement therapy,* Elsevier Pub., Vol. 1, Issue 3, pp. 260–266.

[10] Srinivasan, J. and Venkatesh, Balasubramanian (Dec. 2005). "Muscle Fatigue Analysis for Bicyclist Using sEMG" *12th International Conference on Biomedical Engineering (ICBME), Singapore.*

[11] Stulen, F.B. and Deluca, C.J. (1981). "Frequency Parameters of the Myoelectric Signal as Measure of Muscle Conduction Velocity", *IEEE Transactions on Biomedical Engineering,* col. BME-28, 1981, No. 7, pp. 515–523.

[12] Van, der Hoeven J.H., Weerden, T.W. Van and Zwarts, M.J. (1993). "Long-Lasting Supernormal Conduction Velocity after Sustained Maximal Isometric Contraction in Human Muscle", *Muscle Nerve*, 16, pp. 312–320.

A Study of Musculoskeletal Health of the Workers of Jewellery Manufacturing in India

U. Dhar, R. Salve and Amitabha De

National Institute of Industrial Engineering (NITIE)
Vihar Lake, Mumbai–400 087, India

ABSTRACT

Work Related Musculoskeletal Disorder (WRMSD) is one of the major contributors that affect the workers' health and productivity as a whole. There are several factors that may cause WRMSD. Repetitiveness of the job, high exposure to force, long duration of exposure, awkward posture, contact to the sharp edge and hard surface are such factors. Jewellery manufacturing is one of the important economic activities in terms of its contributions in the Indian economy. The jewellery manufacturing activity like any other sedentary assembly jobs requires repetition, force, awkward (non-optimal) posture, static posture, contact stress, extreme temperatures, vibration and may result in work related musculoskeletal disorders. The present study was conducted to find out the status of the musculoskeletal health of the workers of the jewellery manufacturing sector in India. Modified Nordic musculoskeletal disorder survey and physical examination of posture of the participants of this study were conducted to fulfil the objective. The study was carried out in Mumbai, on the workers of the jewellery manufacturing sector (N = 358). The results showed that the leading musculoskeletal problems were in the region of neck (35.29%), low back (65.27%) and knee (75.35%) in comparison to shoulder (4.75%), elbows (8.38%), wrist (4.47%), upper back (8.66%), hip/thigh/buttock (0.28%), ankle/feet (2.23%). posture analysis showed that 49.01% were normal. Among the others protruded neck with rounded shoulder, only Protruded neck, drooping posture and forwarded posture, others are 24.90%, 10.28%, 3.56%, 1.58%, 5.93%, respectively.

Keywords: MSD, Jewellery Manufacturing, Posture, Modified NMQ.

INTRODUCTION

Musculoskeletal Disorder is a widely accepted term in the field of occupational health hazards. An injury to musculoskeletal system is called Musculoskeletal Disorder (MSD). It has few synonyms, such as Repetitive Strain Injury (RSI), Repetitive Motion Injury (RMI), Cumulative Trauma Disorder (CTD), etc. CTD is a more commonly accepted synonym. The abbreviation CTD can be defined as: Cumulative signifies the injury develops gradually over periods of week, months and year as result of repetitive stresses on a particular body parts, trauma indicates the bodily injury from mechanical stresses and disorder refers to the physical abnormal condition.

Musculoskeletal disorders were recognized as having occupational etiology factor at the beginning of the 18th century (NIOSH report). World Health Organization (WHO) has characterized 'work related' diseases as multifactorial to indicate that a number of risk factors (*e.g.*, physical, work organizational, psychosocial, individual and sociocultural) contribute to causing these diseases (WHO, 1985). Work and workplace hazard is closely related to the development of the CTD. However, the prevalence of CTDs are not only related to work and workplace conditions, many other activities, such as golf playing, tennis playing etc., those require highly physical and physiological demand on musculoskeletal systems can cause CTD. There are few occupational and non-occupational factors that influence or affect the CTD. Few non-occupational factors are age, gender, acute trauma, chronic disease, use of birth control pill, pregnancy, menopause etc. The occupational factors are repetition, force, awkward posture, static posture, contact stress, extreme temperature, and vibration.

In 2003–04 it was estimated that about 1.0 million people in Great Britain were suffering from MSD caused by work, accounting for 50 per cent of all occupational ill health. It was estimated that about 2,00,000 new cases of WRMSD occurs each year; and that around 11.6 million people lose their working days each year due to WRMSD.

Jewellery forms an integral part of India's history, culture and economy for over five thousand years. It has been used as an investment, an art form, and as an important trade. Jewellery manufacturing is one of the important economic activities in terms of its contributions in the Indian economy especially in terms of export earnings and job creation (GJEPC report, 2006). The Indian market size of jewellery manufacturing in 2005–06 was more than USD 13.00 billion (indianembassy.org.ua, 2007) of which the export accounted for USD 3.524 billion (www.indiainbusiness.nic.in/trade-india/trade.htm, 2007). About two million highly skilled workforce are engaged in this sector, out of which one million is engaged in export production (www.gjepc.org, 2007).

The jewellery manufacturing activity like in any other sedentary assembly jobs, requires repetition, force, awkward (non-optimal) posture, static posture, contact stress, extreme temperatures, vibration (Hales, 1991) and may result in work related musculoskeletal disorders (Kuorinka *et al.*, 1995; Latko *et al.*, 1996; Bjorksten *et al.*, 2001; Guo *et al.*, 2004, Kim *et al.*, 2004). A questionnaire survey by NIOSH investigators in a jewellery manufacturing unit in USA indicated that 66 per cent employees reported work related upper extremity disorders. In another study carried out by NIOSH (1997) in more than 200 jewellery manufacturing organization, it was observed that there were 750 occupational health and safety violations. Visual problems among the jewellery workers are also not uncommon; in a study carried out among the jewellery workers in Thailand, Untimanon *et al.* (2006) reported 52 per cent of subjects took part in the study had vision problems that affected their work performance. Studies carried out in India have also shown similar results. In a study to find out the body part discomforts as reported by the employees ($n = 206$) of the jewellery manufacturing organizations in Mumbai, Manna *et al.* (2004) found out that, 32 per cent, 45 percent, 52 percent of the respondents had body ache, stiff shoulder and pain in the wrist; 57 per cent of the employees had reported eyestrain. The above problems are also being recognized by the jewellery manufacturing organizations in India. They felt the need to incorporate ergonomic issues such proper workstation, proper work posture and comfortable physical working environment in their business excellence model (Rio Tinto Diamonds, 2006). Keeping the above in mind this study aims at the status of the musculoskeletal health and its relation with the duration of exposure to the job of the workers of jewellery manufacturing sector.

METHODS

Subjects

Three hundred fifty young male subjects voluntarily participated in this study. Signed informed consents were taken. The inclusion criteria for subjects were minimum employment for at least one year with same kind of nature of work.

Local of the Study

The study was carried out in Zaveri Bazar region of Mumbai.

Tools Used

Musculoskeletal disorder survey was carried out using modified Nordic Musculoskeletal Questionnaire (Dickinson *et al.*, 2002) on the experimental group. The questionnaire was used for identifying the presence of musculoskeletal problems in various body parts (neck, shoulders, elbows, wrists/hands, upper back, lower back, both hips/thighs/buttock, both knees, ankle/feet). The questionnaire was also used to find out the reasons for the musculoskeletal problems and the level or degree and duration of pain, ache or discomfort.

Study Protocol

The participants were interviewed by a trained person for the modified Nordic Musculoskeletal Disorders at the participants' leisure time.

Evaluation of Postural Alignment

A qualified physical therapist performed the posture evaluation of the participants using method described by Kendal.

Statistical Treatment

The descriptive statistics were performed with the data collected through the study.

RESULTS

The subjects of the study were the workers of the jewellery manufacturing sector. The mean age, weight and height of the participants were 22.98 (±4.40), 63.58 (±10.03) and

170.88 (±6.32) respectively. The participants who were having less than one year of work experiences were excluded form the study. The demographic data of the subjects are presented in Table 1.

Table 1: Demographic Data of the Subjects

Age, years	22.98 (±4.40)
Weight, kg	53.46 (±7.67)
Height, cm	162.42 (±6.15)

The result of the Nordic Musculoskeletal disorder questionnaire showed that the MSD problems are specific to the neck, lower back and knee region (Table 2). While questionnaire study it was found that 13.65% workers of the jewellery manufacturing sector were having sensation of tingling throughout the vertebral column.

Table 2: The Results of the Nordic Musculoskeletal Disorders Survey

	Body Parts								
	Neck	*Shoulder*	*Elbows*	*Wrist*	*Upper Back*	*Lower Back*	*Hip/Thigh /Buttock*	*Knees*	*Ankle/ Feet*
No. of Complains	139	28	21	33	33	264	2	300	9
Percentage	34.32	6.91	5.19	8.15	8.15	65.19	0.49	74.07	2.22

The results of the posture analysis showed that 49.01% participants were having normal alignment of the spine (Table 3). Only 24.90% participants were having protruded neck with rounded shoulder.

Table 3: Results of the Posture Analysis

Posture Type	*Normal*	*Protruded Neck, Rounded Shoulder*	*Protruded Neck*	*Drooping*	*Forwarded*	*Others*
Results in Percentage	49.01	24.90	10.28	3.56	1.58	5.93

DISCUSSION

Exposure to fixed posture for long duration causes pain. Lack of change of posture, *i.e.* maintaining static posture for a long time period causes ligament elongation with a reduction of spine stability (Renney, 1997). Frequently the cause of low back pain is not known. Some authors have attributed low back pain, which is due to spinal joint dysfunction to homo sapiens' bipedal posture (Cailliet, 1968). Poor posture, in which the head is thrust forward with excessive spinal curves in the sagittal plane, sloping or hunched shoulders, protruding abdomen and hyper extended knees (Garlick, 1990) may be habitual or occupational and can be related to poor muscle tone. The incidence of low back pain in the sector of jewellery manufacturing can be attributed to the result of another study, which shows that the workers of the jewellery manufacturing occupy non-optimal posture (Dhar *et al.*, 2004).

Neck has two groups of muscles, one is intrinsic, *i.e.* for neck itself and the other is extrinsic. Extrinsic muscles connect both the neck and upper limb and are important for stabilizing the head while the gaze is fixed. These extrinsic muscles suffer from pain and fatigue during immobile condition for a long duration. From Table 2 we can see that the 35.29% workers were suffering from neck pain. Discomfort in the low back region was 65.27%. Both these problems are assumed to be because of continuous working schedule in non-optimal posture. Although the cause effect relationship needs a detailed study.

The major complain of the musculoskeletal health of this sector is discomfort in knee. Studies have shown that in the workers of forest department the problems in knee was only 20% and 5% (Kaewboonchoo *et al.*, 1998). As the working posture of the workers is squatting, the muscles of the lower limb remain in constantly stressed condition. The workers in squatting position flex their knee for whole of the working hour. This may cause impaired blood flow resulting in muscle pain.

The workers of the jewellery manufacturing sector occupy the forward bending position while working. Long term exposure to such posture may lead to change in posture of spine. 24.90% of the workers are having protruded neck with rounded shoulder. This type of posture may cause weakness of neck muscle and may lead to neck pain. 3.56% People are having drooping posture. Drooping posture can lead to forward bending of the trunk and may lead to back pain.

CONCLUSION

From the above result and discussion it is clear that the musculoskeletal discomfort/ problems (neck, low back and knee) are present in this working population. But the root cause analysis only can identify the generic risk factors of this jewellery manufacturing.

ACKNOWLEDGEMENT

We are thankful to National Institute of Industrial Engineering (NITIE) and TIFAC for providing the necessary finance and support for the project. We are also thankful to the association of jewellery manufacturing units for allowing the research team to conduct the study.

REFERENCES

[1] Bjorksten, M.G., Boquist, B., Talback, M. and Edlind, C. (2001). "Reported Neck and Shoulder Problems in Female Industriall Workers: Importance of Factors at Work and Home", *International Journal of Industrial Ergonomics*, 27, 159–170.

[2] Cailliet, R. (1968). Low Back Pain Syndrome, 2nd ed. Philadelphia: F.A. Davis Co.

[3] Dhar, U., Mutakekar, M.V., Ghosh, S., Savanur, C. and De, Amitabha (2004). "Work Related Musculoskeletal Disorder in Organized and Unorganized Sector of Jewellery Manufacturing in India." *A national conference on Humanizing Work and Work Environment held during 22nd–24th April, 2004.*

[4] Dickinson, C.E., Campion, K., Foster, A.F., Newman, S.J., O'Rourke, A.M.T. and Thomas, P.G. (1992). "Questionnaire Development: An Examination of the Nordic Musculoskeletal Questionnaire." *Applied Ergonomics*, 23,197–201.

[5] Garlick, D. (1990). "The Lost Sixth Sense." University of New South Wales.

[6] Guo, H.R., Ching, Y.C., Yeh, W.Y., Chen, C.W. and Guo, Y.L. (2004). "Prevalence of Musculoskeletal Disorders among Workers in Taiwan: A Nation Wide Study." *Journal of Occupational Health*, 46, 26–36.

[7] Hales, T., Grant, K.A., Daniels, W. and Habes, D.J. (1991). Hazard Evaluation and Technical Assistance Report: FL Thorpe & Co., Inc., Deadwood, SD. Cincinnati, OH: U.S. Department of Health and Human Services, Public Health Service, Centres for Disease Control, National Institute for Occupational Safety and Health, *NIOSH Report No. HHE*, 90–273–2130.

[8] Kim, J.Y., Kim Jung, Son J.E. and Yun, S.K. (2004). Prevalence of carpal tunnel syndrome in meat and fish processing plants. *Journal of Occupational Health*, 46, 230–234.

[9] Kaewboonchoo, O., Yamamaoto, H., Miyai, N., Mirbod, S.M., Morioka, I. and Miyashita (1998). The Standardized Nordic Questionnaire Applied to Workers Exposed to Hand-arm Vibration." *Journal of Occupational Health*, 40, 218–222.

[10] Kuorinka, I. and Forcier, L. (1995). Work related Musculoskeletal Disorders (WMSDs): A Reference Book for Prevention. Taylor & Francis, London.

[11] Latko, W.A., Armstrong, T.J., Franzblau, A., Ilin, S.S., Werner, R. and Albers, J.W. (1999). "Cross-sectional Study of the Relationship between Repetitive Work and the Prevalence of Upper Limb Musculoskeletal Disorders." *American Journal of Industrial Medicine*, 36, 248–259.

[12] Ranney, D. (1997). "Chronic Musculoskeletal Injuries in the Workplace", (W.B. Saunders Company), 48–50.

[13] Untimanon, O., Pacharatrakul, W., Boonmeepong, K., Thammagarun, L., Laemun, N., Taptagaporn, S. and Chongsuvivatwong, V. (2006). Visual Problems among Electronic and Jewellery Workers in Thailand. *Journal of Occupational Health*, 48, 407–412.

[14] www.indianembassy.org.ua, 05.01.2007.

[15] www.indiainbusiness.nic.in/trade-india/trade.htm, 05.01.2007.

[16] www.gjepc.org, 05.01.2007.

[17] www.riotintodiamonds.com,05.01.2007.

Relation of Stump Length with Gait Parameters in Trans-tibial Amputees

Prasanna Lenka[1*], R. Kumar[1], A. Roy Choudhury[2] and D. Tiberwala[3]

[1] Dept. of R&D, NIOH (Under Ministry of Social Justice & Empowerment)
B.T. Road, Bon-Hooghly, Kolkata, India
E-mail: *lenka_pk@yahoo.co.uk

[2]Dept. of Applied Mechanics, B.E. College and Science University, Howrah, India

[3]School of bioscience, Jadavpur University, Jadavpur, India

ABSTRACT

This paper evaluates the impact of stump length of unilateral below-knee amputees on the Ground Reaction Force (GRF), energy consumption, velocity and other gait parameters. Amputee gait not only depends on the particular prosthetic component, socket type and suspension but also is highly influenced by the residual anatomy. Nine individuals with unilateral Below Knee Amputation (BKA) were chosen for this experiment and three groups were formed as per the length of the stump: short stump length, medium stump length and long stump length. Each of them underwent a 15s gait analysis test by Computer Dynography (CDG) system to measure the gait parameters in each limb separately and simultaneously. It was found that the ground reaction force was high in the patients with medium stump length whereas the velocity, step length (both of the prosthetic and sound limb) and cadence were high in longer stump length. Statistical analysis showed a significant difference ($p < 0.05$) between the gait parameters of BKA patients with medium and long stump length. The patients with longer stump length were more efficient than medium and short stump patients as they consume comparatively less energy while walking with self-selected velocity and conventional SACH foot.

Keywords: Stump Length, Gait Parameter, Below-knee Amputee.

INTRODUCTION

Optimal rehabilitation depends on good initial healing of the stump, which also needs to be of the correct length and shape (Cosgrove *et al.*, 2002). Incorrect stump length and stumps with excess tissue are significant and can have a permanent influence on the successful fitting of prosthesis (Campbell, 1996). In most of the studies, gait analysis was used as a tool for prosthesis design and also for a proper alignment to the prosthetic devices considering several important gait parameters (Velzen *et al.*, 2005). But very few studies had presented the characteristics of the below-knee amputees with various stump length and their effect in gait. The prosthetic component design, selection, and alignment of amputee's prosthesis are all directed towards obtaining optimal gait (Barth *et al.*, 1992). In this way the residual limb anatomy is a strongly considerable factor in ensuring successful rehabilitation.

A study has revealed that in case of an amputation of lower limb, if the stump has sufficient length, a below-knee amputation has the best gait efficiency and prosthesis fit. If the stump is too short, a lengthening procedure may be beneficial (Mertens and Lammens, 2001). In this study we showed that there were some differences in gait efficiency-regarding the velocity, energy consumption, and with different stride parameters in spite of using a well fitted prosthesis device. With a short trans-tibial amputation, the energy expenditure level increases by 40%, and by 10% if it is too long (Gonzalez *et al.*, 1974).

In case of below-knee amputation, it was recommended that at least 8 cm below the tibial plateau be retained to allow optimal control of the socket and end-bearing (Harnot *et al.*, 2000).

Physiological Cost Index (PCI) is a cardiopulmonary factor as an indicator of energy cost described by MacGregor, 1981 (Mac, 1981). In this study we investigated and quantified the difference in the gait parameters, electromyography characteristics of vastus lateralis muscle as well as PCI data among the patients with different stump length. Sanders and Daly (1993) found that interface force between the socket and stump were highly sensitive to the stump length. Results from this analysis showed a more efficient gait with patients of longer stump length than the patients with medium and short stump length.

METHODOLOGY

Nine traumatic trans-tibial amputees were selected. All were able to walk with the prosthesis, without any additional aid. The descriptive data of subjects were shown in Table 1. None of the subjects had any residual limb pain, sores, swelling and contracture, etc. All subjects were fitted with conventional SACH foot.

The gait analysis system, Ultraflex Infotronic, was used for data collection. Pair of foot sensors called CDG (Computer Dynography) shoes was used to collect force distribution. Disposable surface EMG electrodes were used to record bilateral quadriceps (Vastus Lateralis) EMG. The foot sensor and EMG data were digitally acquired at a sampling frequency of 100 HZ and 1000HZ respectively and stored in portable data logger unit. The ECG system was used to record Heart Rate. The digitally collected EMG raw data were processed using MATLAB 7.0.

Gait analysis was performed with the self selected velocity of the subjects after a short trial period of free walking. Subjects (nine) were divided into three groups as per their stump length normalized with their height: Group I patients with short stump length (range 8–10 cm), Group II patients with medium stump length (range 15–16 cm) and Group III patients with long stump length (range 25–31 cm).

An index was derived from the ratio of difference of active heart rate with rest to the velocity. This index reflects about the gait efficiency of the subject. The other research papers calculated the energy cost by oxygen uptake methods (Barth *et al.*, 1992).

Statistical analysis was done by the help of statistical analysis software MS Excel. Analysis of variance (ANOVA) single factor test, paired t-test were used to analyze the data.

Table 1: Group-wise Descriptive Data of the Subjects

	Group I (short)	*Group II (medium)*	*Group III (long)*	*Total (Gr. I, II, III)*
	Mean ± SD	*Mean ± SD*	*Mean ± SD*	*Mean ± SD*
Age, years	30.67 ± 1.15	36 ± 11.53	24 ± 2.65	29.78 ± 8.45
Height, cm	163 ± 14.8	159 ± 3.61	170.67 ± 2.31	164.22 ± 9.26
Weight, kg	60.74 ± 10.8	65.9 ± 3.92	65.21 ± 10.51	63.95 ± 8.15
Stump length, cm	15.92 ± 0.64	9.89 ± 1.27	28.67 ± 2.84	18.16 ± 8.45
Prosthesis height, cm	47.44 ± 3.15	37.33 ± 14.18	47.67 ± 0.58	44.15 ± 8.89
Prosthesis weight, kg	1.97 ± 0.50	1.71 ± 0.37	2.13 ± 0.36	1.94 ± 0.40

RESULTS

Various gait parameters were considered in all three group of stump length—short, medium and long.

Stump Length and Energy Index

In this particular study energy index of the subjects with long stump length was always less than the subjects with short stump length and medium stump length after the work. Here, the specific work was of walking with a self selected velocity for 15s. The energy index decreased with the increment of stump length. A significant difference was found for this parameter as a result of statistical analysis.

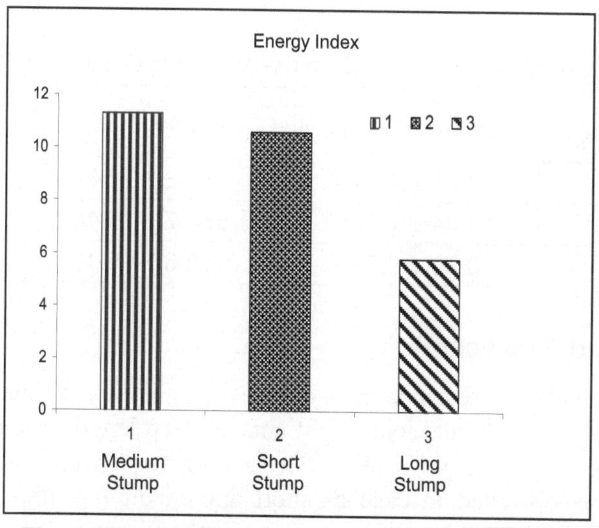

Fig. 1: Energy Index in Various Stump Length Groups

Stump Length and Velocity

Velocity was greater in case of the patients with longer stump length. The velocity linearly increased with increasing stump length.

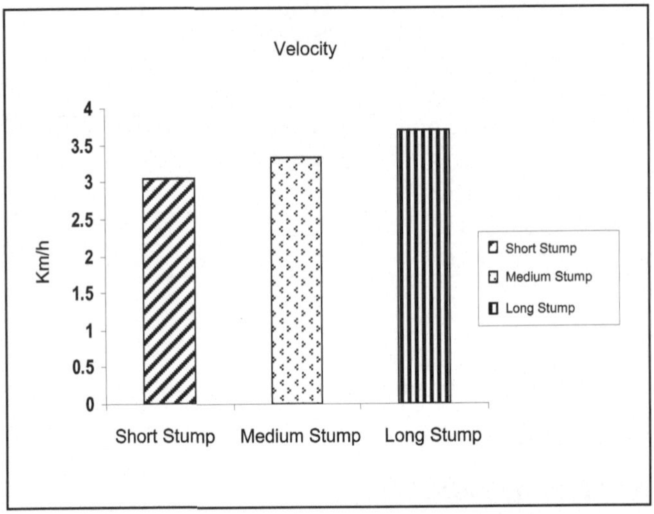

Fig. 2: Velocity in Various Stump Length Groups

Stump Length and Cadence

Correlation among cadence of three different groups was calculated which showed a significant difference with $p < 0.025$.Cadence of the subjects was increased normally with increasing stump length.

Table 2: Relation between Velocity, Cadence and Stride Length with Stump Length

Stump Length, % of Body Height	*Velocity (km/h)*	*Cadence*	*Stride Length, m*
Short	3.04 ± 1.14	90 ± 10.54	1.11 ± 0.32
Medium	3.34 ± 1.47	93.67 ± 15.18	1.17 ± 0.36
Long	3.70 ± 0.74	108.67 ± 5.13	1.13 ± 0.19

Stump Length and Step Length

In a complete gait cycle, the distance from one heel contact to opposite heel contact is the step lengths. During normal gait, equal right and left step lengths are taken. Considering different stump lengths, a significant difference in step length between sound and prosthetic limb was observed in case of short and medium stump length. The graph showed that difference in step length between normal and prosthetic limb was shorter in case of patients with long stump length. Group III with long stump length had a significantly shorter sound limb step length, however Group II, comprising patients with medium stump length had a significantly greater sound limb step length. But the statistical analysis shows that there is no significant difference in the step length of three different groups.

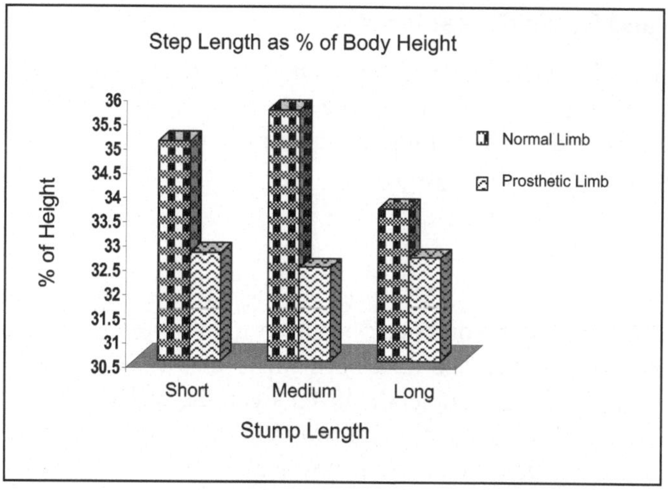

Fig. 3: Stump Length and Step Length Relationship

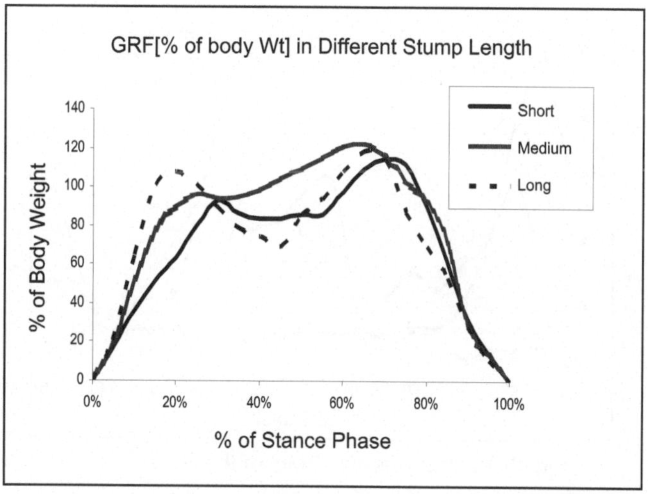

Fig. 4: GRF in Different Stump Length for Prosthetic Limb

Stump Length and Ground Reaction Force (GRF)

Ground reaction forces at each instance of the stance phase were calculated as the percentage of body weight for normalization, while walking at self selected speed with conventional SACH foot. GRF graph proved this fact that a greater peak vertical force in short stump length group was appeared in 80% than 20% of gait cycle. GRF pattern of long stump length patients showed a more similar 'M' shape curve with little bit greater impulse in push-off phase. This indicated a more energy storing gait in long stump length subjects.

Stump Length and Stride Characteristics

A significant correlation was observed between Group II and Group III. In this case the stride duration correlation coefficient between these two groups was 0.4483 ($p < 0.02$). The P-value for stride duration within three groups was $p < 0.03$ which indicated a significant difference among the 3 groups.

EMG and Stump Length

Normally over ground, the mean EMG peak of Vastus Lateralis (VL) activity at natural speed was in early stance. A second peak occurred in late swing in preparation for initial contact (Nymark *et al.*, 2005). The data normalization was carried out by adopting the maximum mean value of each subject's EMG over the stride period as the reference value (100%) (Winter, 1991). Each stride was divided into 10% interval and average peak amplitude of ten strides for each subject was given a value 100%.

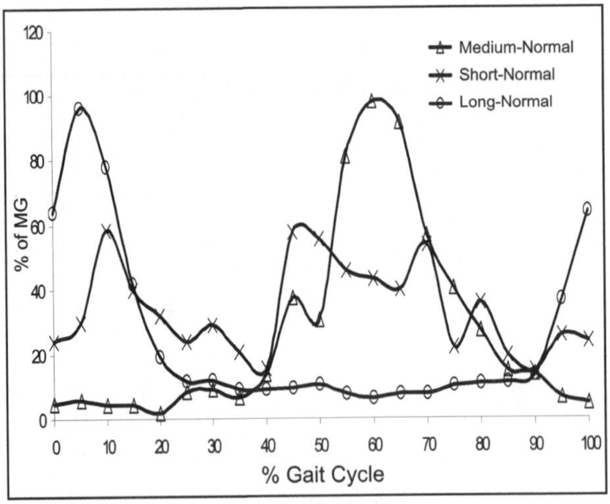

Fig. 5: Vastus Lateralis EMG of the Sound Limb

Each sub phase (10% of Gait Cycle) was expressed as a percentage of mean peak amplitude. The average value of amplitude expressed in percentage of Maximum Gait Contraction (MGC) for all nine subjects was calculated for each 10% of gait cycle. The EMG pattern of the subjects with medium stump length had peaks at 60 to 70% of gait cycle. The EMG pattern of the subjects with long stump length had peaks in the 0 to 10%

region of the gait cycle. These EMG patterns were almost similar in both legs. The only difference was the amplitudes higher in normal limb.

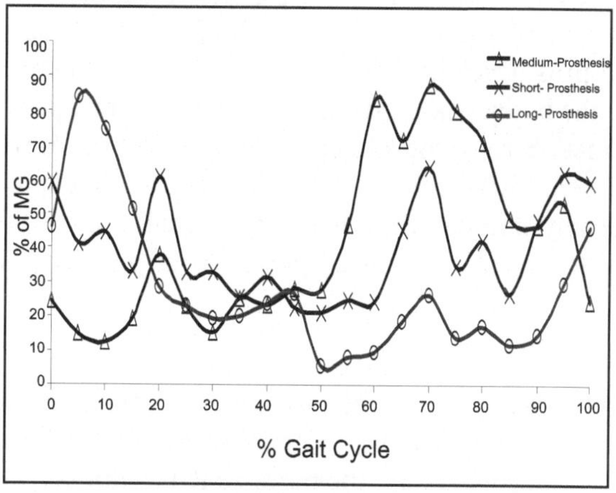

Fig. 6: Vastus Lateralis EMG of the Amputee Limb

DISCUSSION

A previous study had discussed that the successful rehabilitation of the amputee on a prosthetic limb depended on stump quality, thus, this study had only been confined to observe the impact of stump length on different gait parameters and EMG.

The result showed that the energy index decreased linearly with the increase in stump length which agrees with other work (Mertens and Lammens, 2001; Torburn, *et al.*, 1990; Gonzalez *et al.*, 1974).

The ideal mechanical advantage for a lever, like all other Simple machines is the length of effort arm divided by the length of resistance arm. According to this predicted theory below-knee amputees with long stump length where knee joint act as the fulcrum, have a resistance limb length longer than the effort limb length which cooperates to get a greater leverage arc and mechanical advantage with a respective increased ideal mechanical advantage, whereas in case of short stump length patients the mechanical advantage is limited because of the shortening of resistance limb length. It is obvious that a longer stump allows a better cantilever function in lifting the prosthesis, requiring less energy (Mertens and Lammens, 2001).

But velocity increased linearly with the increase in length of the stump. This indicated the more degree of motion with longer stump. Cadence of the individual group had a significant difference ($p < 0.025$) and it increased with the increase in stump length. Similarly, stride length and step length also increased with the increase in stump length whereas stride duration decreased.

Regarding Ground Reaction Force (GRFs) it had been noted that during the heel strike more force were generated with longer stump length. It was observed that the heel strike force increased with the increase in stump length. Whereas, during the period of the Toe-off, an average same force was generated. Here, the EMG study showed the same pattern in both the sound and prostheses limb, only mean amplitude increased with the increase in the stump length for the initial peak and decreased at the second peak. Initial peak indicated early phase of the stance. Strong muscle contraction occurred during early stance phase as stump length increased. The second peak during late swing depicted the preparation of next initial contact. Study showed that there were weak contractions during the above phase, indicated towards less energy consumption for longer stump length.

CONCLUSION

From our present study we can be able to conclude that, in case of a below-knee amputation, the more the stump length, the best is the gait efficiency in terms of less energy consumption, the more velocity, the more cadence etc.

Moreover, in case of below-knee amputation, a longer amputation stump allows the use of a patella-tendon-bearing suction prosthesis instead of a thigh-bearing prosthesis, providing better skin protection and a more efficient gait (Eldridge *et al*., 1990).

Along with other parameters, stump length is one of the considerable parameters in gait efficiency and prosthesis fitting in case of trans-tibial amputee. The longer the stump length, the productive will be the prosthesis efficiency.

ACKNOWLEDGEMENT

We are thankful to National Institute for the Orthopaedically Handicapped, Ministry of Social Justice & Empowerment, Govt. of India and Jadavpur University, for supporting this experimental study.

REFERENCES

[1] Barth, D.G., Schumacher, L. and Thomas, S.S. (1992). Gait Analysis and Energy Cost of Below-Knee Amputees Wearing Six Different Prosthetic Feet. *Journals of Prosthetics & Orthotics,* 4(2): 63–75.

[2] Campbell, W.B. (1996). Complications in Arterial Surgery. London: Butterworth Heinmann.

[3] Cosgrove, C.M., Thornberry, D.J., Wilkins D.C. and Ashley, S. (2002). "Surgical Experience and Supervision May Influence the Quality of Lower Limb Amputation." *Annals of The Royal College of Surgeons of England,* 84(4): 344–347.

[4] Eldridge, J., Armstrong, P. and Krajbich I. (1990). "Amputation Stump Lengthening with the Ilizarov Technique." *Clin. Orthop*, 256: 76–79.

[5] Gonzalez, E.G., Corcoran, P.F. and Reyas, B.L. (1974). "Energy Expenditure in Below-knee Amputee: Correlation with Stump Length." *Arch Phys Med Rehabil,* 55: 111–119.

[6] Henrot, P., Stines, J., Walter, F., Martinet, N. and Paysant, J.B. (2000). "Imaging of the Painful Lower Limb Stump." *RadioGraphics*, 20. pp. S219–S235.

[7] Mac, G.J. (1981). The Evaluation of Patient Performance Using Long-Term Ambulatory Monitoring Tech. in the Doniciliary Environment, Physiotherapy, 67, pp. 32–33.

[8] Mertens, P. and Lammens, J. (2001). "Short Amputation Stump Lengthening with the Ilizarov Method: Risks Versus Benefits"; *Acta Orthopœdica Belgica,* pp. 67–3.

[9] Nymark, J.R., Balmer, S.J., Melis, E.H., Lemaire, E.D. and Millar, S. (2005). "Electromyographic and Kinematic Nondisabled Gait Differences at Extremely Slow Overground and Treadmill Walking Speeds", *JRR&D,* 42(4), pp. 523–534.

[10] Sanders, J.E. and Daly, C.H. (1993). "Normal and Shear Stresses on a Residual Limb in a Prosthetic Socket during Ambulation: Comparison of Finite Element Results with Experimental Measurements. *Journal of Rehabilitation Research,* 30(2), pp. 191–204.

[11] Torburn, L., Perry, J., Ayyappa, E. and Shanfield, S.L. (1990). "Below-knee Amputee Gait with Dynamic Elastic Response: Prosthetic Feet: A Pilot Study." *Journal of Rehabilitation Research,* 27(4), pp. 369–384.

[12] Velzen, J.M.V., Houdijk, H., Polomski, W. and Bennekom, C.A.M. (2005). "Usability of Gait Analysis in the Alignment of Trans-tibial Prosthesis: A Clinical Study." *Prosthetics and Orthotics International,* 29(3), pp. 255–267.

[13] Winter, D. (1991). "Electromyography in Human Gait, in: Winter D. (ed.)", the *Biomechanics and MotorControl of Human Gait: Normal, Elderly and Pathological.* University of Waterloo Press.

[14] Yang, J.F.W. (1984). "Electromyographic Amplitude Normalization Methods: Improving Their Sensitivity as Diagnostic Tools in Gait Analysis", *Arch Phys. Med. And Rehab*, 65, pp. 517–521.

Section–III

Physical Environment

Assessment of Environmental Parameters in Old Age Homes

Shewanti N. Kashyap[1] and P. Sharma

Department of Family Resource Management, College of Home Science
G.B. Pant University of Agriculture and Technology
Pantnagar–263 145, U.S. Nagar, Uttarakhand, India
E-mail: [1]kashyap_sn123@yahoo.co.in

ABSTRACT

The environment provides more than shelter; it should promote continued development, stimulation, and satisfaction to enhance our psychological wellbeing. This is particularly important for the elderly, who spend considerable time in their homes or in a bedroom of an institution and reduce interaction with the larger environment of their communities. To achieve the fullest satisfaction from their microenvironments, the elderly must have various levels of needs met within their surroundings. The purpose of the study is to carry out appraisal of micro-environment, *i.e.* light, sound, humidity and temperature of old age homes. The present study was conducted in two regions of Uttarakhand State, *i.e.* Kumaon-Haldwani block of Nainital district and Garhwal (Haridwar and Dehradun district). The old age homes, *i.e.* Nirmala (Haldwani block); Geeta Kutir and Vridh Sewa (Haridwar), and Prem Dham (Dehradun) were selected randomly for the study. Thermo hygrometer was used for measuring indoor humidity and temperature. Sound was measured in decibel with the help of sound level meter. Lux meter was used to measure the intensity of light available in the room. The finding showed that the environmental parameters were not comfortable especially the illumination levels were far below the prescribed Indian electrical standard that might make elderly people prone to accidents in area like bathroom. Environmental safety could be maximized through sound insulation; improvement in residential temperature control would avert temperature extremes. Maintain a relative humidity of 30–70% at 18–24°C to accommodate a decrease in thermal range tolerance that comes with age.

Keywords: Old Age Home, Light, Sound, Humidity and Temperature.

INTRODUCTION

The environment provides more than shelter; it helps promote continued development, stimulation, and satisfaction to enhance our psychological well being. This is particularly important for the elderly, who spend considerable time in their homes or in a bedroom of an institution and have reduced interaction with the larger environment of their communities. To achieve the fullest satisfaction from their microenvironments, the elderly must have various levels of needs met within their surroundings. Housing for the older persons is one of the most important yet unappreciated issues confronting our society. The older persons spend 80%–90% of their lives in the home environment. The extent of the person-environment fit determines the quality of life of the older people. It provides access to daylight and fresh air. This can help keep a person's normal circadian rhythm functioning properly; improving overall health.□ Air conditioning must be set with caution, because it impedes adaptation to temperatures outside the regulated environment.

The relative humidity should not normally exceed 70% (Galar, 1995). Harold (2000) mentioned an ideal indoor relative humidity from 30% and 60% to minimize growth of allergenic or pathogenic organisms. Housing must provide protection from extremes of temperature as well as other environmental hazards. Elderly people are sensitive to lower environmental temperatures because of their lower body temperature and decreased amount of natural insulation make the elderly people especially sensitive to lower temperature. Mant (1993) investigated that the respiratory and circulatory diseases are most often associated with wintertime deaths. Rates of mortality for these diseases rise as the temperature drops.

Sound produces a variety of physiologic and emotional effects. Many of the sounds we take for granted—television, traffic noise, conversation from an adjoining room, appliance motors, leaking faucets. These can create difficulties for the elderly people. Many elderly already experience some hearing limitation because of prebycusis and need to be especially attentive to compensate for this defect. Environment noise can have a profound effect on quality of life and some circumstances can lead to hypertension. Being hard of hearing is more common than true deafness. Amplification of sound can be helpful, but too much of it can be painful and harmful. Defective hearing is sometimes associated with vertigo (sensation of rotation of the body), although not necessarily the cause (Salmon, 1993).

As the visual system ages, a number of changes in its structure and capabilities occur. The consequences of these changes with age are reduced visual acuity, reduced contrast sensitivity, reduced colour discrimination, increased time taken to adapt to large and sudden changes in adaptation luminescence and increased sensitivity to glare (Boyce, 2003). Little research has been done on the role of the physical home environment in promoting or restricting performance of daily tasks as people grow old. The purpose of the study was to appraise of micro-environment, *i.e.* light, sound, humidity and temperature of old age homes.

MATERIALS AND METHODS

The present study was conducted in two regions of Uttarakhand state *i.e.* Kumaon Haldwani block (Nainital district) and Garhwal (Haridwar and Dehradun districts). The old age homes, *i.e.* Nirmala (Haldwani block); Geeta Kutir and Vridh Sewa (Haridwar), and Prem Dham (Dehradun) were selected randomly for the study. Thermo hygrometer was used for measuring indoor humidity and temperature. Sound was measured in decibel; with the help of sound level meter. Lux-meter was used to measure the intensity of light available in the room.

RESULTS AND DISCUSSION

The observed environmental parameters are given in Table 1 while the recommended levels for those are given in Table 2. The recommended relative humidity should be 60%, not normally exceed 70% for comfortable living. The average relative humidity level of four old age homes was 67.75%; with 2.62 S.D. falls within the recommended value.

The average ambient temperature in all the four old age homes was 27.6°C in the month of August 2006. The recommended room temperature for an older person should not be lower than 24°C. Room temperature less than 21°C can lead to hypothermia in the elderly people. Brain damage can result from temperature exceeding 41°C. Comfortable temperature is 20°C to 23°C; it showed that the temperatures in the room of all the old age homes were not comfortable. The average noise level in all the old age homes was 41.95 dB with 3.19 S.D. and found within the recommended values of noise level for room *i.e.* 30 to 45 dB (WHO). The average illumination levels of four old age homes in the habitat room and bathroom were 35.92 Lux; with 6.11 S.D. and 21.37 Lux; with 3.44 S.D. was found to be far less than the prescribed Indian electrical standard, *i.e.* 50 Lux for habitat room and 100 Lux for bathroom. This showed that the old age homes provided insufficient illumination level to the elderly people may make elderly people prone to accidents in area like bathroom, etc.

Table 1: Environmental Parameters in Old Age Homes

S. No.	*Environmental Parameters*	*Old Age Home*				*Mean*	*S.D.*
		I	*II*	*III*	*IV*		
1.	Relative humidity (%)	68	69	70	64	67.75	2.62
2.	Ambient temperature (°C)	25.7	30.4	26.5	27.8	27.6	2.05
3.	Noise level (dB)	39.1	46.3	40.1	42.3	41.95	3.19
4.	Light level (Lux) (a) Habitat room (b) Bathroom	 45 25	 31.8 17	 32.9 20.5	 34 23	 35.92 21.37	 6.11 3.44

Table 2: Recommended Levels of Environmental Parameters

	Environmental Parameters	*Recommended Levels*
1.	Relative humidity (%)	60% (Not exceed 70%)
2.	Ambient temperature (°C)	20°C to 30°C
3.	Noise level (dB)	30 to 45 dB
4.	Light level (Lux) (a) Habitat room (b) Bathroom	 50 Lux 100 Lux

CONCLUSIONS

The environmental parameters (temperature, humidity, light and sound level) were not comfortable to the elderly people. The illumination level in all the old age homes was found to be substandard. Fluorescent lights are the most bothering because of eye strain and glare. Rapid changes in temperature can also be a cause of deterioration to health and should be guarded against. Environmental safety can be maximised through sound insulation; improvement in residential temperature control would avert temperature extremes. Ideally, noise control begins with the design of the building. Careful landscaping and walls can buffer outdoor noise. Acoustical ceilings, drapes, and carpeting are helpful.

REFERENCES

[1] Galer, I.A.R. (1995). "Applied Ergonomics Handbook", Butterworth, London, pp. 103–119.

[2] Harold, G.L. (2000). ASHRAE Handbook—HVAC Systems and Equipment Atlanta: American Society of Heating, Refrigerating and Air-conditioning Engineers, Inc., pp. 201–256.

[3] Mant, D. (1993). "The Parameters of Health and Housing Research." In R. Burridge & D. Ormandy (Eds.), Unhealthy Housing: Research, Remedies, and Reform London: E. & F.N. Spon, pp. 1–21.

[4] Salmon, G. (ed.) (1993). "Caring Environments for Frail Elderly People." Essex: Longman Scientific and Technical, New York: J. Wiley and Sons, pp. 8–11.

[5] World Health Organization (1987). "Housing: The Implications for Health." Geneva: World Health Organisation.

Effect of Vehicular Vibration on Two Wheeler Drivers—An Ergonomic Study

M. Muzammil[1], Faisal Hasan and Saman Ahmad

Ergonomics Research Divison, Mechanical Engineering Department
Z.H. College of Engineering and Technology, Aligarh Muslim University
Aligarh–202 002, U.P., India
E-mail: [1]mmuzammil18@yahoo.co.in

ABSTRACT

Exposure to whole body vibration may have detrimental effects on driving performance. As the use of two wheelers is increasing day by day it becomes important to investigate the effect of vibration on the driving task performance. The present study was designed to investigate the physiological effect of vibration on users driving two wheelers. Five subjects all males, enjoying a sound health participated in the study. The subjects adopted a pre-specified sitting posture while driving the vehicle for different durations (30, 60 minutes) under varying levels of vibration (0.8, 1.7 and 2.1 m/s^2) in a real life-driving environment. The levels were selected on the basis of a pilot study conducted to determine the vibration levels as per ISO 2631–1 (1997). Heart rate was recorded before and after performing the task to measure the physiological performance using pulse oximeter. A two-way ANOVA was carried out to determine the effects of parameter under investigation. Results of the study indicated that the effect of level of vibration was statistically significant on the driving task performance of the operators. However, the exposure duration and the interaction between exposure duration and level of equivalent vibration were found to be statistically non-significant. The results have been discussed in the light of previous researches carried out on similar topic. It is recommended that manufacturers of two-wheeler should take necessary steps to minimize the generation of vibration by providing good suspension system, shock absorbers and ergonomically designed seats.

Keywords: Vehicular Vibration, Driving Environment, Heart Rate.

INTRODUCTION

The human vehicle system represents one of the most widely used systems that challenges the system designers of tomorrow in designing and developing a model compatible with human capabilities. The two wheelers are being extensively used as a mode of transport by a large population. The frequent use of vehicles has led to question as to whether users are subjected to environmental stressors or not. The human vehicle system environment is one in which continuous manual control needs to be exercised. Driving is directly related to physical and mental capabilities and limitations of the concerned operator. Several researchers (Seidel and Heide, 1986; Burton and Sandover, 1987; Grandjean, 1988) have agreed that physiological damage occurs when the subjects are exposed to vibrations. Though the International Standards Organization (ISO 2631–1997) has prescribed a limit for vibration exposure, it has been observed that these limits are generally not adhered to. According to Dupuis and Zerlett (1986), failing to comply with the ISO standards in any one of the x, y or z directions rendered the whole body vibration exposure severe and capable of compromising worker safety and health. Realizing the seriousness of the vibration exposure several researchers (Chen and Chang, 2003; Barregard, 2003; Ahlin, 2002) had conducted studies on the topic. Rehn (2005) also conducted a study to investigate the variation in exposure to whole body vibration for operators of forwarders vehicles. It was found that the whole body vibration was affecting human performance in a negatively significant way. In a similar study conducted by Sherwin (2004) it was shown that whole body vibration were causing discomfort to the operator. Literature reviewed on the topic revealed that a large number of studies had been conducted in the past on effect of whole body vibration on users employing four wheelers or other heavy vehicles while not much was done in the context of two wheelers. As a major chunk of population across the globe use two wheelers in one way or the other, it becomes imperative to study the effect of vibration on them. Keeping this in view the problem was formulated to study the effect of vibration on users while driving two wheelers. The experimental hypothesis structure was that, the vibration level and exposure duration had a significant effect on heart rate while driving two wheelers. To test the hypothesis the experimental investigations were carried out as detailed below.

METHOD

Fifty-five male subjects who were enjoying a sound health participated in the study. They were all teetotalers and were not consuming any psychotropic drugs. The subjects had normal vision with no previous history of neuromuscular disorders. All the subjects were between 22 and 30 years of age (Mean = 26.4 yrs., S.D. = ± 1.6 yrs.). As far as the task was concerned the subjects sat on the seat of the two-wheeler in a normal sitting position and drove the vehicle for different durations (30 and 60 minutes) under varying levels of vibration (0.8, 1.7, 2.1 m/s^2) in a real life driving environment. The total equivalent vibration of weighed rms acceleration was calculated as per the recommendations of ISO 2631–1 (1997). The levels were selected on the basis of pilot study conducted to determine the vibration levels as per ISO 2631–1 (1997). The vibration level meter (VR 5100, Ono Sokki, Japan) was used to measure the level of vibration. The values of vibration levels under different vehicular speeds were presented in Table 1. All the data were recorded between 9:00 A.M. and 3:00 P.M. Each subject was completely familiarized with the experimental task before performing the actual experiment. A buzzer was used to signal the starting of the task. After the stipulated period of driving the buzzer was operated again to mark the end of the experimental period. Heart rate was recorded before and after performing the task to measure the physiological performance using pulse oximeter. A two-way ANOVA was carried out to determine the effects of parameter under investigation.

Table 1: Range of Equivalent Vibration Levels under Varying Vehicular Speeds

Equivalent Vibration Levels (m/s²)		
X	Y	Z
0.014–0.038	0.23–0.47	0.7–2.1

RESULTS

Results of the present study (Table 2) indicated that the effect of level of vibration was statistically significant on the driving task performance of the operators. However, the exposure duration and the interaction between exposure duration and level of equivalent vibration were found to be statistically non-significant. The graph plotted between the level of equivalent vibration and change in heart rate showed that the heart rate increased with the increase in the vibration level. The best-fit model (Figure 1) along with the correlation coefficient has been presented below.

For 60 minutes exposure duration:
$HR = 6.1V + 8.67, \quad r^2 = 0.99$

For 30 minutes exposure duration:
$HR = 6.2V + 4.00, \quad r^2 = 0.98$

Where, HR is the change in heart rate value and V is the level of equivalent vibration.

Table 2: ANOVA Results when Operators Performed the Driving Task

Sources of Variation	*df*	*MS*	*F-value*	*p-value*
Exposure Duration (A)	1	149.6	2.628	0.1181
Vibration Level (B)	2	378.3	6.64	0.0051
A x B	2	3.1	0.055	0.9466
Error	24	56.9		

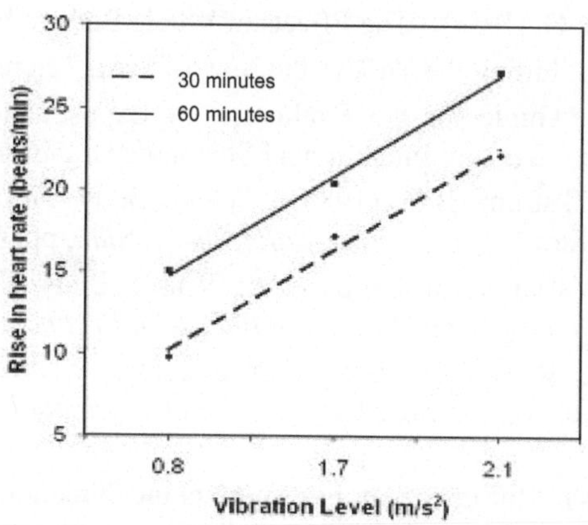

Fig. 1: Variation of Heart Rate with Equivalent Vibration Level for Different Exposure Durations

DISCUSSION

Although, one could hardly find a study involving investigation of the vibration effects on operators of two wheelers, other researchers working on similar topics have shown that vibration affects the operators in a negative significant way (DeWaard *et al.*, 1995; Kjellberg and Wikstrom, 1985). Present study also showed an increase in human fatigue with the increase in the level of vibration. Higher level of vibration requires extra effort to ward off its effect causing operator to get fatigued early. To avoid this situation the equivalent levels of vibration should be kept within the permissible limit. Low back pain, postural stress, etc., are a few manifestations of whole body vibration exposure (Grandjean, 1988; Bovenzi, 1996). Two-wheeler manufacturers should take necessary steps to minimize the generation of vibration by providing good suspension system and shock absorbers. Ergonomic design of seats would also help in reducing them.

REFERENCES

[1] Ahlin, K. (2002). "Relating Road Roughness and Vehicles Speeds to Human Whole Body Vibration and Exposure Limits." *International Journal of Pavement Engineering,* V. 3(4), pp. 207–216.

[2] Barregard, L. (2003). "Short Daily Exposure to Hand Arm Vibration in Swedish Car Mechanics", *Journal of Occupational and Environmental Hygiene*, V. 18(1), pp. 35–40.

[3] Bovenzi, M. (1996). Low Back Pain Disorders and Exposure to Whole Body Vibration in the Workplace." *Semin Perinatol,* V. 20, pp. 38–53.

[4] Burton, A. and Sandover, J. (1987). Backpain in Grand Prix drivers: a found experiment. *Applied Ergonomics.* V. 18, pp. 3–8.

[5] Chen, J.C. and Chang, W.R. (2003). "Predictors of Whole Body Vibration Levels among Urban Taxi Drivers", *Ergonomics,* V. 26(11), pp. 1075–1090.

[6] DeWaard, D., Jessurum, M., Steyvers, F.J., Raggatt, P.T. and Brookhuis, K.A. (1995). "Effect of Road Layout and Road Environment on Driving Performance, Driver's Physiology and Road Appreciation." *Ergonomics*, V. 38, pp. 1395–1407.

[7] Dupuis, H. and Zerlett, G. (1986). "The Effects of Whole Body Vibrations." Springer, Berlin.

[8] Grandjean, E. (1988). "Fitting the Task to the Man (4th ed.)." London: Taylor and Francis.

[9] ISO 2631–1 (1997). "Guide for the Evaluation of Exposure to Whole Body Vibration-general Requirements." Geneva: International Standards Organisation.

[10] Kjellberg, A. and Wikstrom, B.O. (1985). "Whole Body Vibration: Exposure Time and Acute Affects—A review." *Occupational Safety and Health*, pp. 171–184.

[11] Rehn, L. (2005). "Variation in Exposure to Whole Body Vibration for Operators of Forwarder Vehicle", *International Journal of Industrial Ergonomics*, V. 35, pp. 831–842.

[12] Seidel, H. and Heide, R. (1986). "Long term effects of whole body vibration: A critical survey of the literature", *International Archives of Occupational and Environmental Health.* V. 58, pp. 1–26.

[13] Sherwin, L.M. (2004). "Influence of Forest Machine Function on Operator Exposure to Whole Body Vibration in Cut to Length Harvest", *Ergonomics*, V. 47(11), pp. 1145–1159.

Dal Cleaner with Dust Collector: A Remedy to Arrest Dust Pollution in Dal Mills

Nachiket Kotwaliwale[1] **and Krishna Mohan Sahay**

Central Institute of Agricultural Engineering, Bhopal–462 038, India
E-mail: [1]nachiket@ciae.res.in

ABSTRACT

Dust and noise are the two major pollutants in the pulse milling industry. An earlier study revealed that the suspended particle concentration in commercial dal mills was 10 to 100 times the prescribed limit of 5 mg/m^3 for respirable and organic dust. Also the noise level near the machinery reaches to as high as 97 dB (A) against a safe limit of 90 dB (A) for 8 h exposure. Currently, there is no provision for any dust collection in the cleaners being used at traditional dal mills and emitted dust is suspended in the plant environment. Some commercial dal mills have provision of domestic or slightly larger exhaust fans; however, these hardly serve the purpose of clearing the environment inside the plant.

The power operated cleaner developed at CIAE earlier for food grains was modified and a centrifugal aspirator was attached to the grain falling chute. The outlet of this unit was connected to a cyclone separator to collect dust and fines. Therefore, the dust emission during operation is arrested. Capacity of the cleaner is 900 kg/h. Dust emission from the equipment was measured while cleaning milled pigeon pea and chickpea. Grains were milled to separate hull and to break cotyledons using standard method adopted at Agro Processing Centre of CIAE for milling these pulse grains. Suspended particle concentration was in the range of 5.01 to 7.32 mg/m^3 as against 51.94 mg/m^3 without using the dust collection mechanism. Whereas the safe limit for environmental dust pollution is 10 mg/m^3.

INTRODUCTION

Increased awareness for quality product, increased cost of energy, scarcity of energy sources and strict government regulations for pollution control and energy conservation make it necessary to determine the status of technology-cum-management, energy utilization and pollution in the industries. The highly diversified food industry having processing plants of small to high capacity, has different unit operations. It utilizes many types of machineries, power sources like—electrical, thermal, human, animal, solar, fuel, bio energy, etc. and may create pollution to the working or surrounding environment through dust, effluent, noise, etc. The post harvest operations where dust problem exists are cleaning, milling and grinding. Some conditions in which contaminant vapour or dust, may enter the atmosphere include (1) passage of air through the material (2) free fall of material, (3) pouring of the material, (4) agitation of the material, (5) transfer of the material, (6) opening of bags or drum of material, (7) leak of the material, (8) opening up of plant containing material and (9) handling of fabric saturated with the material. In addition the dust generation may occur due to (1) vibration of a powder or (2) a size reduction of material (Lees, 1998). The limits for different types of suspended particle matters in the working environment are shown in Table 1.

An earlier study revealed that the suspended particle concentration in commercial dal mills was 10 to 100 times the prescribed limit of 10 mg/m^3 of total dust and 5 mg/m^3 of respirable and organic dust. It was therefore, considered necessary to develop a cleaner to reduce dust pollution which can be integrated with the existing CIAE dal mill.

Table 1: Recommended Exposure Limits for Dust

	Total Weighted Average		*Peak Limit or Short Term Limit for 15 min*
	Total	*Respirable*	
Dust	10 mg/m^3	5 mg/m^3	–
Organic dust	5 mg/m^3	–	–
Quartz	–	0.1 mg/m^3	
Oil mist, fumes	1 mg/m^3	–	3 mg/m^3

Ref. HSE (1993, HS(G)37).

MATERIALS AND METHOD

Development of the Equipment

A cleaner cum grader for grains already developed in CIAE was modified. The blowing mechanism adopted for the lighter material in the earlier model was replaced with suction mechanism in the modified equipment. To design the suction mechanism, bulk density, true density, moisture content, particle size distribution, settling velocity and stopping distance were considered for the dust emitted from the dal mills.

Testing of the Equipment

Concentration of the suspended particles was measured using a high volume air sampler (Figure 1). The high volume air sampler consisted of an aspirator, a face plate to mount filter paper, a shelter hood, current stabilizer cum timer, manometer for air flow measurement and gas sampling kit. The glass fibre filter paper was exposed to light source and inspected for any damage/pinholes, etc. The filter paper was then weighed to the fourth decimal of a gram. The equipment was dedusted properly prior to its use. The filter paper was mounted under the faceplate and the nuts of faceplate were tightened. The U-tube glass manometer, filled with water, was read, prior to starting the aspirator, for correction, if any. The shelterhood was closed and the unit was run for 5 minutes. The time and date of sampling were recorded and the timer was set for desired duration and equipment was switched on. The manometer reading, *i.e.* water level in both the arms, was observed at 1 minute interval and noted. At the end of experiment, the filter paper was removed, folded and taken back to laboratory. The filter paper was kept in hot air oven, maintained at 105°C, without unfolding for one hour. It was then cooled in desiccators for 15 minutes at room temperature. It was then again weighed to fourth decimal of gram. The average pressure measured by the manometer was used to find out the airflow rate from the calibration curves provided by the manufacturer of the equipment. The concentration of dust, g/m^3, was measured using following equation.

$$D_c = \frac{W}{V.T}$$

Where,

W is difference in initial and final weights of filter paper, g
V is Average air flow rate, m^3/min
T is operation time, min.

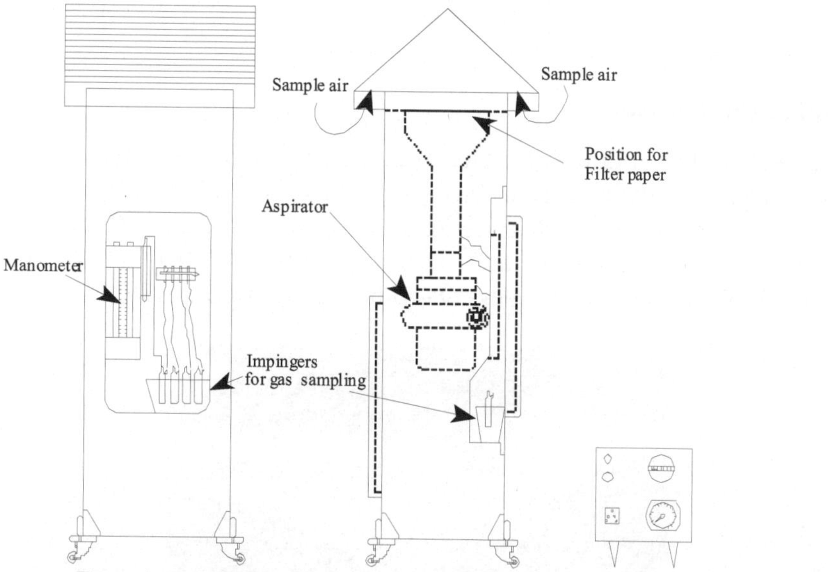

Fig. 1: High Volume Air Sampler

RESULTS AND DISCUSSION

The results for Dal mill dust sample pertaining to chickpea milling are given in Table 2.

Table 2: Density and Moisture Content of Dal Mill Dust

Sample	*Sample Code*	*Bulk Density, kg/m^3*	*True Density, kg/m^3*	*Moisture Content, % (db)*
Milling section dust	MSD-CP	367.85	1285.62	10.27
Cleaning section dust	CSD-CP	349.17	1303.94	9.85

The dust is soluble in strong acid or alkali solutions, *i.e.* pH < 4 or pH > 10. Dust is not soluble in petroleum based solvents *viz.* Toluene, etc. Particle size analysis of the dust collected from milling and cleaning sections of a Dal mill revealed that all the particles in the of milling section were of the size less than 425.3 μm whereas, 90% of the particles were smaller than 182.31 μm. These values for cleaning section dust were observed to be 363.1 μm and 104.46 μm, respectively.

The BOHS guide gives the following values for settling velocity and stopping distance

Particle Diameter (mm)	*Settling Velocity (m/s)*	*Stopping Distance (m)*
0.01	7.0×10^{-8}	6.8×10^{-8}
1.0	3.5×10^{-5}	3.6×10^{-5}
100	0.25	0.127

The Modified Dal Cleaner

A power operated cleaner/grader for cleaning of pulses and splits was developed. The equipment is equipped with suitable ducts, a forward curved centrifugal aspirator and cyclone separator to arrest dust emission during the operation (Figure 2). Major specifications of the equipment are: input capacity 800–900 kg/h, total power requirement 500 W, screen effectiveness 68–87%, seed purity 98.5–99%.

Fig. 2: Dal Cleaner with Dust Collector

Dust emission from the equipment was measured five times; three times while cleaning milled pigeon pea and two times while cleaning milled chickpea. Grains were milled to separate hull and break cotyledons using standard method adopted at Agro Processing Centre of CIAE for milling these pulse grains. Suspended particles in the air while the machine was in operation are shown in Figure 3.

It is noteworthy that the suspended particle concentration is quite low compared to the one reported earlier for grain cleaner with provision of dust collection, which was 51.94 mg/m^3.

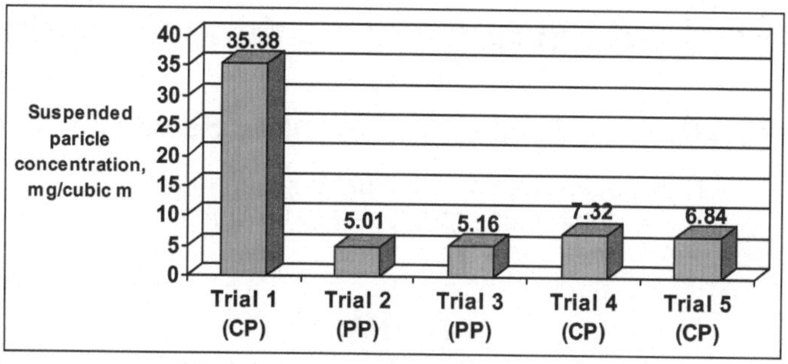

CC- Chickpea milling; PP – Pigeon Pea milling

Fig. 3: Suspended Particle Concentrations Observed during Different Trials of the New Grain Cleaner

Suspended particle concentration in air was observed quite higher than the other trials during one trial. This was due to high dust emission during refilling of hopper and because the dust collector-bag of the cyclone opened accidentally during this trial. Later experiments were conducted for cleaning only one hopper full material, so that error in experiment due to refilling of hopper is eliminated. It is again noteworthy that suspended particle concentration dropped drastically due to this change.

Recommended worker exposure limits for suspended particles in the air are 10 mg/m^3 for total dust and 5 mg/m^3 for respirable dust. It can be noted that all the values except for Trial 1 are much below the limit set for total dust concentration. In fact for pigeon pea grain these values are close to the limit set for respirable dust exposure.

Dust concentration was observed to be higher while cleaning chickpea. This was also reported in an earlier survey work conducted by this investigator. This is probably because of particle size difference between the two grains. Chickpea milling probably generates much smaller particles than pigeon pea milling.

CONCLUSIONS

- The new cleaner had a suction provision for dust rather than blowing the lighter particles.
- Dust emission from the new cleaner was much lesser than the cleaner without dust suction provision. Average suspended particle concentration in ambient air was 5.09 and 7.08 mg/m^3 for pigeon pea and chickpea, respectively. While this value for the cleaner without dust suction provision was 51.94 mg/m^3.
- Suspended particle concentration in ambient air for this new cleaner was below the safe exposure limit specified for total dust concentration (10 mg/m^3) and very close to the safe exposure limit specified for respirable dust concentration (5 mg/m^3).

REFERENCES

[1] HSE (1993), HS (9) 37.

[2] Lees, Frank P. (1998). Loss prevention in the process industries Vols. I, II & III. Butterworth-Heinemann, Oxford. UK.

A Study of Fatigue among School Children

C.S. Savanur, U. Dhar, S. Ghosh, S. Wani, R. Salve and Amitabha De

National Institute of Industrial Engineering (NITIE)
Vihar Lake, Mumbai–400 087, India

ABSTRACT

Fatigue is defined as a relatively temporary inability or decrease in ability or strong disinclination to respond to a situation. The common causes of fatigue are mental workload, prolonged physical exertion, adverse environmental condition, and disturbance in sleep. Fatigue can be divided into three types: General, physical and mental. General fatigue represents feeling of dull, drowsy and exhaustion in the body. Physical fatigue is specific to sensory and nervous (dysfunction of autonomic nervous systems). Mental fatigue is purely mental, having no physical properties. The current study was taken up with an aim to understand if there were any relationships with the classroom factors such as responses for illumination, chalkboard visibility, thermal level, noise level, bag weight and external factors such as physical activity, sleep duration, wake up time and TV viewing duration. The study was carried out on fifty-four students of grade six to grade eight in a school in Mumbai. A questionnaire was administered in the morning on arrival of the students and after three hours of exposure during their recess. The results for percentage students reporting fatigue were; 43.15% group A (general); 47.96%-group B (mental) and 44.62% group C (physical). Significant relationships were established between physical activity and general fatigue, before exposure ($r = 0.295$, significant at 0.032); response for thermal level, sweat and general fatigue after exposure, ($r = 0.351$ and 0.274, significant at 0.012 and 0.049 respectively); visibility of chalkboard, ventilation and mental fatigue after exposure, ($r = 0.332$ and 0.314, significant at 0.015 and 0.023 respectively).

Keywords: Fatigue, Children, Classroom Environment and Physical Activity.

INTRODUCTION

Fatigue is often used with different implications, for instance physiologists consider fatigue as a decrease in physical performance, psychologists consider it as a condition that affects mental process, including factors such as feeling of fatigue, motivation and resulting deterioration of performance, ergonomists and physicians lay more emphasis on consequences of fatigue. The different interpretations of the word 'fatigue' have been caused by the fact that the word does not have a specific scientific meaning (McFarland, 1975). However, there are many definitions for fatigue one of which is that fatigue is a temporary inability, decrease in ability, or strong disinclination to respond to a situation, because of previous over-activity, either mental or physical (OSH, 2003).

In a tired condition, the subjective sensation of fatigue predominates. One feels not only tired or dull in bodily parts and clumsy in motion, but also feels hampered and inhibited in doing either physical or mental work; activities are reduced till given up. Desire for either physical or mental work does not remain. The feeling of fatigue consists of such types of unpleasantness as aversion to work, desire to rest, impatience, and physical, mental and neuro-sensory feelings of incongruity (Hasuo, 1963). The sensation of fatigue is not unpleasant when one is able to rest. But it is painful if rest is not permitted. The sensation of fatigue has protective function and forces to avoid further stress and allows recovery. Its causation is cumulative effect of various factors such as monotony, surroundings *i.e.,* illumination, climate, noise, intensity and the length of manual and mental work, psychic factors *i.e.,* responsibilities, worries, conflicts, illness, pain, eating habits and disturbance in sleep (Grandjean and Kogi, 1975). Unless duly eliminated, fatigue has a tendency to grow into clinical or chronic fatigue. Fatigue can be divided into three types: General, physical and mental. General fatigue represents feeling of dull, drowsy and exhaustion in the body. Physical fatigue is specific to sensory and nervous (dysfunction of autonomic nervous systems). Mental fatigue is purely mental, having no physical properties (Yoshitake, 1975).

Fatigue in school children is not much reported in literature. However, our previous study (Savanur *et al.*, 2006) showed that there was a correlation between test performances of children and fatigue reported by them. Based on the findings of previous study the current study was undertaken. The aim of the current study was to understand if there existed any relationships between fatigue and the classroom factors (such as responses for illumination, chalkboard visibility, thermal level, noise level, bag weight) and also with external factors (such as physical activity, sleep duration, wake up time and TV viewing duration).

METHODS

Subjects

The study was carried out on fifty-four students of grade six to grade eight. The students belonged to the age group of 11–14 years.

Locale of the Study: A School in Mumbai

Tools

Fatigue Questionnaire: The fatigue questionnaire developed by Yoshitake, 1975 was administered in the morning on arrival of the students and after three hours of exposure during their recess. The questionnaire consists of three parts *viz.* group A (general), group B (mental) and group C (sensory/physical). The questionnaire had 30 items (ten items in each part). The responses were put in the formula given below to obtain percentage.

$$\frac{\text{Gross number of the complains in a group}}{\text{Number of items} \times \text{number of each group members}} \times 100\%$$

Classroom Factors: Responses regarding classroom factors such as thermal level, noise, ventilation in the classroom, sweat, illumination on the desks, visibility of the chalkboard, bag weight were obtained from the students.

External Factors: Responses regarding external factors such as physical activity, time taken to travel from home to school and back, sleep duration, wake up time and TV viewing duration were also sought from the students.

Statistical Treatment: The classroom factors and the external factors were correlated with group A (general), group B (mental) and group C (sensory/physical) fatigue, both before exposure and after exposure.

RESULTS

Fatigue

The demographic results that are presented in the Table 1 feature the percentage of students reporting fatigue.

Table 1: Percentage of Fatigue Reported by Students

Fatigue	*Students Reporting Fatigue after 3 hours Exposure (%)*
Group A (general)	43.15
Group B (mental)	47.96
Group C (sensory/physical)	44.62

Classroom Factors

The responses of students with respect to the classroom factors were collected. The problems indicated by the students (in terms of percentages) are presented in the Figure 1.

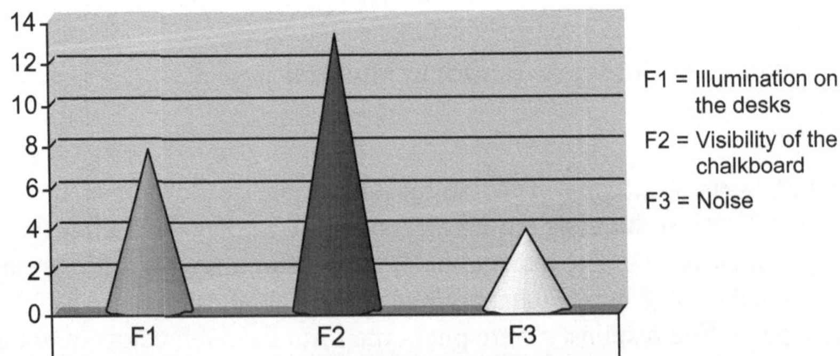

	F1	F2	F3
Problems indicated by students (%)	7.7	13.2	3.8

Fig. 1: Percentage of Students Indicating Problems in the Classroom Factors

External Factors

Nearly 30% of the students took more than 10 minutes to travel, both, from home to school and back, around 49% students slept for less than 8 hours duration. 19.7% students woke up before 6.00 am in the morning and 71.7% students' TV viewing duration was less than 2 hours.

Relationship of Fatigue with Classroom Factors

The students' response to classroom factors such as thermal level, sweat, noise, ventilation in the classroom, illumination on the desks, visibility of the chalkboard, bag weight was correlated with responses for fatigue before and after exposure. Significant relationship was not found between noise, illumination on the desks and bag weight with both fatigue experienced before and after exposure. Significant relationship was established between response for thermal level and general fatigue (after exposure) (Table 2).

Table 2: Relationship between General Fatigue after Exposure (group A) and Classroom Factors

Classroom Factors	*r-value*	*Significance Level*
Thermal level	0.351	0.012

Significant relationship was also found between the visibility of chalkboard and mental fatigue (after exposure) (Table 3).

Table 3: Relationship between Mental Fatigue after Exposure (group B) and Classroom Factors

Classroom Factors	*r-value*	*Significance Level*
Visibility of chalkboard	0.332	0.015

Relationship of Fatigue with External Factors

The students' response to external factors such as physical activity, time taken to travel from home to school and back, sleep duration, wake up time and TV viewing duration was correlated with the responses for fatigue before and after exposure. Factors such as time taken to travel from home to school and back, sleep duration, wake up time and TV viewing duration did not show any significant relationship with fatigue, both before and after exposure. However, significant relationship was established between physical activity and general fatigue, before exposure ($r = 0.295$, significant at 0.032).

DISCUSSION AND CONCLUSION

Nearly 50% of the students under this study had reported fatigue both before exposure and after three hours of exposure, in all the three categories *i.e.,* general, mental and physical/sensory. This calls for concern. Moreover, the responses given by the students regarding the classroom factors such as thermal level was significantly correlated to general fatigue. Responses regarding visibility of the chalkboard were significantly correlated to mental fatigue. These findings are in line with the causes of fatigue suggested by Grandjean and Kogi (1975). Among the external factors, physical activity has shown significant relationship with general fatigue. Although there are significant relationships found between fatigue and both classroom and external factors, the results of this study cannot be generalized as it is based on a small sample size. An extension of the present study may be taken up with larger sample size to be able to generalize the findings.

ACKNOWLEDGEMENT

We are thankful to National Institute of Industrial Engineering (NITIE) and TIFAC for providing the necessary finance and support for the project. We express our sincere gratitude towards the school authority and teachers for allowing the research team to conduct the study. We are also thankful to the students for actively participating in the study.

REFERENCES

[1] Grandjean, E. and Kogi, K. (1975). "Introductory Remarks in Methodology in Human Fatigue Assessment", Hashimoto K., Kogi K. and Grandjean E. (eds), Taylor and Francis Ltd.

[2] Hasuo, T. (1963). "Industrial Fatigue and its Counter Measures." *Management Psychology Series,* Vol. 3 (Tokyo: DAIYAMMONDA-SHA).

[3] McFarland, R.A. (1975). "Understanding Fatigue in Modern Life. In Methodology in Human Fatigue Assessment", Hashimoto K., Kogi K. and Grandgean E. (eds), Taylor & Francis Ltd., pp. 1–10.

[4] OSH (2003). "Healthy Work, Managing Stress and Fatigue at Work Place", Published by the Occupational Safety and Health Service, Department of Labour, Wellington, New Zealand.

[5] Savanur, C.S., Dhar, U., Ghosh, S., Wani, S., Salve, R. and De, Amitabha (2006). "A Study of Classroom Factors Influencing the Performance in School Children." *Paper presented in the 17th Annual Conference of the Physiological Society of India held during December 8–10, 2006* at Presidency College, Kolkata.

[6] Yoshitake, H. (1975). "Relation between the Symptoms and the Feeling of Fatigue. In: Methodology in Human Fatigue Assessment", Hashimoto K., Kogi K. and Grandgean E. (eds), Taylor & Franciss Ltd., pp. 175–185.

Assessment of Available Dust Protectors for Protection to Combine Harvester Operators

N.K. Chhuneja

Department of Farm Power & Machinery, Punjab Agricultural University, Ludhiana–141 004, Punjab, India

ABSTRACT

Six different types of available personal dust protector were evaluated for combine harvester operators against dust hazard. Out of these three were of mask type and remaining three of disposable type. These were assessed for their performance as per BIS standards IS 9473–1980. All the selected dust protectors failed in respect of filtration efficiency tests. The filtration efficiency of the dust protectors varied between 65.5% (single tissue paper) and 92.2% (cloth). The dust protectors were subjected to suitability tests of 1 hour each on 20 combine harvester operators during wheat as well as paddy harvesting season. The suitability performance was recorded in terms of breathing resistance, dust filtration performance, leakage of air from sides, feeling of tightness on face, rate of sweating, and wear & tear of dust protectors while in use. The dust protector of disposable cloth material was noted to have the highest acceptability during wheat as well as paddy harvesting season. Thus, cloth type dust protector provided the highest dust protection. Besides, it was a cost effective, easily available and adaptable solution to the operators.

Keywords: Dust Protection, Dust Hazard, Combine Harvester, Farm Machinery.

INTRODUCTION

Wheat and rice are the most important and major cereal crops of India. Need of timely harvesting of crops sown in large areas and to allow timely sowing of subsequent crop have resulted in introduction of combine harvesters in a large number. Combine harvesters are bulky, heavy and have to work in rough terrain of farm. Heavy dust concentration acts as one of the major health hazards to the operators. The ideal concept for the safeguard for operator is to provide a cabin with a small over pressure, so that the particles cannot penetrate into the workspace. But this involves increase in cost and hence is not being adopted. The economical method for protection of operator of the combine harvester is by use of personal dust protectors.

METHODS

Six different types of commercially available dust protectors were identified and procured for their evaluation. All of the selected dust protectors had different types of material of filter elements. Out of these three were mask type and remaining three disposable types. The purchase price of those ranged between Rs. 2.50/- and Rs. 58.50/- (Table 1).

Fig. 1: Dust Concentration as a Health Hazard to Operators of Combine Harvesters

All the selected dust protectors were got tested from Central Labour Institute, Mumbai as per test procedure in IS 9473–1980 (Specification for filter type particulate matter respirators) for their filtration performance. In addition, all the dust protectors were subjected to suitability tests of 1 hour each on 20 combine harvester operators during wheat as well as paddy harvesting season. The suitability performance was recorded in terms of breathing resistance, dust filtration performance, leakage of air from sides, feeling of tightness on face, rate of sweating, and wear-tear of dust protectors while in use.

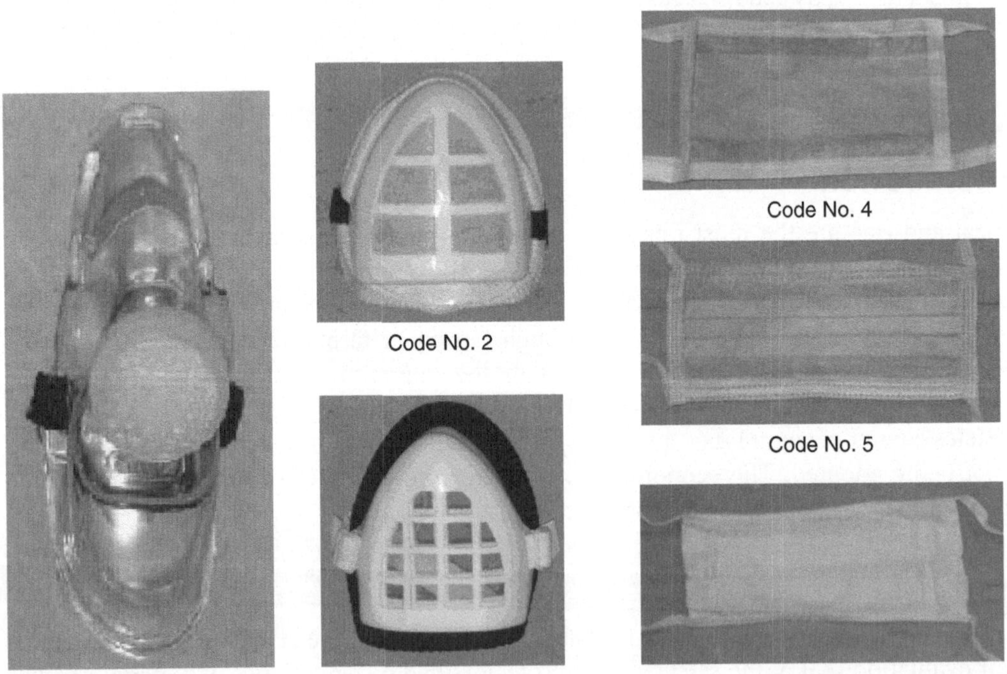

Fig. 2: Selected dust Protectors for Their Comparative Assessment

Table 1: Various Dust Protectors Selected for the Study

Type	*Code No.*	*Filter Material*	*Cost (Rs.)*
Mask type	1	Foam and plastic net	58.50/-
	2	Tissue paper and foam	25.00/-
	3	Plastic net	21.00/-
Disposable type	4	Tissue paper	2.50/-
	5	Tissue paper (2 layers)	2.50/-
	6	Cloth	7.50/-

RESULTS AND DISCUSSION

Conformity of Performance of Dust Protectors with the Standards

All the selected dust protectors were got tested as per IS 9473–1980, which deals with classification of respirators, design requirements, and test procedures. Various tests and their acceptable limits are given in Table 2 and the performance obtained for the dust protectors is given in Table 3.

Table 2: Various Tests and Acceptable Performance as per IS 9473–1980

S. No.	*Tests*	*Acceptable Limit*
1.	Inhalation and exhalation resistance	Maximum breathing resistance for inhalation before silica dust test shall not exceed 30 mm of water column and after silica dust test for 90 minutes shall not exceed 50 mm of water column. Maximum breathing resistance for exhalation shall not exceed 20 mm of water column.
2.	Efficiency of the filter against free silica dust	Efficiency of the filter against free silica dust shall not be less than 98.5% or the leakage shall not be more than 2.0 mg.
3.	Exhalation valve test	Air leakage shall not exceed 30 ml in one minute.

Out of the six selected dust protectors, only one (code no.1) had exhalation valve but it did not conform to the acceptable limit. However, its breathing resistance was within acceptable limit. Again, it was the only selected dust protector that failed conformity test for inhalation resistance. All the remaining five dust protectors confirmed inhalation resistance. All the selected dust protectors failed in respect of filtration efficiency tests. The filtration efficiency of the dust protectors varied between 65.5% (single tissue layer) and 92.2% (cloth). Thus the filtration efficiency was highest, *i.e.* 92.2% in case of cotton cloth disposable dust protector.

Table 3: Test Results of Dust Protectors as per BIS Procedure

Dust Protector Code No.	*Breathing Resistance*		*Filtration Efficiency*		*Exhalation Valve Tests*
	Inhalation	*Exhalation*	*Per cent*	*Leakage*	
1.	Before test 35 mm After test 55 mm	20 mm	86.0	22.1 mg	30 ml in 15 s
2.	Before test 5 mm After test 25 mm	–	81.2	27.9 mg	–
3.	Before test 5 mm After test 10 mm	–	68.9	46.2 mg	–
4.	Before test 4 mm After test 10 mm	–	65.5	54.4 mg	–
5.	Before test 5 mm After test 20 mm	–	79.7	31.7 mg	–
6.	Before test 4 mm After test 38 mm	–	92.2	12.2 mg	–

Suitability Testing of Dust Protectors

The selected dust protectors were tested in field for their suitability to operators of combine harvesters. The operators were found to be educated up to 10th standard. Their age ranged between 20 and 40 years and they were 2–10 years for operation of combine harvesters. All the combine harvesters under study were engine operated and had cutter bar of size 13–16 feet. The suitability test consisted of various performance indicators as observed by the operators. The frequency of operators providing feedback for various performance indicators is given in Table 4.

From the available data, it was evident that the operators felt low to medium breathing resistance for all the dust protectors except the first one that had medium to high breathing resistance. The dust filtration performance of all the dust protectors varied between low and medium as felt by the operators during wheat as well as paddy harvesting season. Air leakage from sides was reported up to low level for all the dust protectors. The operators felt low to medium tightness on the face especially during use of mask type dust protectors. Similarly they felt low to medium rate of sweating for all the dust protectors except the first one that had medium to high rate of sweating. No wear and tear was reported in any of the dust protectors during use.

Comparative Acceptability of Dust Protectors

Each of the operators was asked to give a comparative rank to all the dust protectors as per his preference for acceptability. The comparative rank was converted to comparative acceptability score by giving weightage of 6 to rank 1, 5 to rank 2, 4 to rank 3, 3 to rank 4, 2 to rank 5 and 1 to rank 6. Thus, a total acceptability score was obtained for the selected dust protectors (Table 5).

Table 4: Frequency of Operators Providing Feedback for Various Performance Indicators

Performance Indicator	*Level*	*Dust Protectors Code Number*					
		1	*2*	*3*	*4*	*5*	*6*
During wheat harvesting season							
Breathing resistance	Low	3	17	16	16	17	18
	Med	3	2	1	3	3	0
	High	14	1	3	1	0	2
Dust filtration performance	Low	14	17	18	12	17	16
	Med	3	3	2	8	2	0
	High	3	0	0	0	1	4
Air leakage from sides	Low	20	18	17	19	19	20
	Med	0	1	1	1	1	0
	High	0	1	2	0	0	0
Feeling of tightness on face	Low	11	17	15	14	12	18
	Med	5	3	5	6	6	1
	High	4	0	0	0	2	1
Rate of sweating	Low	1	16	16	14	10	16
	Med	5	4	4	5	8	2
	High	14	0	0	1	2	2
Wear and tear during use	Yes	0	0	0	0	0	0
	No	20	20	20	20	20	20
During paddy harvesting season							
Breathing resistance	Low	0	16	16	16	19	19
	Med	20	4	4	4	1	1
	High	0	0	0	0	0	0
Dust filtration performance	Low	20	17	18	13	15	20
	Med	0	3	2	7	5	0
	High	0	0	0	0	0	0
Air leakage from sides	Low	16	18	14	13	11	20
	Med	4	2	6	7	9	0
	High	0	0	0	0	0	0
Feeling of tightness on face	Low	1	10	7	20	4	20
	Med	19	10	13	0	15	0
	High	0	0	0	0	1	0
Rate of sweating	Low	1	12	9	5	4	12
	Med	10	8	10	11	15	8
	High	9	0	1	4	1	0
Wear and tear during use	Yes	0	0	0	0	0	0
	No	20	20	20	20	20	20

Table 5: Acceptability Score for the Selected Dust Protectors

Dust Protector Code No.	*Acceptability Score*		
	During Wheat Harvesting Season	*During Paddy Harvesting Season*	*Total*
1.	36	24	60
2.	82	85	167
3.	94	71	165
4.	54	60	114
5.	58	73	131
6.	104	109	213

It was evident from the data that disposable dust protector (code no. 6) of cloth material had the highest acceptability (well ahead of all others) during wheat as well as paddy harvesting seasons. The mask type dust protectors having no exhalation valve (code no. 2 and 3) ranked second. The disposable dust protectors of tissue paper material (code no. 4 and 5) scored a total of 114 and 131, respectively. The dust protector having both inhalation and exhalation valve (code no. 1) was the most lacking for acceptability of the operators of combine harvesters.

CONCLUSIONS

The dust protector having both inhalation and exhalation valve (code no. 1) was the most costly and also it performed badly not only as per BIS standards but also as per acceptability of the operators of combine harvesters. The cloth type dust protector provided the highest dust protection (92.2%). Besides, it provided a cost effective, easily available and most acceptable solution to the operators of combine harvesters.

REFERENCE

[1] Anonymous (1980). Specification for Filter Type Particulate Matter Respirators. IS: 9473–1980. Bureau of Indian Standards, p. 16.

SECTION–IV

Cognitive and Design Ergonomics

Understanding Features of Form in Fine Jewellery, by Overt Eye Movement Studies—Catch People Watching Jewels

Parag Vyas

IDC, IIT Bombay, India

ABSTRACT

The study was to understand the overt eye movements that took place during observations of a variety of rings. It was to examine the behaviour of the eye during the free scan and the changes that took place when the scan was task driven. This was done by presenting a question, to motivate subjects, to make them find visual clues in the image present.

The experiment was aimed at getting an insight into those features of form, in a piece of jewellery, which has an ability to draw and hold observer's attention. These factors play a very important role in the process of appreciation and appraisal (decision of selection or further search). If jewellery successfully attracts and holds the attention adequately, it results in a purchase decision (*ceteris paribus*).

Eye movement recorder RED by Senso Motoric Instruments, Germany was used to trace corneal reflex. The instrument bounces an infrared beam off cornea and reflections are traced for overt eye movements.

In the experiments subjects were asked to observe five pieces of jewellery and their eye movements during free examination of object was recorded. They were then asked two motivating questions and individual task driven eye movements recorded.

These observations were analyzed using software B-Gaze for drawing inferences.

Keywords: Overt Eye Movement, Eye Behaviour.

HYPOTHESIS

The null hypothesis taken for the project is "Irrespective of likes, preferences, gender and knowledge of value of the piece being observed, the overt scan paths from eye movement recordings are uniform and are not effected by form, colour or type of setting of piece of jewellery"

Objective

The experiment is aimed at getting an insight into those features of form, in a piece of jewellery, which has an ability to draw and hold observer's attention. These factors play a very important role in the process of appreciation and appraisal (decision of selection or further search). If jewellery successfully attracts and holds the attention adequately, it results in a purchase decision.

Faces, human or animal, particularly the eyes and lips give out critical information about scene or portrait being observed (Yarbus, 1993). The objective here is to see forms in jewellery and investigate if there are such features which attract attention more than others.

If we know why we react and what the factors are, it might help us make jewellery more pleasing and thereby more successful (whether we're going for 'like' or 'dislike'). We have evolved to like the shininess of eyes, teeth, water, and that translates to other things like metals, jewels, silk. And translucent—skin, flesh, teeth, semiprecious stones, frosted glass, flowers, cloudsand similar things (Pagal-Theisen, 1993).

The analysis of eye motion on jewellery gives a more complete understanding of user attention (Li, 2001) and (Salvuchi, 1999).

Visual Attention

The idioms describing the selective nature of attention are the "what" and "where" as per the studies in eye movement, (Duchowski, 2002). Fixation of attention directed towards an element of a stationary object is accompanied by fixation of gaze. Special attention or indifference to the elements of a picture is not due to the number of details composing the elements. Colours that have significance are perceived differently and outlines and contours are important for the appearance of the visual image.

The observer's attention is usually held by only certain elements of the picture that gives information allowing the meaning of the picture to be obtained. Eye movements reflect the human thought process; so the observer's thought may be followed to some extent from the record of eye movement (the thought accompanying the examination of the object). The focus is often on elements that are unusual in particular circumstances, unfamiliar, incomprehensible and so on.

Observers differ in the way they think and therefore differ also to some extent in the way they look at things. Observers familiar with the object would examine the picture differently from people seeing it for the first time and unfamiliar with the background it represents.

Free Examination and Motivated Scan

Eye movements after an instruction are interesting because they help in the analysis of the significance of the eye movements during the free examination of the picture; they show clearly that importance of the elements providing information is determined by the problem. Facing the observer, this importance may vary within extremely wide limits.

EXPERIMENTAL SETUP

The experimental setup consisted of two computers and one eye movement recording camera. One was a subject computer, for the presentation of visuals to the subject and the other for operator to calibrate and record the eye movement. Two computers were connected to each other through network of appropriate cables for the purpose of data recording, transfer and control.

Eye movement recording camera was placed on the side of the subject computer screen and recorded eye movement when the person was looking at the screen that was presenting the images to be observed by the subject. Once focused camera was capable of tracking the corneal reflex taking care of minor head movements. A chin support helped in stabilizing the head. The camera moved in vertical and horizontal arcs for alignment, focusing was inbuilt and had a facility to zoom in and out. It was connected to the computer through a network of cables and operated from the operator computer. For focusing and tracing the eye movement recorder bounced an infrared beam of light to the eyeball.

After search, images were taken by judgement/purposive sampling method. Care was taken to choose images that were not seen by the subjects earlier. Images fulfilling following criteria were selected for display:

- Good figure and ground contrast between the image and its background.
- Material contrast within themselves, as metal to gemstone and different gemstone to gemstones.
- Tentacle-like shapes, forms with acutely changing curves, textural variations, projections and peculiar forms were taken.
- Forms that were neither too complex nor too simple were taken.
- Images were similar in size and visual weight covering similar areas of screen.
- Solitaires with a single confidently placed diamond, gemstones and having simple forms were taken.
- A view presenting maximum details was taken.

Images of a model wearing the jewellery were avoided to keep attention on jewellery. To counter the effect of anticipating motivation questions before they were presented, sequence was randomized by fishbowl draw. To neutralize afterimage and rest the eye, before the next image was presented, a separator slide (a dot with a hollow point on a dark background, placed at bottom corner of the slide on a dark background) was used. This also was a common entry point to the scan paths.

Subjects were taken from age group 20 to 35, both married and unmarried. Group was heterogeneous and constituted of both the genders equitably. Subjects with normal eyesight and moderate numbers were taken. Extreme cases where the camera could not focus because of long black eyelashes and certain types of spectacles could not be included in the study. The dominant eye was determined before the experiment and camera placed accordingly. An attempt was made to minimize the ambient disturbances by carrying out the experiment in evening or night.

A proxy question was used for the question pertaining to estimation of price that proved to be too complex (association of high, medium and low price was proxy by a suggestive approximate sum of money in Indian Rupees). A moderately large sample was taken and answers were recorded on the spot on the question sheet. Relevant information like the subject's eyesight, age was recorded in case follow up was needed. Ideally the question must be presented from the front, since that was not possible, the questions were presented from the side delicately and subjects were advised not to move. They were presented verbally during the spacer slide to achieve desired effect.

Only one type of jewellery, rings, was taken to have a common denominator. Rings were found to be appropriate being the most common item.

EXPERIMENTATION

A brief about the subject's name, age, eyesight and background and knowledge about jewellery was taken to warm up and develop a repertoire. The subjects were briefed about the procedure and made to feel relaxed. Chin rest and height of chair was adjusted to make the posture as comfortable as possible.

The camera was focused on the dominant eye cornea and the settings were saved so the camera could trace corneal reflex. Once the basic setting was achieved, the calibration was done by using the software 'win cal' and strong or medium calibrations were chosen. The subjects were then presented the images for duration of 15 seconds. A free scan of overt eye movement was recorded for total five images.

The subjects were then asked first and second motivating questions and overt eye movements were recorded for the images, which appeared in random order.

Data Analysis

Data was taken to the software 'Be Gaze' for analysis. The software offered interesting and different ways to look at data. Two of them were found to be appropriate for the purpose. The raw data could be seen in one way or a combination of ways for the interpretation.

Heat Map

The heat map gives a visual representation of the areas where the eyes have visited during the appearance of image. The least visited appear in a shade of cool blue and as the visits and attention increase the colour changes to yellow, orange and red. This superimposed on image indicates the areas which are more 'hot' and draw attention for longer duration and the areas that are not visited or visited for short duration.

Scan Path

The scan path is the visual representation of the path of the eye taken during the process of scanning. Fixations are represented by circles that increase with the increase in the duration. The software also allowed for the image to be marked with AOI (areas of interest) and this made it possible to see how often different areas were being visited. A comparison between two such marked areas was made possible.

A bar chart was printed to see the comparison of two or more such AOI's and inferences drawn. These were listed at the inferences section (The scan path were seen in slow motion for detailed analysis).

Questionnaire

Two motivating questions were asked and the answers noted. First question was directed to assessment of the price of the ring in subject's opinion. Second question was to assess like and dislike about the ring. A three-point scale was taken for measurement.

OBSERVATIONS

Eyes tended to move directly to the gemstones rather than the shiny metal parts like the highly polished shanks presenting large metal areas and pave beadwork.

Eyes visiting the gemstones shuttled between the metal prongs that hold the gemstone in place, or the part of the shank nearby, trying to interpret it in its context. The metallic surroundings provided the context for interpretation.

Overall form of ring was not visited uniformly and the focus of attention was gemstone and its immediate surrounding. Type of shank or amount of information it provided seem to play little role, although it may present larger surface area.

Solitaires as single gemstones attracted most attention, although minimal details were present.

After a short while, when the principle features were taken into account regressions took place to reconfirm and appreciate.

Important features were visited in the first 6 to 10 seconds, and potentially this was the time where the decision took place.

Simple forms were liked more than the complex forms. Choice (like, dislike) was independent of the perception of wealth.

INFERENCES

Affinity of the eyes to the gemstones is independent of view of presentation.

Competing edges become important and draw attention at the border of gemstones, and metals.

Among the major gemstones diamond attracts the attention and more referencing takes place in border areas and adjoining metal.

Large sized diamonds (which may not present as much surface visibility) are located and visited in priority than the ones that are closer from the point of entry but smaller in size (which present larger surface visibility).

Subject A (Female)

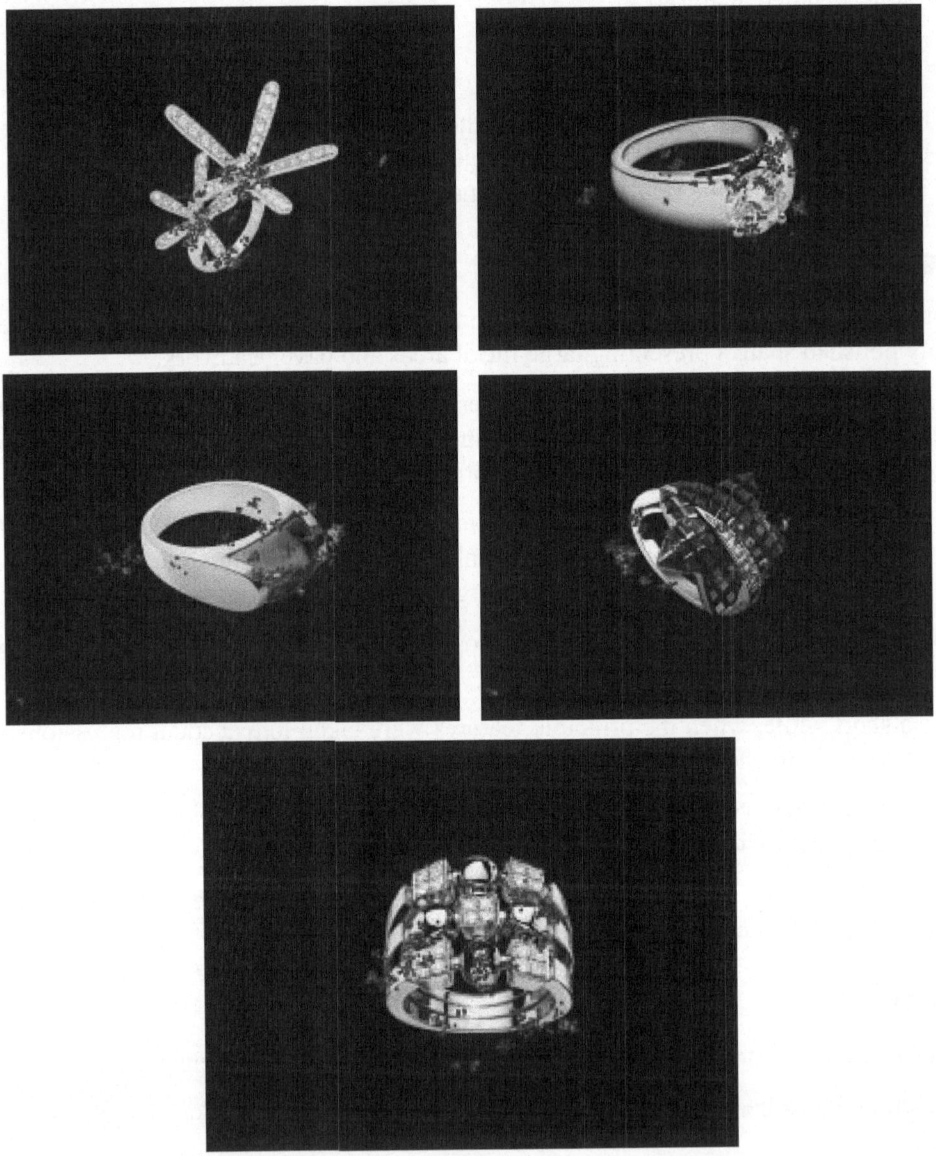

Subject B (Male)

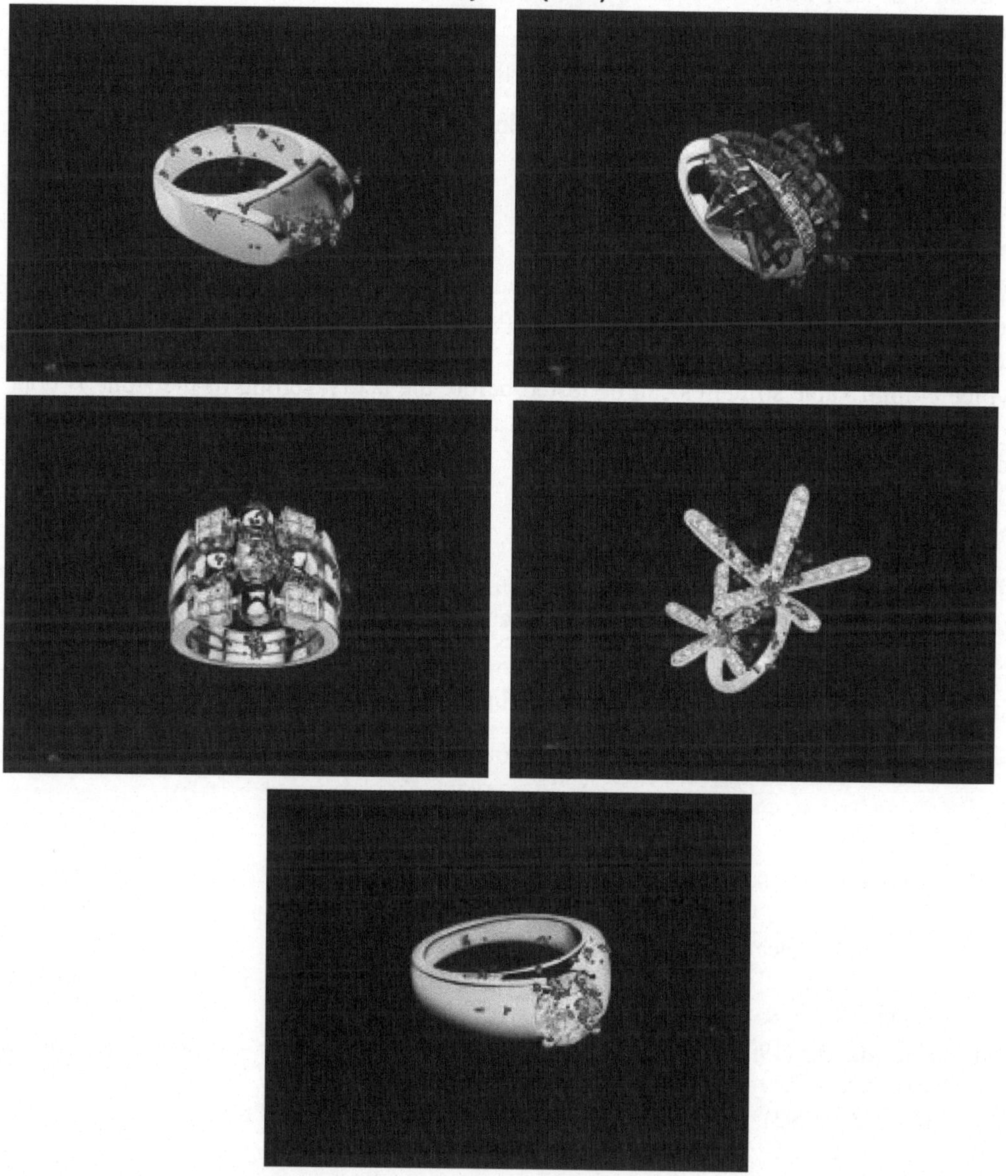

Differences between Male and Female Observers

A trend was seen that men were prone to take a broader area into the field of their visual exploration of the scene. In comparison women remained focused in key areas and covered less area in heat maps.

A trend was seen where women tended to look more on focus areas and dwelled on finer details.

Men tended to go more on the extremities of the form than the women.

Women remained attentive in key areas looking at the contents more in depth.

SUGGESTIONS AND CONCLUSION

The result is expected to help jewellery designers and manufacturers than the by revealing features of form that have a better aesthetic appeal and are liked better as compared to others. It will allow particular features to be incorporated in the jewellery design and to appeal to senses in a more effective way. Very expensive jewellery, typically high value solitaires can be tested before the actual setting of the stone is done by presenting a photo realistic rendering and taking proof of concept through eye movement recorder.

GLOSSARY OF TERMS

Gemstones: A crystalline rock (mineral origin) that can be cut and polished for jewellery, a precious or semiprecious stone that can be cut and polished for use as a gem.

Prong: Jewellery. A tapering metal projection, usually heavier than a claw, rising from the base of a jewellery setting and used to hold a stone in position as needed.

Shank: Jewellery. The part of a ring that surrounds the finger; hoop.

REFERENCES

[1] Duchowski, A. (2002). *Eye Tracking Methodology, Theory and Practice,* London: Springer.

[2] Li, Michael W. (2001). *In Vision: a System Using Vision Patterns to Understand User Attention,* Massachusetts, MIT.

[3] Pagal-Theisen, V. (1993). *Diamond Grading ABC,* Handbook For Diamond Grading, Belgium: Rubin & Son.

[4] Salvuchi, D. (1999). *Mapping Eye Movements to Cognitive Processes, Pittsburgh:* Department of Computer Science Carnegie Mellon University.

[5] Yarbus, A. (1993). *Eye movement and Vision*, New York: Plenum Press.

Ergonomics of Call Centres and Psychosomatic Problems among Employees

Indra Nanda

Dept. of Home Sc., V.M.L.G. College
Dayanand Nagar, Ghaziabad, India

ABSTRACT

A call centre creates employments and working nights. Although it is financially rewarding, but at the cost of mental and physical well being of employees. When the emotional differences are expressed through body systems innervated by the autonomic nervous systems, resulting conditions are called psychosomatic problems. The main objective of this research paper is to study the relationship between ergonomics and psycho-somatic disorders of employees working in night shifts. The sample drawn was 100 respondents belonging to age group of 20 to 35 years working in call centres at NCR. It is a co-related study. Ergonomic parameters *i.e.,* light intensity sound level, temperature, humidity, infrastructure and safety were measured weekly. An interview schedule was formulated to measure physical problems of respondents and the Occupational Stress Index was administered to calculate the level of stress. The results of the study reveals that the ergonomics of call centre was highly correlated with occupational stress index and was significant at 0.01 level. More than half the respondents (53%) were facing the problem of eye strain followed by 51% suffering from gastrointestinal disorders. 17.5% of the respondents complained about hearing problems due to long hours working in night shifts. The result of the study proves that continuous usage of electronic gadgets adversely affected the user's health and general well being.

INTRODUCTION

India has all the ingredients to become a major hub for outsourced services. Despite all challenges Indian BPO service providers are extremely optimist over industry prospects. The study conducted by NASSCOM shows a projection of about 11,000,000 employees in 2008–2009 from 23,000 in 1998–1999 and an increase in money gain from $224 million in 1998–1999 to $18617 million in 2008–2009. This projection was welcomed by the economists but ergonomic is a much better technique for obtaining balance between economic growth and humanity. The office itself, the architectural feature together with temperature, humidity, lighting, space and air quality—all have an influence on task performance and working long hours in front of computer can affect a worker's visual musculoskeletal and psycho-social comforts. Studies indicate that for extended exposures temperature between 20°C and 25°C are more acceptable. As humidity increases, discomfort will be felt. Humidity below 70% is preferable in summer. Illumination more than 200 lux is recommended as the minimum luminances for workstation where work is executed for long periods. Noise level is one of the main aspects while designing all ergonomically designed workplace. People in noisy environment suffer from lack of sleep, irritability, severe tension, illness, annoyance, anxiety, tension and sudden unexpected shock.

Ear Diseases take place and at above 120db noise painful effects occur. It is important to evaluate the heat stress on the human body as it comes from environmental heat (external) and metabolic heat (internal). A study by Ritvanen *et al.* (2006) reports that psychosomatic systems are significantly more common during the high stress work periods both in younger or older workers. Stress is a situation of threat, frustration or conflicts that strains a person's physical and mental abilities. In order to maintain level of growth as well as care of the worker ergonomic should be given vital importance to make work and worker better, safer, efficient and more productive. High work stress has been associated with increased risk of suffering from cardiovascular disease. Psychosocial stress is related to pain in the neck and shoulder region and affects health more than the mental or physical demands occurring during work. Keeping in view all problems related to work, worker and work environment the study was conducted with following objectives.

- To investigate the perceived physical problems among employees working in call centre at NCR (National Capital Region).
- To evaluate the level of occupational stress on employees working in night shifts at call centres.
- To study the effects of ergonomics of call centre and Psychosomatic problems among employees of call centres.

Hypothesis

There would be positive relationship between ergonomic of call centres and psychophysical problems perceived by employees working in night shift.

There would be a positive relationship between psychological (occupational stress) and physical problem among employees especially working in night shifts.

There would be positive relationship between ergonomics and psychosomatic problems among employees.

METHODOLOGY

The present study is a co-relational to investigate the hypothesis. A sample of 100 respondents working in night shifts belonging to the age group of 25–35 years was collected for investigation from call centres at NCR (Delhi, Noida, Gurgaon, and Faridabad).

Tools

A self-designed questionnaire (non-medical parametric) was formulated to evaluate the perceived physiological problem among employees.

The ergonomic environment of the call centres such as temperature, humidity, lighting, noise level, space and air quality along with architectural features was monitored with required tools fortnightly.

Occupational Stress Index (OSI) by Shrivastava and Singh (1984) was administered to measure the level of stress among employees. The index consisted of 46 items each to be rated on the five point scale. All items were almost relevant components of job attitude and behaviour which causes stress. The main twelve factors in the scale were-role overload, role ambiguity, role conflict, unreasonable group and political pressure, under participation, powerlessness, poor peer relation, intrinsic impoverishment, low status, strenuous working condition, unprofitability.

Procedures

Weekly visits were made to meet respondents. They were given self formulated questionnaire to be answered in yes or no formation to acknowledge the common physiological disorder perceived by them, *e.g.* pain, eye strain, sleeplessness digestive disorder, *etc.* During visits to the centres temperature, humidity, light and sound levels were recorded with the help of thermometer, barometer, hygrometer, sound level meter, *etc.*

The questionnaires of OSI were distributed and related instructions were given to the respondents. Filled in sheets were collected following the instructions of manual of the author. Data were processed into percentage from the scores of true keyed items and false keyed items were calculated into total scores and then level of stress was calculated on the basis of total score.

RESULTS AND DISCUSSIONS

Perceived physical problems by male and female respondents are shown in Table 1. The table signifies that eye strain was the most persistent complaint made by respondents (53%), digestive disorder (51%) second and neck and back pain being the third highest problem (50%), Pain in lower limbs (40%) and so on above problems were reported. Thus beside good ergonomically designed work stations above complaints persist among employees working in night shifts, may be due to various other factors.

Factors Affecting Level of Occupational Stress

The OSI (Occupational Stress Index) purports to measure the extent of stress which employees perceive arising from various consistent and conditions of their job. The scale consists of 46 items all are relevant components of job life which often cause stress. Table 2 shows the level of occupational stress among those employees who work in call centres. The ergonomic environment of call centres can affect the working condition.

The data in the table indicates that except factor (unprofitability) level of stress remained of moderate level. The percentage of the most of the component, which causes stress, is moderate. In the factors (powerlessness, poor peer relations, low status) the high level stress was nil Unprofitability (85%) and role ambiguity (50%) were the factors of high level stress. Powerlessness (65%) and strenuous working condition (55%) were the factor of lower stress. Low status (85%) poor peer relation and intrinsic impoverishment and role conflict (70%) were the highest level as the factor of moderate stress. All respondents felt stress due to one or other factor. On the basis of data we can substantiate the fact that these factors increase the level of stresses among employees.

Table 1: Physical Problems Analysis

Physical Problems	*Female n = 38*		*Male n = 62*		*N = 100 Total %*	
	Yes	*No*	*Yes*	*No*	*Yes*	*No*
Loss of Appetite	19 (50%)	19 (50%)	14 (22.6%)	48 (77.4%)	33	67
Gastro-intestinal problems	18 (47.3%)	20 (52.6%)	33 (53.3%)	29 (46.7%)	51	49
Pain in lower limbs	16 (42.1%)	22 (57.3%)	24 (38.7%)	38 (61.2%)	40	60
Eye strain	21 (55.2%)	17 (44.7%)	32 (51.6%)	30 (48.3%)	53	47
Headache	17 (44.7%)	21 (55.2%)	22 (35.4%)	40 (64.5%)	39	61
Sleeplessness	13 (34.2%)	25 (65.7%)	24 (38.7%)	38 (61.2%)	37	63
Hearing problems	7.5 (19.7%)	30.5 (80.2)	10 (16.1)	52 (83.8)	17.5	82.5
Change in body weight	6 (15.78)	32 (84.2)	12 (19.3)	50 (80.6)	18	82
Neck and back pain	24 (63.1)	14 (36.8)	26 (41.9)	36 (58.0)	50	50
Skin Problems	4 (10.5)	34 (89.4)	6 (9.6)	56 (90.3)	10	90

Multiple responses
Figure in parentheses indicate percentages.

Table 2: Level of Occupational Stress (N = 100)

Factor/Components	*Low Stress (%)*	*Moderate Stress (%)*	*High Stress (%)*
Role overload	20	50	30
Role ambiguity	20	30	50
Role conflict	20	70	10
Unreasonable group and political pressures	15	60	25
Responsibility for persons	10	65	25
Under participation	28	48	04
Powerlessness	65	35	00
Poor peer relations	20	80	00
Intrinsic impoverishment	00	80	20
Low status	15	85	00
Strenuous working conditions	55	25	20
Unprofitability	10	25	65

CONCLUSIONS

The study can be concluded with the note that newly set up call centres are maintaining a satisfactory & comfortable ergonomic environment, temperature. Humidity light, sound, infrastructures are well maintained to make employees feel very comfortable. The approaches to perceive the working conditions and style of working may be different as Longnarus stated "a phenomenon becomes stressful only when perceived it as stressful". Thus very less or nil fluctuations were noted in ergonomic environment. However, still physical problems and occupational stress persist but the level being moderate in the percentage. The causes of stress may vary but if ergonomic is not maintained the condition will never be favourable and occupational stress will boost up.

REFERENCES

[1] Ritavanen, T., Louhevaara, V., Helin, P., Vaisanen, S. and Hanninen, O. (2006). "Responses of the Autonomic Nervous System during Periods of Perceived High and Low Stress in Younger and Older Female. *Applied Ergonomics*, 37, pp. 311–318.

[2] Sensharma, S.K. (1999). Heat stress in industry Indoshnews July September.

[3] Srivastava, A.K. and Singh, A.P. (1984). "Manul of Occupational Stress Index." Manovigyanik Parichhan Sansthan. Varanasi.

Analyzing Factors Affecting Driving Comfort and their Inter-relationship in a Passenger Car Using Analytical Hierarchy Process

R.G. Rajhans[1], N.R. Rajhans[2] and B.B. Ahuja[1]

[1]Tata Technology Ltd., Pune, India

[2]Department of Industrial Engineering, VIT, Pune, India

ABSTRACT

Driving an automobile (passenger car) in a crowded street is a complicated task. Many a time people complain about exhausted feeling after driving a car even after 30 to 40 minutes. Driving an automobile especially in India requires the full range of human capabilities, including perception, decision making and motor skills. These skills must be performed in a highly coordinated fashion often under stressful conditions.

Driver's comfort depends on many factors and a qualitative analysis of these factors is generally done. It is, therefore, necessary to quantify these factors and identify the factors which contribute maximum towards perceived driving comfort.

An attempt has been made to quantify the factors using Analytical Hierarchy Process (AHP). This analysis will provide inter-relation of these factors and their relative importance in driving comfort. More than fifty factors such as—Anthropometric data, driving posture, seat design, environmental factors, energy consumption, biomechanical as well as psychological perception are considered for quantification and four-tire hierarchy is constructed based on these factors.

The hierarchy is generated based on vehicle component systems, subsystems and human factors which after the study reoriented based on its relative importance for logical conclusion. Relative importance ratings were given and the ratings were confirmed using Saaty's procedure. Percentage contribution of each factor was calculated. This relative rating will help the designers to identify the most important factor from drivers' point of view.

Keywords: AHP, Driving Comfort, Perceived Comfort, Relative Importance.

INTRODUCTION

An automobile designer concentrates generally on engineering parameters like engine, suspension, brakes, cabin, wiring harnesses *etc*., while designing a car. In initial phase of designing/conceptualization designers consider human placement in a vehicle layout. These layout considerations are always compromised for one/many engineering parameters or marketing requirements. Although vehicle designing is upgraded over the years with engineering innovations and technological breakthroughs, market feedback shows that numbers of accidents are also on the rise. Engineers and technicians have tried analyzing these accidents and also have found active as well as passive safety devices like seat belts, air bags, anti brake system and so on to prevent or reduce the life risk.

Most of the drivers narrate driving as a tiring experience. It is a multitask activity. Driving on a crowded road during peak hours is an experience in itself. Driver has to be very alert on controls, surroundings (relative vehicle speed, distance, *etc.*) as well as unexpected incidences (like some animal running across the road or unpredicted behaviour of other drivers).

Designers/testing engineers after designing a new vehicle take physical tests on prototypes to get comfort assessment of new vehicle with reference to benchmark vehicle. Many times designers as well as test engineers also ask drivers to rate benchmark vehicles for subjective comfort rating to identify most comfortable vehicle. As this information is at macro level, one can hardly use it directly while designing a new car. Even if somebody duplicates driving position/posture from most comfortably assessed car it may not result in high comfort rating for a new car.

Comfort is a relative term and is an overall effect of many factors. Tangible and intangible factors related to driving comfort were identified through brainstorming sessions. These factors need to be prioritized based on their relative importance for assessment in perceived comfort. Analytical Hierarchy Process (AHP) is a logical approach which is used to analyze these prioritized factors.

Analytical Hierarchy Process (AHP)

Saaty (1985) has evolved AHP which enables decision makers' to develop a hierarchical structure for the factors which are explicate in the given problem; and to provide judgment about the relative importance of each of these factors specify a preference for each decision alternate with respect to each other, providing a prioritized ranking order of preferences for decision alternatives.

The general approach of AHP consists of three stages:

1. Determination of self importance of an attribute.
2. Determination of relative importance of each of the alternative with respect to each other.
3. Overall priority weight determination of each of these alternatives.

METHOD

Factors affecting perceived driving comfort were collected through brainstorming, books, journals and user magazines. Further these factors were grouped and regrouped to generate hierarchy chart considering their interrelationship considered as main criteria. A group of 9 drivers/test engineers participated in this study. The selection was based on the number of vehicles they had driven and years of car/long driving experience (Falou *et al.*, 2003). All the group members were explained meaning of every term/factor used in this study. This helped in developing same understanding about the terms across every group member. Rating for various factors was confirmed using Saatys' procedure.

Next step was to select most important factor from group of factors at each level as well as sub group and the quantitative rate with reference to competing factors. For this standard rating scheme as described by Saaty was used (Table 1). All the group members were asked to debate within themselves before giving importance rating to one factor over other. Group members were allowed to change their rating in case the result was erratic or not as per their perceived expectations.

Rating of level 2 factors was multiplied with level 1 parent group rating to get absolute importance rating number and percentage importance across level 2 factors, *e.g.* anthropometry aspects rating of level 2 was 43% which multiplied by parent group rating (related to Human) 56% gave overall rating number '24', which was 25.2% at level 2. Same method was extended for getting overall percentage ratting of factors across level 3 and level 4.

Table 1: Saaty's Relative Importance Rating Criteria

Intensity of importance	*Definition*	*Explanation*
1.	Equal importance	Two activities contribute equally to the objective
3.	Weak importance of one on other	The judgement is to favour one activity over other but it is not conclusive
5.	Essential or strong Importance	The judgement is strongly in favour of one activity over other activity
7.	Demonstration importance	Conclusive judgement as to the importance of one activity over another
9.	Absolute importance	The judgement in favour of one activity over another is of the highest possible order of affirmation
2, 4, 6, 8	Intermediate values between the two adjacent judgements	When compromise is needed

If the activity '*i*' has one of the above non-zero numbers assigned to it when compared to *j*, then j has the reciprocal value when compared with '*i*'.

RESULTS AND DISCUSSION

Human factors dominate comfort assessment. This shows capacity of human brain and body sensitivity to discomfort. This also could be due to compromises done for human factors against engineering parameters while designing vehicle layout. Next important factor is related to vehicle compared to environment. This could be due to vehicle systems development to overcome day to day environmental variation issues like HVAC for climate control, multi-speed wiper system for various rain patterns, *etc.*

Level 1

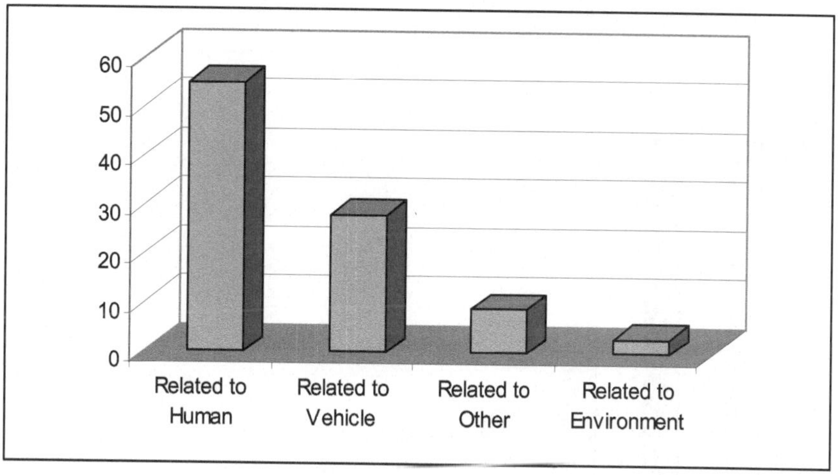

Anthropometric aspects as well as visual aspects dominate level 2 factors. This is also reflected in accident statistics, published data. Majority accidents occur due to human errors, for example poor visibility or poor perception of vehicle speed. Due to stringent safety (crash) norms many vehicles are designed for 95th percentile man, compromising 5th percentile woman. Next factor is seat design. Designer can work on them as major component which is very well being perceived by users in comfort assessment. Due to its high rating, it has been studied in detail in level 3 and level 4 assessments. Vibrations and Biomechanics aspects have lesser priority as vehicles used for this study were of similar category (passenger cars).

Level 3 shows similar results as level 2. However, after anthropometry, visual, seat design factors, one can see driver's age, suspended body weight, whole body vibration as well as traffic congestion which could be next factors affecting comfort assessment. Surprisingly instrument visibility, monotonous or new path, comfort expectations from vehicle type had very low rating.

As seat being seen as major component in comfort assessment in level 2 and 3, it was specifically studied for its sub factors for their relative importance in comfort assessment. The study shows that seat stiffness/rigidity followed by foam contours have major contribution to comfort rating followed by proper/scientifically designed lumber support and good mounting strategy. Surprisingly seat foam stiffness shows less priority. Probably people are not so sensitive to difference made in foam stiffness in benchmark cars or the range is narrow not be detected. Longitudinal adjustment also shows low priority as every vehicle provides this and perceived as granted feature.

Level 2

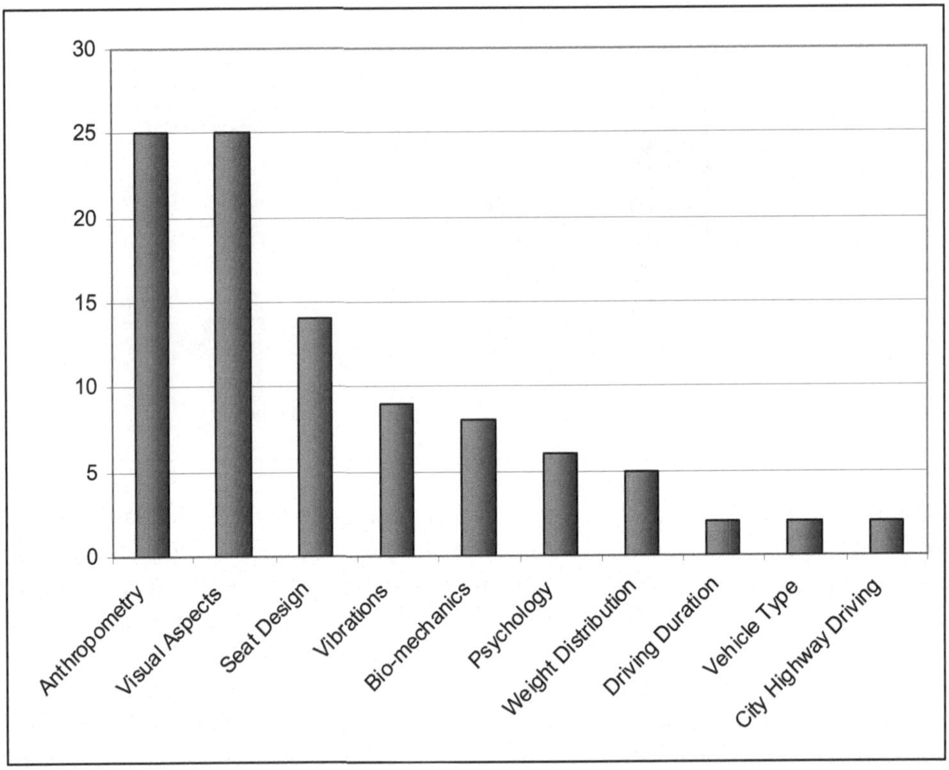

CONCLUSIONS

AHP was found to be very simple but effective method for assessing decision making process. It also provided quantitative feedback connected to qualitative judgement.

The methodology used in this paper could be further detailed for getting three-axis assessment methodology changing fixed factors.

At the time of interior layout conceptualization/finalization designers come across best compromises to be done for engineering or packaging constraints. This study gives a good guideline for deciding which factors should not be compromised based on their relative customer perception for driving comfort assessment.

ACKNOWLEDGEMENT

The work was supported by Engineering Research Centre, Tata motors Ltd, Pune, and Production Engineering Department, College of Engineering, Pune.

REFERENCES

[1] Falou, Wassim El, Jacques Duchene, Michel Grabisch, David Hewson, Yves Langerson and Frederic Lino. (2003). Evaluation of driver discomfort during long duration car driving. *Applied Ergonomics*, 34(3), pp. 249–255.

[2] Nishiyama S., Uesugi, N., Takeshima, T., Kano, Y. and Togii, H. (2000). "Research on Vibration Characteristics between Body and Seat, Steering Wheel and Pedals (effect of seat position on ride comfort)" by, *Journal of sound and vibration*, 236 (1), pp. 1–21.

[3] Park, Se Jin, Kim, Chae Bogk, Kim, Chul Jung and Lee, Jeong Woo (2000). "Comfortable Driving Postures for Koreans." *International journal of industrial ergonomics,* 26(4), pp. 489–497

[4] Saaty, T.L. (1980). Analytical Hierarchy process, McGraw Hill.

Effect of Posture, Length of Hammer-handle and Frequency on Discomfort

Abid Ali Khan[1] **and Mohd Shakeb**

Ergonomics Laboratory, Department of Mechanical Engineering
ZH College of Engineering & Technology, Aligarh Muslim University
Aligarh–202 002, India
E-mail: [1]abidak71@yahoo.com

ABSTRACT

Present study was to investigate the effects of hammer length, body posture with frequency of exertions. The experiment was performed for repetitive hammering for upsetting a fixed size of mild steel rod as done in case of conversion of rod to a bolt. Eight right-handed male subjects were randomly selected from the population of the college students and staff for this experiment.

Two body postures (standing and sitting) × two lengths of hammer handle (203 mm and 279 mm) × two levels of frequency of exertions (10 and 20/min) × eight subjects full factorial design were used for this experiment. The dependent variable was subjective discomfort on a 100 mm Visual Analogue Scale (VAS) at the end of each three minutes of exertion. Symptoms of discomfort included aching, fatigue, soreness, warmth, cramping, pulling, numbness, tenderness, pressing or pain. Analysis of covariance (ANCOVA) was performed only height was found to be significant covariate ($p = 0.046$) with $R^2 = 0.191$ (Adjusted $R^2 = 0.073$). The result of the ANCOVA showed that the frequency of exertion and the length of the hammer handle were not significant ($p = 0.138$ and 0.579, respectively) but the whole body posture was found significant at $p = 0.03$. There were not any significant interactions of the main factors on discomfort. However, pair wise t-test showed that the frequency of exertions was significant for 203 mm length of hammer handle ($p = 0.03$) but was not found significant for 279 mm length of the hammer handle ($p = 0.63$). The reported discomfort was mostly in the region of shoulder and hand/palm as per body discomfort map. It was concluded that the risk of injury could be reduced if the bending in the different parts of the body is minimized, because the discomfort was significantly higher for sitting than standing.

Keywords: Posture, Hammer Handle, Discomfort.

INTRODUCTION

Work-related MusculoSkeletal Disorders (WMSDs) are injuries and disorders of the muscles, nerves, tendons and supporting structures of the body (Bernard, 1997). MSDs can be directly and indirectly related to the work or the work environment known as risk factors. When an MSD is caused by occupational task, it is referred to as a WMSD. There is strong evidence for the relatedness of work design factors with WMSD (Bernard, 1997). WMSDs can be caused by repeated efforts, uncomfortable posture, static work, continuous loading of the tissue structures or lack of recovery time trigger or by a pathological process that then manifests itself as a WMSD (Hagberg *et al.*, 1995).

Industrial survey was carried out to find out the most common task of the forging industry. There were many sorts of tasks, carried out, which were seemed to be responsible for WMSDs. However, hammering was seen as very uncomfortable task in the small scale forging industry around Aligarh and in many places in India. Further, survey was carried out to find the most frequently employed postures and movements of upper limbs while performing hammering operations. These postures showed most frequent deviation of trunk, elbow, forearm, knee and shoulder joint. The possible bad postures observed for the workers involved in the hammering task were tilted back, abduction of upper arm, elbow extension/flexion, radial deviation of wrist, knee bending, deviation of shoulder joint, etc. There were different types different sizes of hammers used in the industries surveyed. It was also observed that the size of hammer is also an important factor responsible for the WMSDs. Literature shows that the diameter and length of the hammer is an important factor to control WMSDs.

Hertzberg (1973) reported grip strength of 43 kg for a 40 mm grip, 65 for a 65 mm grip, 48 for a 100 mm grip, and 36 for a 125 mm grip. Ayoub and LoPresti (1971) used electromyography and reported 40 mm was preferred to 50 and 65 mm. However, no study was found to report the combined effect of whole body posture and hammer size on discomfort. Keeping this in view and the factors responsible for WMSDs, present study was designed to investigate the effect of whole body posture, frequency, length and diameter of the handle of the most commonly used hammer (Ball Pin Hammer) in forging industry. The experiment was performed for repetitive hammering for upsetting a fixed size of mild steel rod, similar to the task of experiment one. The aim was to focus more on the postural change of whole body and the length of the hammer handle for the only extreme levels of the hammer length; those were selected for the experiment one. The dependent variables were discomfort score recorded on 100 mm VAS. Null Hypothesis for the present study was as follows:

"The length of the handle of hammer, body-posture, frequency and their interactions were not having significant effect on discomfort."

METHOD

Two whole body postures (standing and sitting) × two lengths of hammer handle (203 mm, and 279 mm) × two levels of frequency of exertions (10/min and 20/min) × eight subjects; full factorial design was used for this experiment. Eight right-handed male subjects were randomly selected from college population. The subjects signed the consent form, before the starting of the experiment. Additionally subject's demographic and relative anthropometric data (palm length, palm width, mass, age and height) were recorded. To obviate order effects, a full Latin square was employed to decided experimental orders. The dependent variable was subjective discomfort on a 100 mm Visual Analogue Scale (VAS) at the end of each three minutes of experimental exercise. Subjects were told that symptoms of discomfort included aching, fatigue, soreness, warmth, cramping, pulling, numbness, tenderness, pressing or pain (Lin *et al.*, 1997).

As per the treatment number, each subject was asked to sit/stand parallel to the work platform. The body posture of the subjects was controlled so as to minimize the effect of other postural deviations on WMSDs. The experimental setup is shown in the Figure 1. The hammer was kept above shoulder and striking task was started when the stop clock showed the duration of six seconds and repeated thereafter at an interval of six seconds until the 30 impacts were completed for the treatment of the frequency of 10 exertions/min and 60 impacts for the frequency of 20 exertions/min. The repetitive exertion of hammer for upsetting of mild steel rod was of diameter (10 mm) and length (100 mm). This experimental task was the experimental simulation of the part of the square bolt-making job. This simulation was also conducted in the Smithy Shop of ZH College of Engineering and Technology, Aligarh, India. The total time for the set of experiment was set at three minutes. The level of discomfort reported by the subject was recorded on the VAS and points of discomfort were marked on the body map. The intensity of discomfort at different points of hand arm and wrist system was separately noted on the sheet. A rest of at least five minutes or until the subject felt no discomfort was given between two blocks, assuming no cumulative fatigue (Carey & Gallwey, 2002). A trial of the experiment was run to get subject familiar with the task of the experiment.

Fig. 1: Experimental Setup (standing & sitting postures)

RESULTS

Discomfort Score

The mean values of the discomfort score recorded were summarized in Table 1. Before performing the Analysis of Variance (ANOVA) data was tested for normality. To verify the normality of the data Levene's test was conducted, it showed that the data was not from different population groups at $p = 0.924$. The distribution based on histogram and normal score plot showed that the data was close to normal distribution. Further ANOVA was performed using SPSS11.0.1 for discomfort score as dependent variable, Body Posture (BP), Frequency of Exertions (FR) and the length of hammer handle as fixed independent variables, while subject was random factor because of the availability of subject randomly. The results of ANOVA showed that the main factors; body posture, frequency, and the length of hammer handle were not significant on discomfort (p = 0.076, 0.107 and 0.442 respectively). There were also not significant interactions of these variables. Only it was noticed that the 'p' value for frequency and posture were close to 0.05.

Table 1: Summary of the Data of the Discomfort Score Recorded on 100 mm VAS for all Treatments of Experiment

Independent Variables		*Dependent Variable (Discomfort Score on 100 mm VAS)*		
FR (exr./min)	*HL (mm)*	*BP*	*Mean*	*Std. dev.*
10	203	Standing	41.88	9.69
20	203	Standing	51.88	17.84
10	279	Standing	40.50	13.22
20	279	Standing	47.25	16.38
10	203	Sitting	51.00	17.38
20	203	Sitting	57.38	16.91
10	279	Sitting	53.50	13.62
20	279	Sitting	52.63	14.83

In the preliminary data collection height, weight, and palm size were also recorded to be possibly considered as co-variates while the ANOVA was performed using each of the demographic variables as covariate. Only height of the subject was found to be significant covariate (p = 0.046) with R-Squared = 0.191 (Adjusted R-Squared = 0.073). The result of the ANOVA is presented in Table 2. The frequency of exertion and the length of the hammer handle were not significant (p = 0.138 and 0.579 respectively) but the whole body posture was found significant at p = 0.03. There were not any significant interactions of the main factors on discomfort.

Interaction Plots/Graphical Analysis

The profile of the discomfort score with respect to the length of the hammer handle is shown in Figures 2 and 3. This shows a decrease in the value of mean discomfort score for the frequency of exertion 20/min. The pair wise t-test was performed to verify whether the discomfort for 279 mm length of hammer-handle was significantly less compared to 203 mm length of hammer-handle for the frequency of 20 exertions/min. The t-test showed that these were not significantly different at p = 0.209 and t = 1.312. However, the little increase was noticed for the change in hammer length at 10 exertions/ min. The pair wise t-test found no significant difference (p = 0.881 and t = 0.152) in the hammer length for the frequency of 10 exertions/min. For the change in frequency from 10 to 20/min, the increase in discomfort was more for the 203 mm length of hammer handle compared to 279 mm length of the hammer handle. Further pair wise t-test was performed for the difference of discomfort for the change in frequency of exertions from 10 to 20/min at 203mm hammer length and 279 mm hammer length. The results of the t-tests showed that the difference of discomfort scores was significantly different for the change in frequency if hammer length was 203 mm ($p = 0.03$ and $t = 2.396$), but this difference was not significant for hammer length of 279 mm ($p = 0.627$ and $t = 0.495$).

Table 2: The Results of the ANOVA Considering Subject's Height as Covariate (Dependent variable: discomfort score on 100 mm VAS)

Source	*Type III SS*	*df*	*MS*	*F*	*Sig. (p-value)*
Subject height (covariate)	912.793	1	912.79	4.175	0.046
Frequency (FR)	495.063	1	495.06	2.264	0.138
Hammer Length (HL)	68.062	1	68.06	.311	0.579
Whole Body Posture (WBP)	1089.000	1	1089.00	4.980	0.030
FR * HL	110.250	1	110.25	.504	0.481
FR * WBP	126.562	1	126.56	.579	0.450
HL * WBP	14.063	1	14.06	.064	0.801
FR * HL * WBP	16.000	1	16.00	.073	0.788
Error	12026.21	55	218.66		
Corrected total	14858.00	63			

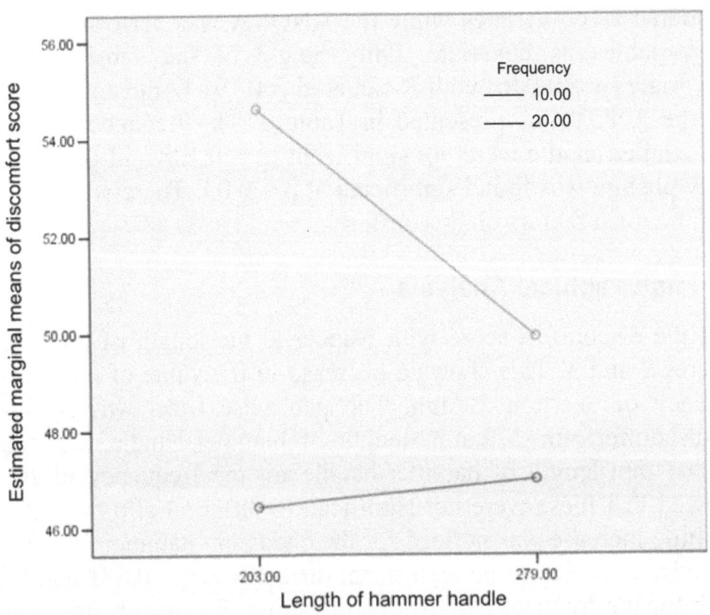

Fig. 2: The Mean Discomfort Score for the Length of Hammer *vs.* Frequency

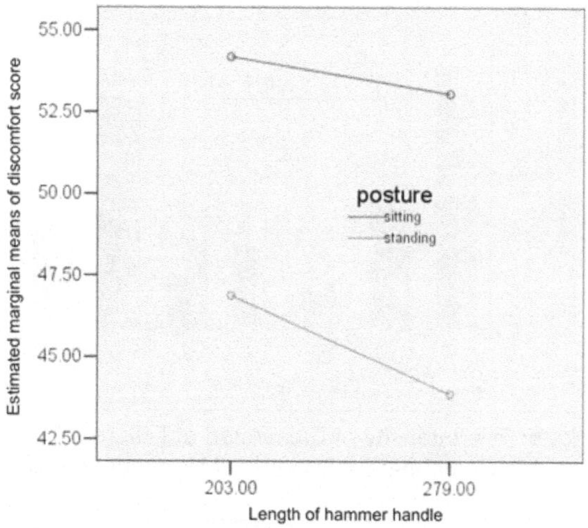

Fig. 3: The Mean Discomfort Score for the Length of Hammer *vs.* Body Posture

In Figure 3 the change in the discomfort for both the postures selected for the experimental simulation is shown. The difference noticed was of about 7 units for the hammer length of 203 mm but this difference was little more (approximately 8 units) for the hammer length of 279 mm. This difference of about 7 to 8 units was little compared to the size of discomfort of 100 units.

Body Discomfort Map

The bar charts were drawn based on the marking of intensity of discomfort on body discomfort map (Figures 4, 5 and 6). The bar charts showed that the discomfort was mostly reported in the shoulder, upper arm and palm regions. The difference of sitting and standing posture was also noticed in shoulder and upper arm. As well as the differences due to hammer length and frequency were noticed in the bar carts as shown.

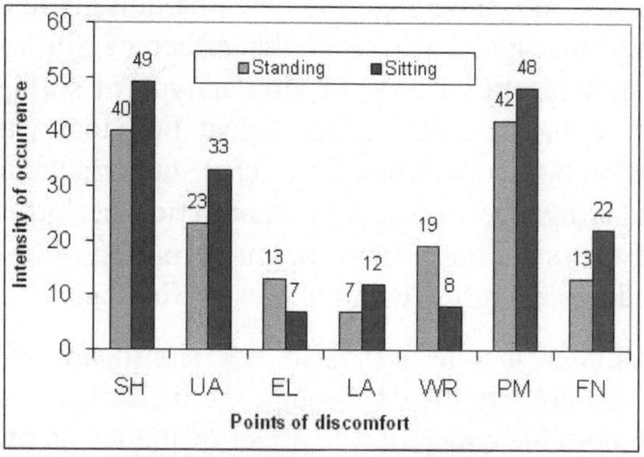

Fig. 4: The Intensity of Discomfort at Different Points of the Body Map for Standing and Sitting Posture

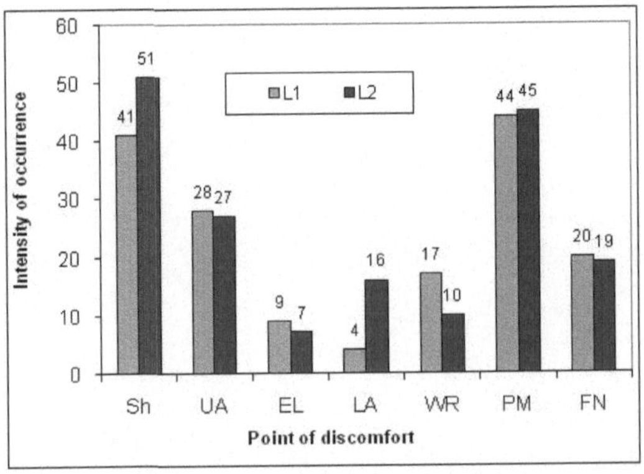

Fig. 5: The Intensity of Discomfort at Different Points of the Body Map for Lengths of Hammer Handle

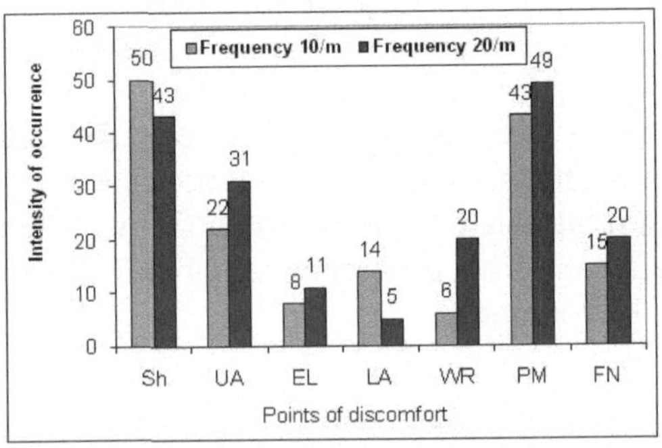

Fig. 6: The Intensity of Discomfort at Different Points of the Body Map for Both Frequencies

DISCUSSION

The sitting posture contributed more discomfort than standing posture. Stal *et al.* (1999) studied wrist positions and movements during machine milking to compare tethering and loose-housing system. This study also find the effect of sitting posture for milking operation that was also a repetitive task for similar type of sitting posture. Karwowski *et al.* (2003) found that hammering surface orientation that was affecting the body posture was significant on task completion time. That supports the finding that the change in body posture makes a significant effect on hammering performance. MacKinnon and Stone (2003) reported that standing posture might be preferred to the sitting posture for pulling task. These findings are inline with the present findings.

The length of the hammer handle was also not significant in the findings of this experiment too, as it was not significant through the investigations of the experiment of the present study. The possible cause of this effect of the levels of hammer length is the range of investigation which was low at the larger end out of the selected lengths of the hammer handle. The frequency of exertion was not significant but was seen different for both levels of the frequency of exertions. The p value for frequency was 0.107 which was not very high for being not significant. The possibility might be the duration of each treatment. The duration of each treatment was not kept more than three minutes due to the risk of permanent soft tissue injury. The results of these tasks of small durations can be predicted for longer durations as per reported symptoms.

Meagher (1986) reported that size, shape, handle texture, and length substantially affect WMSDs for the repetitive use of hand tools. However, present investigations could not contribute to the complete design of hammer. If the other factors such as angle of handle, handle diameter and texture are considered further, the studies can give complete guidelines to design a hammer for different purpose with the reduced risk of WMSDs.

CONCLUSIONS

- The height of the subject was found significantly covariate for discomfort score ($p = 0.046$).
- Sitting without support is more uncomfortable than standing for Hammering.
- The frequency of exertions was significant for 203 mm length of hammer handle ($p = 0.03$) but was not found significant for 279 mm length of the hammer handle. Therefore, it is required to investigate the effect of frequency for longer durations.
- The length of hammer was not significant on discomfort score for light upsetting task as was selected in this experiment. Though the values were found to be higher for minimum length of hammer handle and lower values of discomfort for larger length of the hammer handle.
- The reported discomfort was mostly in the region of shoulder and hand/palm as per body discomfort map.

REFERENCES

[1] Ayoub, M.M. and LoPresti, P. (1971). The determination of optimum size cylindrical handle by use of EMG. Ergonomics, 14(4): 509–518.

[2] Bernard, B.P. (1997). Musculoskeletal disorders and workplace factors, National Institute for Occupational Safety and Health, Cincinnatti.

[3] Carey, E.J. and Gallwey, T.J. (2002). "Effects of Wrist Posture, Pace and Exertion on Discomfort", *International Journal of Industrial Ergonomics,* 29: pp. 85–94.

[4] Hagberg, M., Silverstein, B., Wells, R., Smith, M., Hendrick, Carayon, P. and Perusse, M. (1995). Work related musculoskeletal disorders: A reference book of Prevention, Taylor & Francis, London.

[5] Hertzberg, H. (1973). "Engineering Anthropometry", In *Human Engineering Guide to Equipment Design*, VanCott, H. and Kincaid, R. (Eds).

[6] Karwowski, W., Mattila, M., Quesada, P.M., Kwon, Y.G. and James, D. (2003). "The Effects of Hammer Design and Hammering Surface Orientation on Task Performance", *Occupational Ergonomics,* 3, pp. 45–54.

[7] Lin, M.L., Radwin, R.G. and Snook, S.H. (1997). "A Single Metric for Quantifying Biomechanical Stress in Repetitive Motions and Exertions", *Ergonomics*, 40(5): 543–558.

[8] MacKinnon, A.N. and Stone, G.W. (2003). "Effects of Standing and Sitting Postures on an Isoinertial Pulling Task", *Occupational Ergonomics*, 3, pp. 55–64.

[9] Meagher, S.W. (1986). "Hand Tools: Cumulative Trauma Disorders Caused by Improper Use of Design Elements", In: *Trends in Ergonomics/Human Factors*, (Vol. III), Karwowski, W. (Ed.), Elsevier.

[10] Stal, M., Hansson, G.A. and Moritz, U. (1999). Wrist Positions and Movements as Possible Risk Factors during Machine Milking", *Applied ergonomics*, 30, pp. 527–533.

Occupational Self Efficacy among Employees Working in BPOs at National Capital Region

Kusum Tyagi

Department of Psychology

VMLG (PG) College, Ghaziabad, Uttar Pradesh, India

ABSTRACT

The study has its focus on role of ergonomics of BPOs and occupational self-efficacy among male and female employees. The objective of the study is to reveal the totality of occupational self efficacy through OSE Scale which has calculated the factors (Confidence, Command, Adaptability, Personal effectiveness, Positive attitude and Individuality). The sample of the study was 100 employees belonging to the age group of 25–30 working in BPOs. The design of the study was stratified randomized group and the statistical tools were t-test and correlation. The findings of the study have proved significant difference among male and female respondent regarding self-efficacy. A positive correlation at 0.01 levels was found between ergonomics of BPOs and self-efficacy. The general information by the employees indicated an increased number of female employees in BPOs and decreased level of higher education among all respondents. More than 61% of employees were undergraduate. These results may help the youth in keeping physical as well as mental fitness to lead a healthy life.

Keywords: Self Efficacy; BPO, OSE Scale.

INTRODUCTION

Ergonomics of any organization is one of the important factors to influence the motivation level of its employees that can change the economy of a developing country. It is a multidisciplinary science which applies principles based on physical and psychological capabilities of employee's to the design or modification of jobs, equipment, products and workplace.

The goal of ergonomics is to decrease risk of injuries and illness to improve workers' performance, to decrease workers' discomfort and to improve the quality of workers' life. Ergonomics also covers all aspect of a job from the physical stresses it places on joints, muscles, nerves, tendons, bones, to environmental factors which can affect hearing vision and general comfort and health.

Today Business Process Outsourcing (BPOs) is the expression of an ideal organization with a good ergonomics to achieve the development of international trade by providing cost effective output in globalized world. This service provider owns entire responsibilities for the customer's operations and manages the operation in totality. BPO is the fastest growing sector in the Information Technologies Enabled Services (ITES) industry and a biggest source of employment for youth. The Indian information technologies and BPO sector, both in exports and domestic business, are expected to register a growth of 28% in 2006–07 and revenues are likely to exceed $47.8 billion (Modi 2007). The direct employment offered by the sector is likely to cross 1.6 million. The Indian BPO sector is expected to maintain the current export momentum and grow by 35–40% in this fiscal year (2006–2007) to achieve $8.85 billion agencies $6.3 billion in the previous fiscal year (2005–2006).

Besides, Bangalore, Chennai, Pune, Mumbai, Hyderabad, Kolkata, NCR (Gurgaon, New Delhi, Faridabad, Greater Noida) are tier cities that are leading IT Cities in India which have setup of BPOs. The major part of the economic growth rate of India is being covered by the service sector rather than traditional sectors of manufacturing and agriculture. Thus it has been a long time that we are enjoying the fruits of globalization and India is emerging as an economic power gradually on the world map. .

Occupational self-efficacy among employees is motivational and strong determinant and predictions of the level of accomplishment that individual finally attain at his work place. It has important influence on our abilities to withstand stress occupation. Employees with high self-efficacy completes task more successfully than people of comparable ability, but lower self-efficacy. Subject with high occupational, self-efficacy expectations shows lower emotional arousal as they work, allowing them to maintain more of a task orientation. They regulate problem solving behaviour more effectively and bounce back more readily from failure. In these ways it seems that life's challenge may be less stressful for them. Occupational self-efficacy also refers to a global confidence in one's coping ability across a wide range of demanding or novel situations. Beliefs are multifaceted contextual (Locke *et al.,* 1984) but the level of generality of the efficacy items with in given domain of functioning varies depending on the degree of situation resemblance and the ability to foresee of task. Thus self is on the verge of dominating the field of motivation (Pajares, 1996). Hence by keeping all these points in consideration following objectives have been formulated for the present study.

Objectives

- To study the level of self efficacy among male and female employees working in BPOs.
- To study the correlation between ergonomics of BPO's occupational self efficacy among employees in BPOs.

Hypothesis

In respect of occupational self-efficacy there would be a significant difference between male and female employees working in BPOs.

There would be a positive effect of ergonomics of BPOs on occupational self-efficacy among employees.

METHODOLOGY

Design

The study is based on stratified randomized group design. The ergonomics of BPOs and gender effect have been treated as Independent Variable (IV) whereas occupational self-efficacy is as Dependent Variable (DV).

Sample

The sample of 50 male and 50 female employees belonging to the age group of 25–30, working for six months to one year, was constituted for investigation from multinational companies at NCR (Gurgaon, Greater Noida, Faridabd and Delhi).

Tools

Occupational self-efficacy scale was used for measuring the level of occupational self-efficacy. It is a scale having 50 items measuring six factors (confidence, command, adaptability, personal effectiveness, positive attitude, and individuality) that express the total occupational self-efficacy of the employee. Self-prepared questionnaire-related ergonomics of the organization was administered to measure the attitude of employee.

Procedure

The scale was administered to the subjects individually. Directions and instructions given by the authors of the manual were followed. The response sheets were scored as per instructions and raw data was quantified and statistically processed by t-test.

INTERPRETATION AND DISCUSSION

Table 1: Factor I of O.S.E.

Subject	*N*	*Mean*	*SD*	*t-Value*	*Significance Level*
Male	50	18.98	4.65	2.42	.05 *
Female	50	17.38	3.22		

As shown in above table t-value is significant at .05 levels. It makes clear that the confidence level of male employees is higher than female employees.

Table 2: Factor II of O.S.E.

Subject	*N*	*Mean*	*S.D.*	*t-value*	*Significance Level*
Male	50	13.20	2.48	1,88	.05*
Female	50	14.12	2.36		

Result in the above table makes clear that ability to command is higher in male employees than female employees.

Table 3: Factor III of O.S.E.

Subject	*N*	*Mean*	*S.D.*	*t-Value*	*Significance Level*
Male	50	13.96	2.47	5.11	.01**
Female	50	11.20	2.87		.05*

The t-value in the above table is significant at .01 level as well as at .05 levels. Both show a significant difference between male and female employees in respect of adaptability.

Table 4: Factors IV of O.S.E.

Subject	*N*	*Mean*	*S.D.*	*t-value*	*Significance Level*
Male	50	18.50	3.23	0.17	Not Significant
Female	50	18.40	2.65		

Above table indicates that t-value 0.17 is not significant at .01 level and .05 level both mean scores also indicate minor difference that is (0.10) not relevant. So, it can be explained that there is no significant difference in personal effectiveness among male and female employees in BPOs.

Table 5: Factor V of O.S.E.

Subject	*N*	*Mean*	*S.D.*	*t-value*	*Significance Level*
Male	50	13.90	2.89	2.19	.01**
Female	50	12.72	2.38		.05*

Table 5 indicates that t-value 2.19 of positive attitude is significant at both .01 level and .05 levels. It shows that male employees have more positive attitude towards their work and working conditions than female employees at their work place.

Table 6: Individuality

Subject	*N*	*Mean*	*S.D.*	*t-value*	*Significance Level*
Male	50	8.90	1.84	0.70	Not significance
Female	50	8.72	1.80		

Results in the above table show that there is no significant difference between male and female employees in respect of individuality that indicates independent thinking in making decisions.

Table 7: Total Results of O.S.E.

Subject	*No. of Sample*	*Means*	*S.D.*	*t-Value*	*Significance Level*
Male	50	87.84	15.29	1.76*	.05
Female	50	83.00	11.69		

As results shown in above Table 7, the t-value 1.76 is significant at .05 level. It indicates that occupational self-efficacy is higher among male employees than female employees working in B.P.O's at National Capital Region (NCR).

CONCLUSION

Thus overall occupational self-efficacy scores of male employees are higher than the score of female employees. The observed difference may be due to only personality factors, gender difference and style of working. As result shows, the confidence depends on one's own ability, adaptability and positive attitude. The ability to evaluate optimistically is higher in males than females. In respect of command that is a sense of control over the situation, females are higher than male.

The results also show that both male and female do not differ in respect of individuality (as factor VI) that is independence in making decision and setting standard for performance and personal effectiveness (as factor IV), *i.e.* inclination towards continuous development.

Graphical presentation (Figure 1) also makes it clear that mean scores of male employees are higher than female employees in respect of confidence, adaptability personal effectiveness, and positive attitude as factors out of six identified factors of occupational self-efficacy.

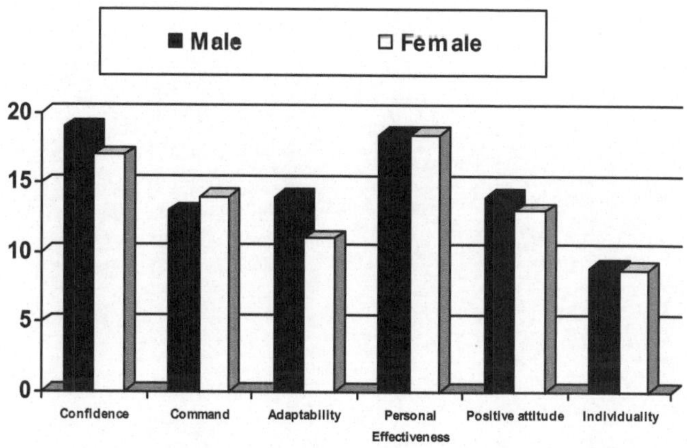

Fig. 1: Graphical Presentation of Mean Score and Factors of Occupational Self-efficacy Showing Total Results

Information collected from male and female employees through self-prepared questionnaire, related to ergonomics of BPOs, indicates that the ergonomics of organization is very positive and motivating. Though the work environment is the same, their approaches towards work environment may be different.

REFERENCES

[1] Locke, E.A., Frederick, Lee, C. and Bobko, P. (1984). Effect of Self Efficacy, Goal and Task Strategies on Task Performance. *Journal of Applied Psychology*.

[2] Modi, Y. (2007). Insight of Souvenir.

[3] Pajares, F. (1996). Directions in Self Efficacy Research: *Self Efficacy Paper Presented at Annual Meeting of the American Educational Research Association New-York*, from the World Wide Web.

Human Errors in Medical Field

Avinash Konkani

Department of Biomedical Engineering, Trident Academy of Technology
Chandaka Industrial Estate, Bhubaneswar–751 024, Orissa, India
E-mail: avinashkonkani@gmail.com

ABSTRACT

Human errors are unavoidable but can be minimized to reduce the accidents or injuries. In the field of medicine where the patients undergo treatment to get fast recovery, these human errors create difficulties in the process of diagnosis, treatment, surgery and/or recovery of the patient. A major human error may cause the death of the patient also. Therefore the risk involved with a human error in the medical field is very high. This article focuses on the causes of these errors and how can we minimize the errors by applying Human Factor principles in order to have a safe, effective and quality health care environment. Some of the case studies will be discussed.

Keywords: Human Error, Medical Error, Patient Safety, Medication Errors and Humans in Complex System.

INTRODUCTION

Three large studies conducted in United States and Australia revealed that adverse events occur during 3%–17% of all hospitalizations, a large portion of which resulted from medical errors and could have been prevented (Inoue and Koizumi, 2004)). Human error is an inappropriate or undesirable human decision or behaviour that reduces, or has the potential for reducing effectiveness, safety or system performance (Sanders and McCormick, 1993).

This definition of human error implies that a human error is that action which has the potential to affect harmfully the system or human criteria. For example if I forget to get my lunch box to my office, it will not be considered as a human error as compared with if I forget to wear lead apron before entering an operating x-ray room, because this action will cause harm to me and is not a safe act.

The Human errors can be classified in different ways; some of them are as follows (Sanders and McCormick, 1993):

Discrete-action Classification

Depending on the action performed, the errors can be classified as:

- *Error of Omission:* Error due to failure to do some action.
- *Error of Commission:* Error due to performing an act incorrectly.
- *Sequence Error:* Error due to performing an action in wrong sequence.

- *Timing Error:* Error due to failing to complete the task in given time (too slow or too fast action).

Information Processing Classification

These errors depend on the type or level of behaviour involved. They are execution of procedure errors.

- *Skill-based:* These are primarily errors of execution.
- *Rule-based:* Errors due to failing to apply the correct rules.
- *Knowledge based:* Errors due to inadequate analysis or decision making.

Causes of Human Errors

The human error is very complex subject and the above classifications will not cover the whole complexity of human errors. These classifications give us an idea about how human errors can occur. Now that we have gained some knowledge about human errors and its causes we look into the human errors in medical field. In the medical field, "the human errors" are considered as "the medical errors". According to the study conducted by Zhang *et al.*, 2002 the different levels at which these medical errors occur in the system are:

- *Individual level:* Cognitive factors of individual are important at this level.
- *Interaction between individual and technology:* Man-Machine interaction level.
- *Distributed system:* Interactions between groups of people and complex technology.
- *Organizations structures:* Lack of co-ordination and communication, standardization of work process.
- *Institutional functions:* Inaccuracy in making policies, guidelines, etc.
- *National Regulations.*

Most of the human errors can be controlled at individual level, interaction with technology level and/or distribution system level. But the root cause of the medical/human error can go up to the National Regulations level depending on the medical error occurred.

The results of an Institute of Medicine–led committee on quality of health care in America indicated that medical error is the fourth largest killer of Americans. Using medical error data from Colorado, New York, and Utah hospitals, it was estimated that at least 44,000 Americans die each year from medical error (Gawron *et al.*, 2006). This data shows us the impact of medical errors on patient-safety and the importance to reduce these errors. Gawron *et al.*, 2006 states that the most common cause for human error is leaving out a necessary step in a procedure. They also list the following factors as the sources of medical error:

- Errors occur due to poor readability of instructions.
- Use of technical language in instructions for user.

- New technology and automation surprises.
- Stress and work load.
- Human-human coordination breakdowns.
- Poor organization of procedures/work methods.
- Patient-related factors such as misleading normal findings.

"Errors, Stress, and team work in medicine and aviation: cross sectional surveys"—a study done by Helmreich, Sexton *et al.*, 2000 at University of Texas indicates the following responses from the medical professionals towards the human errors:

In general, a minority of respondents agree that stress has effect on performance, most of them think they handle the situation well under stress.

Teamwork and communication between doctors and nurses has effect on errors. 77% of intensive care doctors have agreed with high level of teamwork with nurses whereas only 40% of nurses agreed to that they have high level of teamwork with doctors.

From this survey we can conclude that some people do not accept openly that they do not commit mistakes under high stress. Highly skilled professionals can handle the situations very well in stress but still there are possibilities of committing an error by them because they are also humans. But the probability of committing a mistake may be very low with them. Teamwork with improper communication with team members plays a role in causing an error. If the team works with the team spirit and can ask for the help of other team members whenever necessary then the errors can be reduced.

Another study states that a major contributor to the occurrence of accidents related to anesthesia is Human Error (Weinger, 1999). But the root causes for these human errors are defective organizational structure, poor equipment design, complexity of the work area, and lack of knowledge of technology in the user. We can conclude that the human errors occur due to humans, machines and work environment.

"Errors in the Intensive Care Unit (ICU)" a study concludes that errors with administration of medications and infusions were the most frequent one, and only a minor part was concerned with technical equipment (Flaatten and Hevroy, 1998). Medication errors are very important because these can cause adverse effect on patient's health. Reducing these medication errors will improve patient safety.

Reason's organizational accident causation model states that human decision and actions play an important role in all accidents which are contributed by Latent and Active failures. Latent failures are as a result of faulty decision by higher management and these will have effect in long terms. Whereas the Active failures are unsafe acts or omissions committed by end operators or users. These actions will have quick adverse effect. The active failures mainly depend on the individual performance, increased workload, fatigue and poor training (Taylor-Adams *et al.*, 1999).

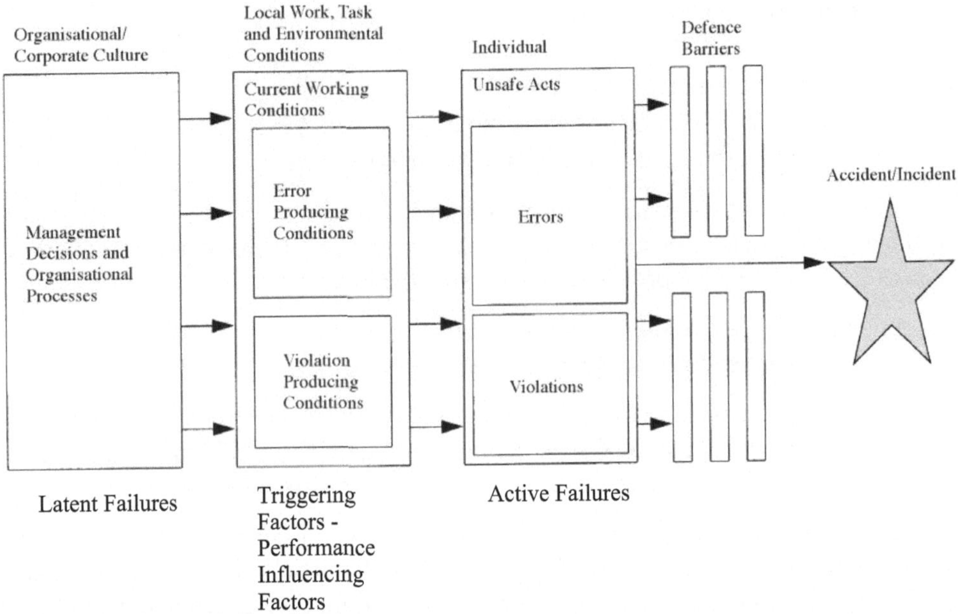

Fig. 1: Reason's Organizational Accident Causation Model (Reason, 1993)

Applications of Human Factors Methodologies to Reduce Human Errors

Dr. Alphonse Chapanis defines Human Factors Engineering (HFE) as: "It is the application of human factors information to the design of tools, machines, systems, tasks, jobs, and environment for safe, comfortable, and effective human use."

"Medical Error: Is the solution Medical or Cognitive?"—A study conducted at University of Texas Health Science Centre at Huston and Columbia University, New York concludes that the medical errors are human errors in healthcare and are domain subject of Cognitive and Human Factors Engineering. The solution to medical error is cognitive. The better person to solve these errors is a double expert having knowledge in Cognitive science, which is part of Human Factors Engineering and understanding of healthcare system.

The first Institute of Medicine report To Err is Human—Building a Safer Health System stated that the application of human factors in other industries has successfully reduced error. Health care has to look at medical error not as a special case of medicine, but as a special case of error, and to apply the theory and approaches already used in other fields to reduce errors and improve reliability (Schneider, 2002).

From the above reports we understand that the applications of HFE have a vital role in reducing the human errors in medical field. The following are Human Factors (HF) Methodologies that can be applied in order to reduce the human errors.

Schneider (2002) in his report "Applying Human Factors in improving medication—use safety" states the following HF Methods:

- *Root Cause Analysis:* It is a method of finding the root cause of the error and suggesting ways to eliminate these kinds of causes.
- *Failure Mode Effect and Criticality Analysis (FMECA):* In this method a process is analyzed to determine the points at which failure may occur by a panel of experts. This point of failure is known as failure mode. These failure modes are evaluated and ranked as most critical to least critical depending on severity, occurrence and detection.
- *Usability Testing:* This technique is applied in the design of the equipment or a procedure to get the feedback from the end user in order to improve the deficiencies of the design or the procedure.
- *Heuristic Evaluation:* It is a checklist approach to evaluate procedures, software and devices for the ease of effective use. It helps in identifying general problems related to physical user interfacc with machines, Graphical User Interface (GUI).
- *Evaluation of Medical Devices:* Using the HF principles a Human Factors expert will evaluate the medical device for its safety and poor design related factors.

Sanders and McCormick (1993) states that the following are the tools to reduce human errors:

- *Selection:* Selecting people with the capabilities and skills required to perform a job will result in fewer errors.
- *Training:* We can reduce number of errors by training the people in their job.
- *Design:* Design of equipment, procedures and environments can improve the performance of people.

Some of the studies show that the real culprit in most errors or accidents involving complex systems is, almost always, poor design (Weinger, 1999). Therefore a user-friendly design of medical equipment is necessary to reduce the human errors. The designer of the medical equipments should keep in mind that the end user should be able to use the equipment with minimum steps and without making an error. To achieve this kind of a design the involvement of the user in order to give the right feedback to the designer is very important.

The design of the human-machine interface plays a critical role in operating any medical equipment and this is designed by a human factors expert who has knowledge of medical equipment. The best way to reduce the human errors is by simple and effective design of the medical technology, procedures and tasks. This can be achieved by Human Factors Engineering.

CONCLUSION

With the help of HF methodologies we can find the root causes of the errors and provide an appropriate solution to overcome the error causing factors. Human Factors Engineering was not much applied in health care systems but now the time has come to concentrate on the applications of HFE in health care system to improve and provide a safe, effective and comfortable work environment to reduce the stress on the medical professionals so as to provide a high quality health care system to our society.

REFERENCES

[1] Flaatten, H. and Hevroy, O. (1998). "Errors in the Intensive Care Unit (ICU)", *Acta Anaesthesiol Scand,* 43, pp. 614–617.

[2] Gawron, V.J., Drury, C.G., Fairbanks, R.J. and Berger, R.C. (2006). "Medical Error and Human Factors Engineering: Where are we now?" *American Journal of Medical Quality,* 21(1), pp. 57–67.

[3] Inoue, K. and Koizumi, A. (2004). "Application of human reliability analysis to nursing errors in hospitals." *Risk Analysis,* 24(6): 1459–1473.

[4] Reason, J.T. (1993). "The human factor in medical accidents." In: Vincent, C. (Ed.), *Medical Accidents*. Oxford Medical Publications, Oxford, pp. 1–16.

[5] Sanders, M.S. and McCormick, E.J. (1993). "Human Factors in Engineering and Design." McGraw-Hill, Inc.

[6] Schneider, P.J. (2002). "Applying human factors in improving medication-use safety." American society of health-system pharmacist, Inc, 59: 1155–1159.

[7] Sexton, J.B., Thomas, E.J. and Helmreich, R.L. (2000). "Error, Stress, and Teamwork in Medicine and Aviation: Cross Sectional Surveys." *BMJ*, 320, pp. 745–749.

[8] Taylor-Adams, S., Vincent, C. and Stanhope, N. (1999). "Applying human factors methods to the investigation and analysis of clinical adverse events." *Safety Science,* 31: 143–159.

[9] Wienger, M.B. (1999). "Anaesthesia Equipment and Human Error." *Journal of Clinical Monitoring and Computin,*. 15: 319–323.

[10] Zhang, J., Patel, V.L. and Johnson, T.R. (2002). "Medical Error": Is the Solution Medical or Cognitive? *Journal of American Medical Informatics Association,* 9(6): S75–S77.

Modelling Human Hand for DHM Applications

B. Mukesh, Dibakar Sen and J.S. Vipin

Centre for Product Design and Manufacturing
Indian Institute of Science, Bangalore–560 012, India

ABSTRACT

The present paper develops a human hand model that is kinematically closer to the anatomical hand. This is achieved through the development of a technique to estimate the kinematic parameters of the hand experimentally using photogram metric techniques. The technique involves taking a series of photographs of the hand with markers and determining the joint axis locations by analyzing the markers in different hand configurations. The hand model so developed has 22 Degrees-Of-Freedom (DOF), four DOF for each finger and two DOF for the wrist. The first joint of the fingers have two DOF which are anatomically saddle joints characterized by two mutually perpendicular axes of rotation that are non-intersecting. The wrist is anatomically complex, but functionally it has two DOF. Hence it is also modelled as a saddle joint. The remaining joints are revolute joints. The hand is modelled using the D-H scheme widely used in robotics literature for serial manipulators. Hand being a tree structure rather than a simple serial chain with five fingers connected to the palm, the above scheme was modified. The present implementation incorporated a nominal non-deformable geometry with each segment of the hand for the purpose of realistic visualization and physically meaningful grasp related computations. The hand model implemented in C++ is available as an object class for general purpose usage. A GUI is developed to demonstrate the hand model features using direct kinematics. An application which computes the distance between a finger and an object or another finger, using the hand class, is presented. This is an important step towards the study of adaptive grasp strategies. The result is visualized using the same GUI.

INTRODUCTION

For Ergonomics to be considered in the early stages of the design of product, process or workspaces, a computer aided ergonomic evaluation model could serve as an effective tool for the designers. Usually designers find difficulty in incorporating ergonomic information to the design as they try to meet other design constraints. Digital Human Modelling (DHM) can be used for incorporating human elements in the early stages of design. DHM provides a realistic representation of human beings embedded in the virtual environment of a work scenario. The simulations and graphics provided by the DHM enable the designers to see the potential problems a user would encounter while using his design. DHM is prevalently used to simulate people of extreme sizes for the purpose of providing designs wherein a variety of people can reach, see and manipulate objects (Chaffin, 2002). The study of the effects of wearing protective gadgets, cloths, etc. are useful, especially for soldiers, which can also be done using DHM. The application of DHM will enable designers to have a proactive approach in their design rather than a reactive approach. Some of the commercially available digital human models are JACK, Boeing Human Modelling System (BHMS), SAFEWORKS, etc.

Hands are the vital organs for the human beings for physically manipulating the environment. In order to accomplish almost all day-to-day task people use their hands. Since DHM is used to simulate people and their behaviour/action in real life, in a virtual environment an anatomically closer hand is needed in DHM to accurately model the people's interactions with the objects. Human hand has high degree of dexterity; 29 bones intricately held in place with ligaments and 35 muscles actuating the bones through a network of tendons are responsible for the dexterity and strength it enjoys. Even though there are many Degrees-Of-Freedom (DOF) in human hand, the motions of the fingers are not completely independent.

A simplified form of the human hand model can be used in computer aided product design. Such a model should satisfy the following criteria (Freivalds and Davidoff, 1993): (a) The model should approximate the geometry of a human hand closely enough to be useful for designers. (b) The model should limit motions of joints of the hand to realistic postures. (c) The model should be scalable so that different populations and percentile groups can be represented. (d) The user should be able to check for intersections between solid sections of the hand and between the hand and a solid object being grasped. In the present work, (a), (b) and (d) have been addressed, (c) can also be addressed so far as the kinematics of the hand is concerned.

The primary objective of this work is to create a kinematically accurate model of human hand which could be made to follow configurations, motions, and dimensions of the real hand closely enough for use in designing products. The modelling of bones, tendons and muscles are not considered in this work. A simplified form of the skin representing the physical extent of the hand has been included in the model.

Experimental Determination of Kinematic Parameters

Human hand consists of 29 bones; each finger consists of three phalanges: Proximal, intermediate, and distal, except the thumb, which has only two phalanges. The motion of the fingers is controlled by the following tendons: Extensor (EXT), Interosseus Left (INL), Interosseus Right (INR), Flexor Digitorum Profundis (FDP), and Flexor Digitorum Superficialis (FDS) (Boyd, 1966). EXT, FDP and FDS create motions in the flexion-extension plane, while INL and INR are responsible for abduction-adduction movements, accompanied by partial extension. One end of each tendon connects to hand muscles and the other is attached to the bone structure. Hand-segment movements are usually described by three physiological angles: Flexion/extension (F/E), abduction/ adduction (Ab/Ad) and pronation/supination (P/S). Distal and proximal interphalangeal (DIP and PIP) joints of fingers and interphalangeal (IP) joint of the thumb are usually considered with one DOF (F/E). Metacarpophalangeal (MCP) joints are usually considered as two DOF joints (F/E and Ab/Ad). The movement of the carpometacarpal (CMC) joint of the thumb is more complex, combining F/E, Ab/Ad and P/S.

A goniometer is a simple protractor based device used for measuring joint angles in biomechanical and ergonomic applications. It measures only one joint angle at a time. On the other hand data gloves are strain gauge based instrumented device capable of measuring multiple joint angles of the hand. Conventional anthropometric devices such as sliding or spreading callipers measure the distance between the landmarks of the body. Since, the joint axes cannot be determined from distance between landmarks or joint range of motion, the above devices are not useful for estimating the kinematic parameters of hand. In the following a photogrammetry based technique is described for estimation of joint axis locations; the distance between two consecutive joint axes, in turn, gives a measure of the link lengths associated with a moving segment.

First markers are put on the subject's hand such that every moving segment has two markers on it. The Reference Position (RP) of the finger is fixed such that the palmar side of the finger is in full contact with a table. Subject is asked to slowly move the particular finger under consideration and it is photographed at different configurations (Figure 1). The coordinates of the markers at each configuration are retrieved using an image processing software. These coordinate values are used to find a best-fit circle first for the segment adjacent to the RP. The least square error method is used to calculate the best-fit circle.

$$\text{Error} = \left\{ \sum_{i=1}^{n} (x - x_i)^2 + \sum_{i=1}^{n} (y - y_i)^2 - nr^2 \right\}^2 \quad \dots (1)$$

Where x is the ordinate of the centre of the best-fit circle
y is the abscissa of the centre of the best-fit circle
x_i is the ordinate of the ith point
y_i is the abscissa of the ith point
r is the radius of the circle
n is the number of points

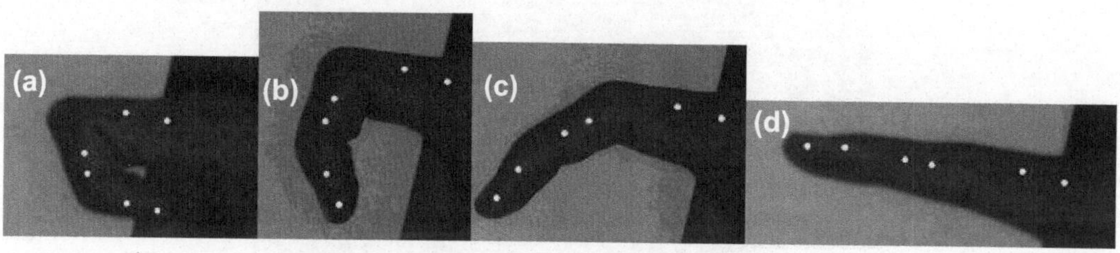

Fig. 1: The Motion of the Finger with Markers

The equation (1) is differentiated with respect to x, y and r and equates it to zero. The three equations so obtained are then solved using conventional technique to obtain the centre and radius of the best-fit circle (Figure 2). The centre of the best-fit circle will give the axis of rotation of the joint. This axis of rotation is very crucial in the kinematic modelling of the human hand as D-H parameters are defined with respect to axes of rotations.

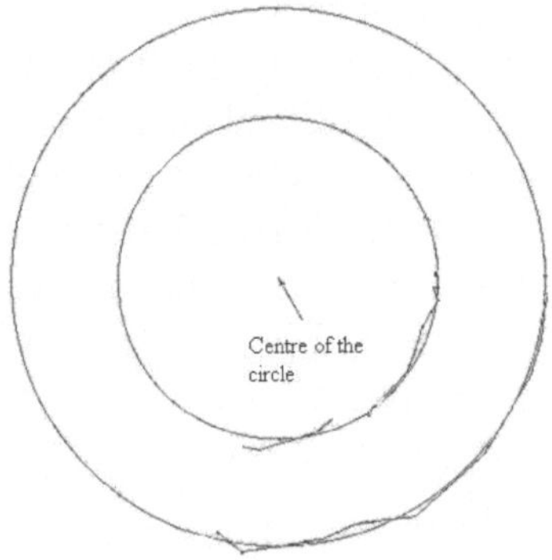

Fig. 2: Estimated Best Fit Circle

For a distal segment, first the coordinates of the markers on it in different configurations are computed in the moving frame of the adjacent proximal segment using the axis location computed as above. This is done as follows: Angle of rotation, say θ, of the proximal segment in the global frame is computed as the angle between the lines through the two markers on it in two configurations of interest. Here we take the straight finger configuration (Figure 1(d)) as the reference configuration. Coordinates of the markers on subsequent segments are changed by rotating them with respect to the axis computed by an amount (–θ). When this is done for all configurations and the above circle fitting is performed we get the new joint location. This is illustrated in Figure 3(a)&(b). The recursive application of the procedure for all the segments we get the kinematic model of the finger (Figure 4). When the above procedure is followed for all the fingers as well as the palm we get the kinematic model of the hand (Figure 5).

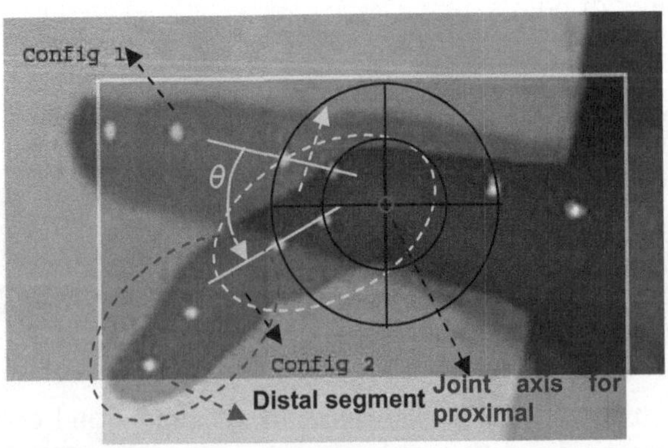

Fig. 3(a): Original Configurations of the Finger

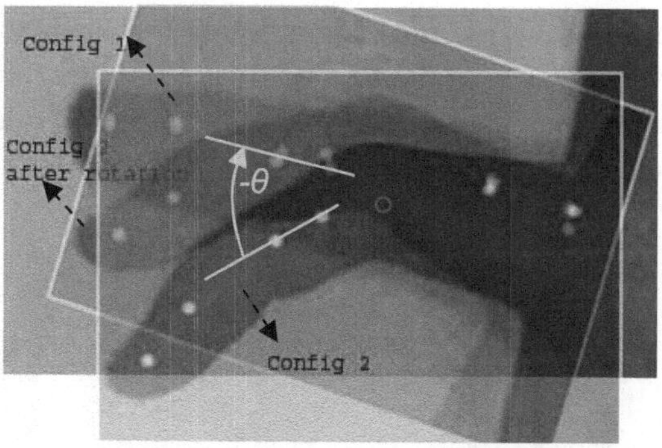

Fig. 3(b): Configurations w.r.t. Proximal Segment

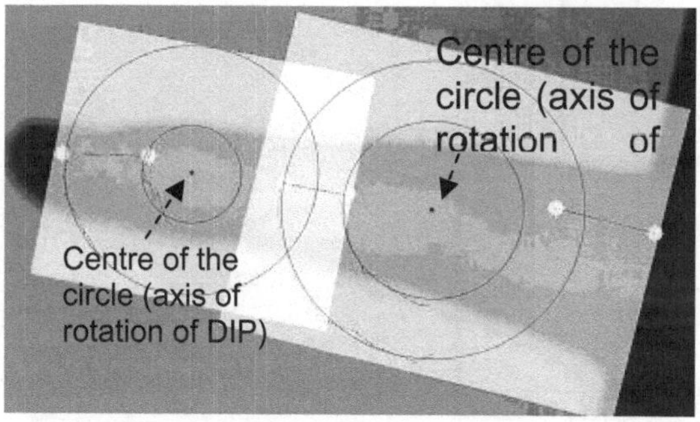

Fig. 4: Best-fit Circles Superimposed on Finger

Kinematic Modelling of Hand

The articulating bones of MCP (except thumb) and CMC of the thumb have saddle shape, *i.e.*, opposing surfaces are reciprocally concave-convex, thus giving two DOF for the joints. The motion of CMC joint of the thumb is complex, due to the incongruity between the articulating surfaces of the joint and will produce an additional degree of freedom, *i.e.*, axial rotation. The centre of circles of principle curvature of the saddle will give the axes of rotation (Figure 6). If the centre of the two circles is joined, additional link is obtained which can be used to model the offset between axes at saddle.

Saddle joints (CMC of thumb and MCP of other finger) are modelled as two mutually perpendicular revolute joints and all other joints in the hand are modelled as revolute joints. Lee and Kunii (1995) modelled the hand by assigning local frames to every joint and represented any joint rotation by a sequence of rotations occurring about the three mutually perpendicular axes of local coordinate system. Yi *et al.* (2005) have modelled hand with 16 joints and attached a local coordinate system to each joint. Hand developed has been modelled using Denavit-Hartenberg (D-H) parameters (Craig, 2004). In literature, D-H parameters are widely used for modelling serial manipulators. Human hand has got a tree structure, necessitating a modified scheme for D-H parameters. The local coordinate systems for wrist and finger joints are such that z axis of the joint coordinate system is coincident with the axis of rotation at that joint and x axis along the segment and y axis direction is determined by right hand thumb rule (An *et al.,* 1979).

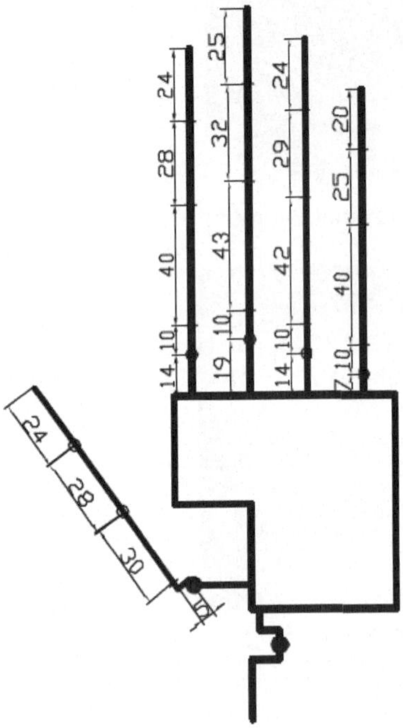

Fig. 5: Kinematic Model of Human Hand

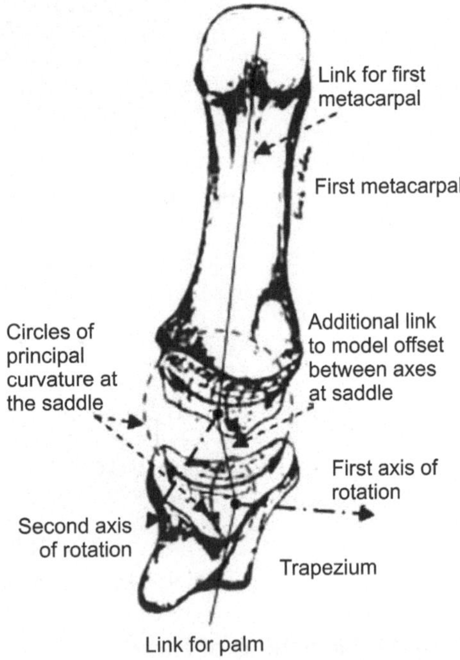

Fig. 6: Modelling Saddle Shape CMC of Thumb

Modelling Coupled Motion in Fingers

Human hand has got many DOF, and finger's motion is also not independent. Angle of each finger joint is influenced by the adjacent joint or by the neighbouring finger joints (Yasumuro *et al.*, 1999). Although people display differences in hand constraint magnitudes, all natural movements observe the same pattern (Yi *et al.*, 2005). Two types of coupled motions can be observed: Inter-finger and intra-finger constraints for the dependency of motion between two adjacent fingers and that between two segments in the same finger respectively.

Inter-Finger Constraint

Owing to the like structure between the two neighbouring fingers, if one finger get flexed from MCP joint beyond some value, the neighbouring finger also get flexed (except thumb). The finger which is causing motion is called the active finger and the finger following the motion of the active finger is called the passive finger. The influence depends on the active finger's flexion angle and the proximity of the passive finger to the active one (Yi *et al.*, 2005). In the following, the constraint is modelled such that it is active only for flexion and is passive for extension. The motion of the fingers considered for modelling the constraint here is only in one direction. This constraint is modelled as an inextensible string connected between the two fingers (Figure 7). Let the maximum flexion (θ) of the active finger, achievable without disturbing the other neighbouring fingers in their nominal configuration, be called its critical flexion angle. Then the length of the tout string (Ls) between the two anchor points at equal length (l) from the origin of the corresponding MCP joints is calculated as follows,

$$L_s = \sqrt{d^2 + \{l(1 - \cos\theta) + a\}^2 + (l\sin\theta)^2} \quad \dots (2)$$

When the active finger moves beyond the critical value, the correlated movement of the passive finger can be calculated using equation (3); where θ1 and θ2 are the flexion angles of the active and passive fingers respectively. The result is illustrated in Figure 8.

$$L_s^2 = d^2 + \{l(\cos\theta_1 - \cos\theta_2) + a\}^2 + (l\sin\theta_1 - l\sin\theta_2)^2 \quad \dots (3)$$

It is observed that if the active finger extends, the passive finger need not move as per equation (3). The unidirectionality of the constraint is modelled as follows: For any arbitrary configuration for the active finger, define the distance between the anchor points as *La*. It can be appreciated that if *La*<*Ls* the string will be slack; hence the active finger will not be able to pull the passive finger. But if *La*>*Ls*, the passive finger should move at least by an amount so as to make the current *La*=*Ls*. Thus the unidirectionality of the constraint is easily modelled by observing the sign of *La-Ls* for every configuration of the active finger. Equation (3) is used to compute the modified value of the passive finger only if *La-Ls* is negative.

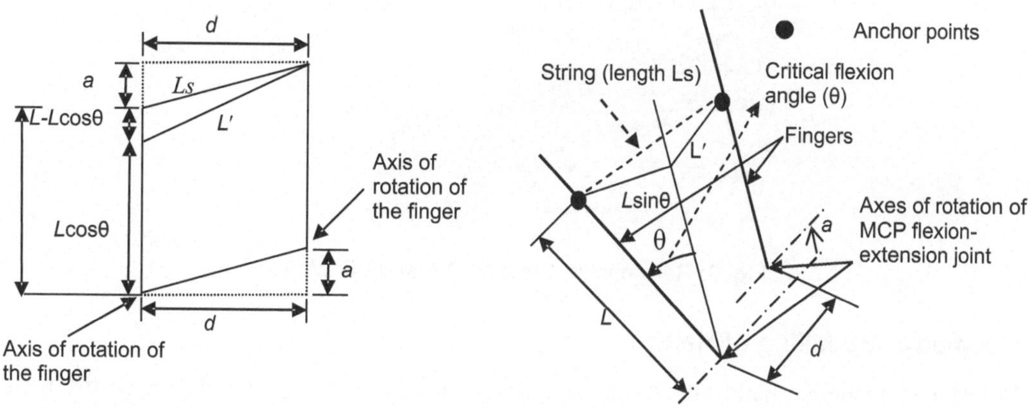

Fig. 7: Coupling between two Neighbouring MCP Joints Modelled Using Strings

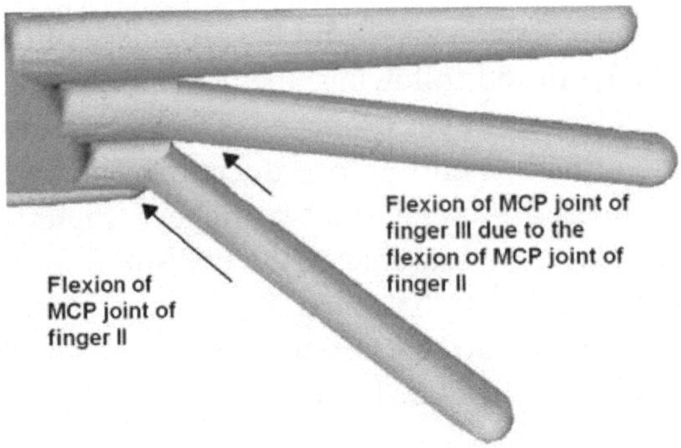

Fig. 8: Effect of MCP Movement of Finger on Neighbouring Finger

Intra-Finger Constraint

Intra-finger constraints are motion dependency between the segments of the same finger. In human hand it is nearly impossible to move DIP without moving the adjacent PIP joint and vice-versa. This dependency is due to the tendon that runs through the finger. If DIP joint is flexed past the maximum constraint limit of the DIP, the flexion of the PIP joint will begin.

The relation between flexion of PIP and DIP joints is given by equation (4) (Lin, *et al.*, 2000). This dependency of PIP and DIP is as shown in Figure 9.

$$\theta_{DIP} = \frac{2}{3}\theta_{PIP} \qquad \ldots (4)$$

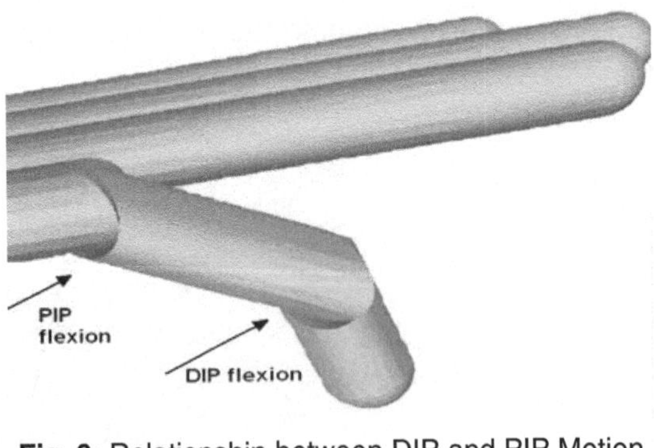

Fig. 9: Relationship between DIP and PIP Motion

Geometric Modelling of Hand

In order to visualise hand and its movement capability, various geometries of the hand segments should be modelled. Appropriate geometries for the hand segments are generated according to the kinematic parameters determined. A segment of the hand between two consecutive axes is modelled as a rigid body. Each finger is modelled using four segments, palm is modelled using 1 segment and wrist is modelled using two segments, as shown in Figure 10. A geometric modelling software is used to model the hand and is saved in VRML format wherein the models are represented using triangles. The hand segments and the corresponding number of triangles in the model used here are given in Table 1. The transformations of various hand segments calculated in the kinematics module is applied to each triangle of the corresponding segment in the geometric module and is rendered using OpenGL for visualisation (Figure 10).

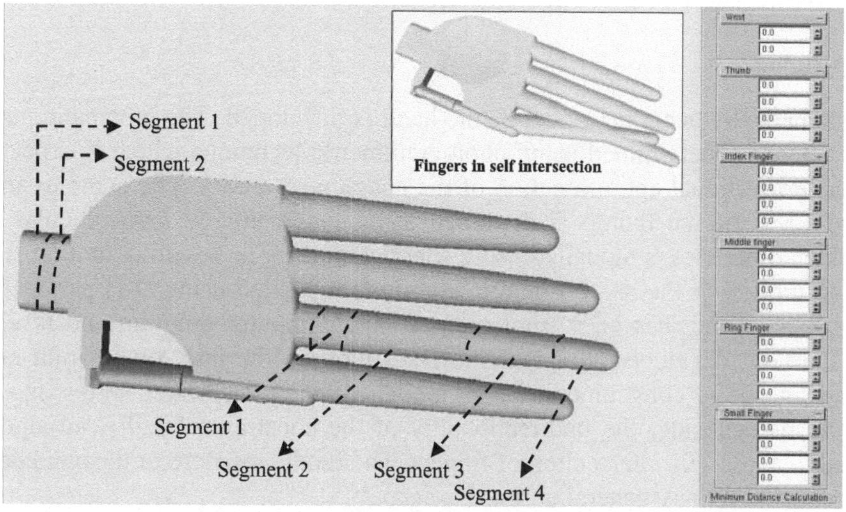

Fig. 10: Visualising the Hand in the GUI with Interactive Direct Kinematics

Table 1: Triangles in Each Segment of Hand

Segments	*1*	*2*	*3*	*4*
Thumb	1024	530	1070	964
Index finger	498	646	672	1008
Middle finger	498	592	650	970
Ring finger	498	616	678	980
Small finger	498	646	738	1094
Wrist	128	254	–	–
Palm	4940	–	–	–

Avoidance of Self-Intersecting Configurations

In the abduction-adduction movement of the fingers, if the motion is not stopped on contact with the neighbouring finger, it will visually penetrate through that finger, which is unrealistic (inset of Figure 10). Similarly, when all the fingers are fully flexed (at the extremities of their individual ranges of motion) it could penetrate the palm. In order to avoid such unrealistic motion of the fingers, the motion is arrested when the distance of the moving finger from other segments is zero. The distance between the segments is computed through the determination of the minimum of the distances among the vertices in one segment with those in other segments of the hand. The same algorithm can be made use for computing contact and collision of hand and an object to be manipulated, to enable free from grasping unknown objects. The movements of hands can be stopped in a systematic way when the palm or the fingers comes in contact with the object, using the same algorithm. This problem is under investigation of the authors.

CONCLUSION

A 3D kinematically accurate model of the hand is developed. The kinematic parameters are experimentally determined using photogrammetric technique which is expected to be a reasonable functional approximation of the anatomical reality. The complex motion of wrist and CMC of the thumb is modelled using two mutually perpendicular revolute joints, which otherwise is modelled using spherical joint. The resulting motion is close to the natural motion of the hand. The articulation is modelled using D-H parameters. The proposed hand model has been implemented in a computer program and is visualised through the GUI developed. For realistic hand motion, the inter-finger joint constraint and intra-finger joint constraints are also modelled using an inextensible string analogy which will also consider the undirectionality of the constraint. Finally, an algorithm is developed to avoid self-intersection of fingers. The modular nature of the hand developed makes it suitable for any general purpose usage.

REFERENCES

[1] An, K.N., Chao, E.Y.S., Cooney, W.P. and Linscheid, R.L. (1979). "Normative Model of Human Hand for Biomechanical Analysis", *Journal of Biomechanics,* 12, pp. 775–788.

[2] Chaffin, D.B. (2002). "On Simulating Human Reach Motions for Ergonomics Analyses, Human Factors and Ergonomics in Manufacturing", Vol. 12(3), pp. 235–247.

[3] Craig, J.J. "Introduction to Robotics Mechanics and Control", Pearson Education (Singapore), Pte. Ltd., India 2004.

[4] Freivalds, A. and Davidoff, N.A. (1993). "A Graphic Model of the Human Hand using CATIA", *International Journal of Industrial Ergonomics,* 12, pp. 255–264.

[5] Lin, J., Wu, Y. and Huang, T.S. (2000). "Modelling the Constraints of Human Hand Motion", *IEEE Proceedings of Workshop on Human Motion*, pp. 12–126.

[6] Yasumuro, Y., Chen, Q. and Chihara, K. (1999). "Three-dimensional Modeling of the *Human Hand with Motion Constraints, Image and Vision Computing*, 17, pp. 149–156.

[7] Yi, B., Harris, F.C., Wang, L. and Yan, Y. (2005). "Real-Time Natural Hand Gestures, Computing in Science and Engineering", Vol. 7, No. 3, pp. 92–96.

Problems Faced and Coping Strategies Adopted by the Employed Couples to Relieve Stress

Neerja Jaiswal[1] and Sarjoo Patel[2]

Department of Home Management, Faculty of Family & Community Sciences,
The Maharaja Sayaji Rao University of Baroda, Vadodara, India
E-mail: [1]neerja_jaiswal@rediffmail.com; [2]sarjoo_patel@yahoo.com

ABSTRACT

Dual career couples experience unique problems but the fact is that the number of couples choosing this life style will continue to increase. The focus here is how work-arrangements at home and at work place affect marital satisfaction and work spouse conflict. The fact that married women are going to work and working more consistently than ever before is undesirable. It is also true that in majority of dual career families, the responsibility for the domestic sphere lies with the wife. Thus, the present study was undertaken to find out the problems faced and the coping strategies adopted by employed couples, to relieve their stress arisen from their dual careers in the family. Descriptive research design was used for the study. Pre-validated interview schedule was used as a tool for data collection. The sample for the study constituted sixty employed couples of Vadodara city, selected purposively by snowball technique. For the purpose of data analysis descriptive as well as relational statistics were used.

The findings of the study revealed that majority of the husbands had longer working hours as compared to their counterparts. The former had 8–10 hours of working as compared to the latter who had 5–6 hours of working. Majority of the couples experienced psychological problems, management problems and children related problems to a moderate extent. Further, it was observed that majority of the couples used coping strategies to a high extent for coping with social problems and communication problems, whereas for coping with other problems they used coping strategies to a moderate extent. The paper also discusses in detail about the nature of coping strategies adopted by the couples to overcome their stress.

Keywords: Employed Couples, Problems, Coping Strategies, Stress.

INTRODUCTION

The number of women in the workforce has increased tremendously during the last decade. These working women have become a part of a relatively new life-style, the dual-career family. The dual-career life-style has resulted primarily from the expanded educational opportunities for women and the increasing job opportunities for these educated, aspiring women.

The time and energy demands of dual-career couples thus have to be balanced with the jobs at the home and outside the home. Even though a slight deviation in these, however, leads to family conflicts resulting in stress for both of them.

Work and marriage are two important part of adult's life. Sometimes the demand of one can interfere with the demand of the other. The focus here is how work arrangements at home and at work place affect marital satisfaction and working spouse conflict.

Rapoport and Rapoport (1976) have defined dual-career family as one in which both heads of households pursue careers and at the same time maintain a family life together. The fact that married women are going to work and working more consistently than ever before is undesirable.

The dual-career life-style of the employed couples has created a unique set of challenges, many of which relate to socialisation and role expectations, work-role conflicts and family role conflicts.

A woman who tries to combine a career and a family is soon reminded that she is flaunting the socially accepted norms. She finds herself in a seemingly no-win-situation. The quality associated with the role of wife-mother (nurturance, emotionality, responsiveness to people rather than ideas) is seen to be incompatible with those qualities associated with independence, rationality and assertiveness.

The man, too, is struggling with his definition of masculine and feminine roles as he has been socialised to understand them. A man grows up expecting to be the head of a household, to be the one who earns the money and has the power. Men are not socialised, nor educated to fill roles calling for skills in child rearing or homemaking. Even if a man has the skills, he may perceive that dovetailing a great deal of time and emotional energy to domestic activities may negatively affect his career, particularly if he is competing with other man who does not have similar family roles.

The home environment is a special challenge for the employed couple as both of them try to meet demands of careers and build a family life together. The fact remains that in majority of dual-career families, the responsibility for the domestic sphere lies with the wife. Even highly educated professional women also retain that responsibility.

Does this family responsibility give rise to the problems? What nature of problems did the employed couples experience? Which coping strategies did they adopt to overcome their stress? Were some of the queries which the investigators felt the need to be explored. Thus, the present study was undertaken with the following objectives:

Objectives

1. To find out the problems faced by the employed couples pursuing dual careers.
2. To explore the coping strategies adopted by them to relieve their stress while managing their dual careers in the family.

METHODS

The research design of the present study was descriptive in nature. The extent of problems faced and the extent of coping strategies adopted were judged on a three-point continuum scale as to great extent, to moderate extent and to no extent, scored as 1, 2, 3 for the former and 3, 2, 1 for the latter. Problems faced were operationally defined as the difficulties experienced by the employed couples related to social, psychological, economical, employment, management of household work, communication and child related aspects. The coping strategies were operationally defined as the actions or behaviours taken by the employed couples to cope up with the problems faced to relieve stress. The data were collected through a survey on a pre-validated interview schedules. The sample size of the study constituted sixty middle class employed couples of Vadodara city. The samples were selected purposively with the help of snowball technique. The data were further coded, tabulated, and analysed by computing descriptive statistics.

RESULTS AND DISCUSSION

Background Information: A comparatively high percentage of the husbands and wives was young, aged between 25–35 years. Majority of the employed couples were highly educated, having their college degree above graduation. Viewing the occupational status of respondents, it was found that a higher percentage of them (husbands 58.33% and wives 41.64%) was engaged in service. A teaching profession was pursued more by wives (35%) than husbands (8.33%). The wives were engaged more in professional consultancy (23.33%) as compared to husbands (11.67%). It is very interesting to note that not a single wife was involved in business as compared to 21.67% of the husbands.

The data on the income of wives and husbands highlighted that the latter were earning more than their counterparts. Majority of the wives had their monthly income ranging less than or equal to Rs. 15,000/- where as more than one-half of the husbands were found earning Rs. 15,000/- and above There were dominance of nuclear family, among the respondents. Most of the couples possessed moderate sized family with 3–5 members residing in it.

While analyzing the data on the working hours of the husbands and wives it is highlighted that majority of the wives had less working hours ranging between 5–7 hours per day whereas majority of the husbands worked for 8–10 hours daily.

Nature and Extent of Problems Faced: The findings of the study further revealed that the husbands and wives both experienced different nature of problems due to their dual careers. The major psychological problem experienced by both of the respondents was their guilt feeling arose from their inability to balance the household as well as their career life ideally. The pressures and demands at the job sometimes became the major cause of stress to experience the psychological problems to a moderate extent.

The employed couple's also experienced social problems to a moderate extent. Taking out time to visit friends and relatives was the major social problem faced by them. It is very interesting to note that not a single wife reported to experience problem of sexual harassment at their place of work. The tension at the workplace and at the home sometimes was the major cause of unhappiness experienced by the employed couples due their employment. Proportionately a very less percentage of them (10%) experienced the employment problem to a higher extent.

The findings on economic problems elicited that a higher expenditure on the paid assistance was reported to be the most dominating one. Since the employed couples belonged to middle class families, both preferred to work outside the home to run their house as wished by them. Only one-fifth of the employed couples experienced economic problems to a higher extent. Majority of the employed couples experienced the problems related to managing their home and career to a moderate extent. Inability to cope up with the household responsibilities and chores was reported to be the major problem faced by them.

The nature of problems related to communication as experienced by the employed couples included negligence of children due to poor communication with them all the time and the restricted communication among the friends and relatives. Most of the respondents faced communication problem to a moderate extent. Of the varied child-related problems faced by the employed couples, loss of childhood at an early age and inability to spare time for them were found to be the most serious ones always. Most of the respondents experience child-related problems to a moderate extent.

Coping Strategies Adopted to Relieve Stress: The investigators were curious to find out the coping strategies adopted by the employed couples in tackling the problems faced by them arisen due to their dual career. The findings on the same highlighted that the nature of coping strategies adopted by them varied with the type of problem. The coping strategies adopted for dealing with psychological problems were related with the relaxation and fitness programmes and consumption of nutritive healthy diet. Most of the employed couples adopted these strategies to a moderate extent.

Coping strategies adopted for solving social problems included sparing extra time for social commitments when actually needed, by compromising and making adjustments at the work place and making friends and relatives realise about their time constraint. Majority of the respondents adopted such coping strategies to a higher extent. The employed couples reported to complete their work at the job on time and switched to small breaks in between the work at the job to relieve stress. Such coping strategies were adopted by them to cope up with their employment problem to a moderate extent.

To cope up with the economic problem, the respondents tried to spend wisely according to their priorities without compromising on the basic necessities and support. This coping strategy was adopted by most of them to a moderate extent. Most of the employed couples also adopted the coping strategies to cope up with their management problem to a moderate extent. The coping strategies adopted by them included the paid assistance rendered by the servants and the utilisation of work simplification methods and labour saving devices.

The coping strategies adopted for solving communication problems, were the employed couples planned outings on weekends to enjoy with their family, friends and relatives. The respondents were also very particular in spending time with their children during holidays and in their free timings as and whenever available while sacrificing their leisure time. Majority of them adopted the coping strategies to take care of communication problems to a high extent. Utilising the services of elders/siblings/servants/neighbours/relatives was one of the most popular coping strategy adopted the employed couples to take care of their children. Some of them resorted to admit their children in day boarding school. Majority of the couples adopted coping strategies to resolve their child-related problems to a moderate extent.

CONCLUSION

The employed couples had to face varied nature of problems at the workplace and at home which ultimately led to stress at both the places. The selection of appropriate coping strategy adopted could be the solution to a stress free and happy married and family life.

REFERENCE

[1] Rapoport, R. and Rapoport, R.N. (1976). Dual-Career Families Re-examined. New York. Harper and Row Publishers.

Aesthetic Functionality and Trust in Product Design: Some Ergonomic Experiences

Debkumar Chakrabarti

Department of Design, Indian Institute of Technology Guwahati
Guwahati–781 039, India
E-mail: dc@iitg.ernet.in

ABSTRACT

Aesthetic functionality ensures products' better acceptance by the intended users. Semiotic inbuilt look and identity features confirm trust for reliable function and safe use of a product or a system. Design features are not only to divulge ease of use and safety but also pleasure of use and enjoyment of the experience. Design ergonomics is devoted to achieve these in a product to call it a good design.

This paper discusses product development strategy using ergonomics design criteria and role of participation of likely to be users and entrepreneurs to take part in the development process. Few product development experiences *e.g.*, indoor planter for ease of operation, green-design idea, which uses locally available raw material bamboo and cane, and making skill that insists users and makers to look for new possibilities in product features; and a shearing device developed for tea leaf plucking where achieving the quantity goes along with the quality of production are cited to demonstrate the process of product development with ergonomics requirements.

Keywords: Aesthetic Functionality, Trust, Design Ergonomics, Design Curriculum, Product Development.

PLEASURE BEYOND FUNCTIONALITY: AN EMERGING ISSUE

Today's design is conceived to cater the mass fulfilling its various context specific human needs. How far this support is justified to meet the need of the appropriate consideration is a challenging task itself. Acceptance and success of the design depends on the ease of decoding its inbuilt features and its utility. Inappropriate or misuse of the product away from its intended use mode and products designed without considering users compatibility features lacks in product reliability and safety which finally land to be trouble makers. Many such examples are reported in various forums and are also individually experienced. Users, while selecting a product analyse its attributes through knowing, doing and feeling which are linked with his cognitive skills, perceptual-motor skills and emotions.

This advocates that though application of best scientific principles and appropriate technologies may generate a design best to deliver its intended function, still its user, the prime system component, ultimately has to feel comfort while using it to qualify the same a good design. Good designs thus follow human, inbuilt as well as acquired, abilities and limitations in terms of physical, physiological and behavioural aspects. Other relevant context specific characteristics should also be taken into consideration while conceptualising a design for human use. A good design aids human functional needs and reduces stress through improved usability elements. It also adds a value to pleasure of using and feeling good to possess. Design should be usable, reliable, fear free, low learning and assuring re-usability. The issue is how we see these in design application; design should be human compatible and context specific.

Ergonomics plays role in framing criteria for good human compatible design. Ergonomics from its equipment-centered approach (traditionally believed to be the domain of design ergonomics), where it experiences human adaptation to the requirements of equipment and work space, has expanded to human-centred practice where equipment is adapted to the requirements of users that ensure aesthetic functionality and trust in product design. Design features, *e.g.,* form and function along with human compatibility features, are not only to divulge ease of use and safety but pleasure of use, enjoyment of the experience of the product or the system. While selecting a design, users normally give a first look on the product, the feeling at first sight, *i.e.*, aesthetic perception of the product, which goes side by side with the usability aspects through judging functional reliability, trust, fear-free use mode and comfortable use. When we judge a product to be beautiful we do not necessarily notice its utility or usefulness at the very first instance. However, it is the utility value that is present along with aesthetics to the extent that we judge the design to be beautiful. That is how we look at human perceptual bias and its relation to our aesthetic preferences in man-made world as well as in nature.

This demands approach of Eco-friendly design. Design is normally created out of local needs, and thus carries an identity. Individuals' labeling of themselves, as members of particular groups and their design of various utility items carries their own identity. Application of ergonomics in design has crossed the limit of usability aspects, and emotion and satisfaction add vital dimensions to the design. Design innovations take traditional elements and modern technological advantages to meet contextual needs and to satisfy intended users. Application of design elements, *e.g.*, use of colour, lines, surface texture, form, shape, etc., creates context specific mood as well as identity of functional utility items. It can be said that a design tending towards organic form and eye soothing overall colour combination is pleasant. These aspects have now become integral component of design ergonomics curriculum.

Ergonomics Design Method: Learning by Doing

Many agencies are imparting ergonomics with specific emphasis on application. Design ergonomics are being taught to undergraduate as well as post graduate level students in design institutes in India. Ergonomics specifications are inbuilt in these course curriculum. Commercial advertisements use ergonomics as a value added term, to express more comfort and safety in use, for commercial benefit. Ergonomics has become a word for business promotion that we hear very often these days particularly in the high-tech industry.

Design methods must stress aesthetic functionality aspect to ensure products' better acceptance by the intended users. Human elements responsible for these needs are to be studied. Semiotic inbuilt look and identity features confirm trust for reliable function and safe use of a product or a system. Design features are not only to divulge ease of use and safety but pleasure of use and enjoyment of the experience. Design ergonomics is devoted to achieve these in a product to call it a good design. Design usability is reflected with various psycho-physiological responses.

As integral component of design curriculum, for both Bachelor and Masters Degrees, Department of Design IIT Guwahati imparts design ergonomics inputs taking references of varied context specific applications. It includes clinical approach for betterment of existing design and new design development to satisfy a need of the time with futuristic anticipation. To acquaint with theory input, students undergo various levels of hands on experiments with projects of various nature real life experiences. It covers identifying users' context specific needs, studying relevant human information, specifying design features and various occupational settings of production process of the conceived design to understand the product development possibilities.

Project assignments and hands on experiments through actually doing and by analysing case studies form integral part of imparting design ergonomics which are normally related to some design themes. Appropriate ergonomics applications in design are learnt and practised through sharing project experiences and case studies and also studying cost benefit effects of such applications. It is stressed that even if a small development were made, considering its large number of beneficiaries, the impact made would be very significant.

Theory Input Linking with Case Study and Hands on Experiment: Examples

To impart practical acquaintances with theory, specific ergonomics aspects is chosen and students carry out projects on a design theme anchoring on that; while undertaking the project relevant ergonomics principles and design criteria applicable to similar context is also discussed. To demonstrate similar ergonomics criteria relevant case studies are presented. This approach of teaching and learning design ergonomics is being used at IIT Guwahati. The below mentioned three themes and their practical acquaintances discuss product development strategy using ergonomics design criteria and role of participation of likely users and entrepreneurs to take part in the development process:

A Piece of Nature at Home, Indoor Planting: Concerning Operational Easiness

Theme taken is natural greenery in interior decoration suitable for high-rise apartments. This has become a hobby-based culture in new apartment residential system. While maintaining live plants some observations and anticipations may be incorporated in design with ergonomic value addition. Decorating interior with houseplants and using varieties of potholders recently draw designers' attention. Normally, earthen pots are used for plantation, which is inserted into decorative outer pots. For nourishing, earthen pots are required to be taken out from their holders, which require clearance in between primary pot and decorative secondary ones, and while keeping these gaps for functional utility, it should not disturb the aesthetic continuity of the form (Figure 1).

Fig. 1: Holding Features are to be Integrated with the Form

For replacing pots or removing them for any other reason a certain amount of gap has to be left, *i.e.*, the outer decorative pot should be of a considerable size or a certain portion may be perturbed out which leaves space between the original pot and the decorative pot rather than both being of almost same size. Finger movement in space in between two pots (Figure 2), holding facility and weight to lift and carry, maintenance easiness and keeping harmonious relationship with the surroundings are to be considered. While conceiving the design for a single item, *e.g.*, for outer decorative pot, a system design approach should be taken. Both the inner and outer pots can be developed as a complementary to each other.

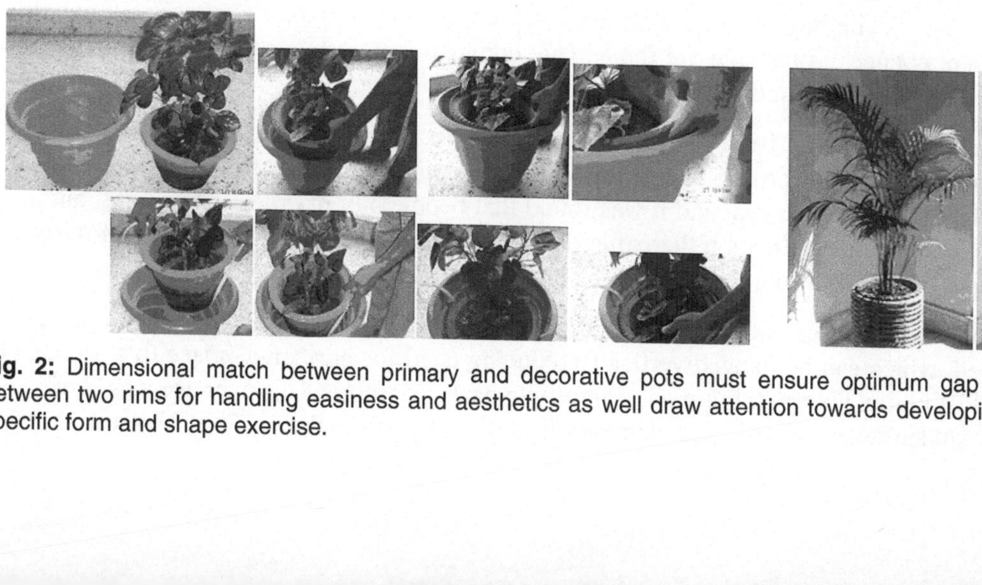

Fig. 2: Dimensional match between primary and decorative pots must ensure optimum gap in between two rims for handling easiness and aesthetics as well draw attention towards developing specific form and shape exercise.

In tune to the theme, another such ergonomics application area can be nourishing tools, *e.g.*, soil tiller which normally consists of a flat trapezoid blade with a handle, which may have one side rounded off so that it would be easy to follow the pot inner curvature to move smoothly (Figure 3), while tilling the soil and to apply manure also. This type of design requires close look of people's behaviour and the way they use the tools. If small developments are done, considering the growing market of indoor plant decoration, a good benefit may be achieved; which would only be possible if aesthetic and usability aspects, with appropriate human information, share equal weight while conceiving design, it would be a pleasure possession.

Fig. 3: Among common hand tools: A, B and C are seen to be used for interior planting, and out of that 'C' appears to be appropriate as it tends to follow the pot curvature, Thus this feature of hand tool specific to use in this context becomes easier and motivating to work with. This may lead to a hand tool development exercise.

Packaging: Design Modifies Conventional Behaviour, Egg Vertical Container

The theme was eco-friendly design and users' behaviour. A green-design idea conceived and developed as student project at IIT Guwahati can be cited here as an example. The concept uses locally available raw material bamboo and cane, and making skill that insists users and makers to look for new possibilities in product features. Eggs are normally placed on a horizontal surface and stalked placing trays of eggs one above another. While buying from shop eggs are either sold on paper trays or in a paper bag where chances to damage are there. This conventional system of raw egg storage mode has occupied our mind.

An experiment has been attempted as student-exercise; whether the system of keeping eggs can be transformed from horizontal to vertical. For this, first common consumer behaviour was observed; and it was found that people buy in order of two eggs, and units of three and multiplication thereafter as half dozen, one dozen or a full crate of thirty in a paper tray. This also goes along with common refrigerator egg tray volume.

Unit of six was considered, a special package using locally available bamboo pieces are used, which can be hanged vertically reducing need for horizontal space to store, can be sold a unit of six eggs, Figure 4, or multiple packs for more. This exercise was carried out as student-exercise till idea generation and prototype making. This idea can bring a change in conventional behaviour to a new one, and yet not completely new that to absorb the design it requires a special training or learning time.

Fig. 4: Bamboo made egg holder unit for six eggs. Mind change is necessary for new design development; a thought with intended use and re-use value.

Design: For Quality and Quantity Production

Hand plucking of tea leaves, two leaves and a bud from tea bushes are an appropriate and conventional image for quality tea. It is also seen that for bulk collection, without much discrimination of plucking two leaves and a bud, apex shoots of tea bushes along with some older leaves are also collected. This very manual operation is done by both hands. Plucked leaves are thrown to store in a basket container, that hangs on the back with a fore-head strap, over the shoulders is a common picture of tea garden operation prevails in our mind.

Another method using shearing operation to clip off the apex shoots with a special device is also introduced by the tea gardeners. This device is basically a scissors where on one blade has a vertical flap and a collection box is fitted on another one. Tea leaves are clipped off by the device with horizontal hand movement over the tea bushes and after each stroke, cut leaves are pushed automatically by the flap to the box. After few times of clippings, collected leaves are released in a bag which hangs on left side of the waist.

Towards improving this operation, with ergonomic requirements, an improved version of this device was developed as a sponsored project at National Institute of Design, Ahmedabad. This gave a new look with improved utility value to the device (Figure 5). It is also observed that, as the shoots are clipped off, every time while releasing the leaves into the bag the pluckers have to open the face of the bag as there is no option for automatic opening. To ease this release operation, the bag should have an optimum opening with a ring fitted on the opening edge of the bag and be carried on a cross belt shoulder strap.

Based on shoots size to be cut, to cover more area of the bush in single stroke, towards ease of releasing the cut leaves into the collecting bag, usage of aluminium perforated sheet for tray and flap to reduce weight and while morning operation to drain out the water droplets, special handle was developed matching with operating hand movement, easy to handle and while shearing hands should remain over the sheared edge of the bush to avoid abrasion of hands have been developed. Both hands equally share the load of operation. While trials were made in tea gardens with the prototype, people opined and other ergonomic evaluation carried out to see whether this device increased productivity and lessened the workload, it appeared to satisfy the need.

Fig. 5: Improved sear device for tealeaf plucking, which has aesthetically appealing and matching user features. Shape is aesthetically appealing, and product geometry and users anthropometry (structural limitations) must fit each other for ensuring safety as well as better function. Size should be handy to operate and easy to maintain. Casing of internal parts should be in such a fashion that if there is any fault in a single part, it could be replaced easily.

Achieving the quantity along with the quality of production is the process of product development with ergonomics requirements. These types of product development processes are shared with the students as relevant case studies to demonstrate ergonomics involvement in design process.

Need for the Day for Design Ergonomics

Ergonomics has advanced its design application aspects from usability to pleasure while interacting with product to ensure joy in use.

Users' needs are to be considered to achieve this:

- Safety and well-being
- Functionality
- Usability
- Pleasure-feeling of satisfaction and joy: sensuous enjoyment as an object of life.

Product should be conceptualised with the below attributes:

- Product
  - Usable
    - Functional
    - Reliable
  - Pleasurable
    - Aesthetic perception
    - Attractiveness
    - Feel good to possess
    - Joy in use

While making the design concept aspects of pleasure, feel good to possess the product and joy in use becomes integral design development philosophy; where appropriate ergonomics studies of context specific application design of elements that influences aesthetic functionality and trust in design are necessary to make a design good.

REFERENCES

[1] Chakrabarti, Debkumar (2001). "Design Ergonomics: Need and Role", *Udyog Pragati-Journal for practicing managers*, 25(2), pp. 37–43.

[2] Chakrabarti, D. (2005). "Ergonomics/Human Factors: Human aspect of Technology, Directions", Indian Institute of Technology Kanpur. Special Issue on Product Design and Media Communication. 7(1), pp. 41–47.

[3] Chakrabarti, D. (2006). "Product Ergonomics", Indo-US Design Workshop, January, IISc., Bangalore.

Analysis of Datamining and Database Creation in Bio-informatics Prostate Cancer: A Case Study

Vasundhara Pandey[1] **and Siddharth Singh**[2]

[1]National Institute of Industrial Engineering, Powai, Mumbai–400 087, India
[2]Bioinformatics Centre, Apticraft Systems, Indore, India
E-mail: amitabhade@nitie.edu

ABSTRACT

The major objective of this research was to retrieve valid, authentic and actionable information for potential prostate cancer biomarkers from published research articles and also to identify novel, potentially useful and understandable correlation and pattern in existing data related to these candidate prostate cancer biomarkers.

A dataset was created having 207 entries, of which 134 entries are for up-regulated candidate biomarkers, 31 entries are for down-regulated candidate biomarkers, 22 entries for differentially expressed candidate biomarkers and 3 entries for candidate biomarkers without any expression level information (about candidate biomarkers, no expression related information was given in concerning research articles). This dataset included information related to 90 different candidate biomarkers and would be useful for research community to find out the genes/proteins which may act as biomarkers in the prostate cancer. It would serve as a crucial information resource to further study these potential biomarkers. Furthermore, data set would be of significant help to the research community to compare some facts of biomarkers and to design a data set of biomarkers for development of specialized micro-array chips. By using real-time qPCR assays or antibody based method such as ELISA and Western blot analysis, it might be useful for an easy and low cost diagnosis for the specific disease.

Keywords: Prostate Cancer, Datamining, Database, Biomarker.

INTRODUCTION

The biomedical literature is an incredibly rich resource for researchers. Information obtained from previous scientific studies helps researchers focus their own efforts. To obtain the maximal benefit from studies in genetics and genomics, there is a need to link this data with the information available in the associated biomedical literature [1]. While the amount of research has increased, the ability to interpret the results has become increasingly difficult. Today's scientific papers are highly focused and require special expertise in the given field for a reader to appreciate the work's significance [2]. To fulfil these entire objectives one can build a database of cancer related information from the existing biomedical literature. Databases are structured to facilitate storage, retrieval, modification, and deletion of data in conjunction with various data-processing operations. A database consists of a file or set of files that can be broken down into records, each of which consists of one or more fields. Fields are the basic units of data storage. Users retrieve database information primarily through queries. Using keywords and sorting commands, users can rapidly search, rearrange, group, and select the field in many records to retrieve or create reports on particular aggregates of data according to the rules of the database management system being used.

For the formation of such database, data mining is first step, which incorporates process of knowledge discovery or retrieval of hidden information from data banks or available literatures [3, 4]. Data mining attempts to discover useful pattern or relationships in a group of data. Initial steps involve selection and pre-processing by an expert in the appropriate knowledge domain [5, 6].

Among different types of cancers, prostate cancer is most frequently diagnosed cancer in the Western male population and is the second leading cause of cancer mortality [9, 11]. There has therefore been large scale search for the potential biomarkers useful in early detection and prognosis of prostate cancer. There has always been a need for the information related to biomarkers for their applications in cancer diagnostic and prognostic studies.

The major objective of this research was to retrieve valid, authentic and actionable information for potential prostate cancer biomarkers [8] from published research articles and also to identify novel, potentially useful and understandable correlation and pattern in existing data related to these candidate prostate cancer biomarkers. ChromSorter PC is such type of database, which provides the information of chromosomes, which are related to prostate cancer [10].

METHODS

Data on prostate cancer biomarkers was collected from published reviews and original papers. PubMed database was also referred for collecting various research papers and review articles. MeSH [Medical Subject Headlines] database was extensively used for Query optimisation.

The resulting downloaded list of references was then manually screened to generate a second list of references containing significant information on prostate cancer biomarkers.

Ten parameters were determined to catalogue the information obtained from the articles. These were biomarker information, disease information, expression details, expression parameter, experimental information, validation of Microarray, model organism details, pathway information, mutation and article information.

The abstracts or full-text articles were obtained from the selected articles and the data was annotated into a simple excel spreadsheet. Information provided by the abstract and full text article of each reference was catalogued across a list of common data fields. Gene names were entered as per the standards present in Entrez gene, *e.g.*, PCA, PSA.

Protein names were entered as per the standards present in Entrez Protein. Names of these molecules were entered as per the standards present in their standard databases.

The name of the disease to which the biomarker was related, was entered as per the standards present in MeSH. The stage of disease was entered depending on the concerning gene/protein/other molecule and might act as a biomarker for the diagnosis of that disease. Name of detection method found in the article was entered as per the standards given in MeSH and NCI [National Cancer Institute]. Preference was however given to MeSh standards, *e.g.*, Blotting, Western, Immunohistochemistry, etc. If a biomarker showed higher level of expression in diseased condition as compared to its expression under normal condition, then expression was entered as up-regulation (not as over expression or other term). If a biomarker showed low level of expression in diseased condition as compared to its expression under normal condition, then expression was entered as down regulation. If a biomarker showed different level of expression in diseased condition as compared to its expression under normal condition then expression was entered as differential expression. If a biomarker expressed in diseased condition but not in normal condition then expression was entered as differential expression. The expression form (cDNA or protein or mRNA) of biomarker, which was detected by detection method in concerned article, was entered as expression type. Cell Line name found in the article was entered as per standards present in American Type Culture Collection (ATCC).

Scientific standards for the relevant information were used. Entrez gene database was utilised as a reference to the names of the genes and proteins and MeSH or NCI database was utilised as a reference to the names of detection method.

After initial annotation, quality assurance of the database was performed by re-review of the entire dataset with the literature, and the data elements were checked for duplications and errors. It was done to test the accuracy of an independent dataset and to assess the sensitivity of the data set. In this manner, standardisation of data entry was established for certain data elements and resulted in a dataset having acceptable entry terms.

An Entity—relationship diagram [7] was made for the graphical representation of the dataset. In order to form this entity-relationship diagram, at first, all entities were determined. Entities were drawn as rectangles and relationships as diamonds. A relationship captured how two or more entities were related to one another. Attributes for these entities were then determined and they were drawn as ovals connected to their owning entity sets by a line.

Data retrieval interface was made to allow the user to find out the information quickly and with a minimal amount of interaction; bad user interface makes it difficult or even impossible to locate and display information stored in database. It was made by including various well known features and various web tools. The search was categorised into quick search and advanced search.

Quick search allows the user to query the database to view all the information distributed in different fields. If the desired biomarker is listed, the user can select that biomarker and will be taken to its full information.

Advance search allows the user to simultaneously search from all determined categories, or to get a filtrated output. To display the information in a systematic and informative way, two interfaces were prepared for the result.

RESULTS AND DISCUSSIONS

The information obtained by curation was annotated into a simple excel spreadsheet. This dataset contains 207 entries, of which 134 entries are for up-regulated candidate biomarkers, 31 entries are for downregulated candidate biomarkers, 22 entries for differentially expressed candidate biomarkers and three entries for candidate biomarkers without any expression level information (about these candidate biomarkers no expression related information was given in concerning research articles. This dataset includes information related to 90 different candidate biomarkers).

The basic idea was the analysis of a series of data mining steps of increasing stringency such that the huge number of entities present in the literature is reduced to a manageable size which could be a useful resource to scientific research community.

This dataset might be useful for the users in following ways:

- To identify the list of the biomarkers with particular expression level. This database would be useful for the user to retrieve complete list of biomarkers associated with particular type of the expression level.
- To retrieve information related to particular candidate biomarker under all determined fields. This type of search has been included under quick search.
- To retrieve the information related to candidate biomarkers in selected fields. This type of search has been included in advanced search.

This dataset contains information related to 90 different candidate biomarkers, 60 of which are up-regulated candidate biomarkers, 8 are down-regulated and 22 are differentially expressed. Figure 1 shows a graph between number of candidate biomarkers and expression level.

During this study it was found that Immunohistochemistry is most common detection method which has been used for the detection of the candidate biomarkers and PCR is second most common detection method. Figure 2 shows a graph between number of citations and detection method.

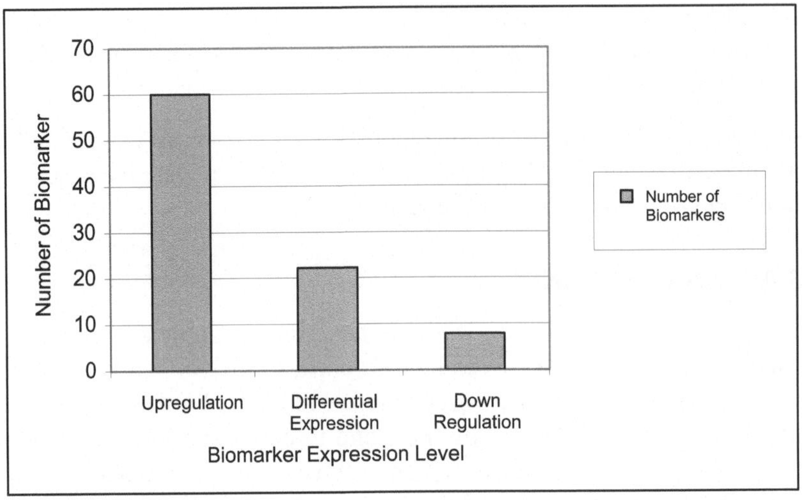

Fig. 1: Relation between Number of Biomarkers and Expression Level

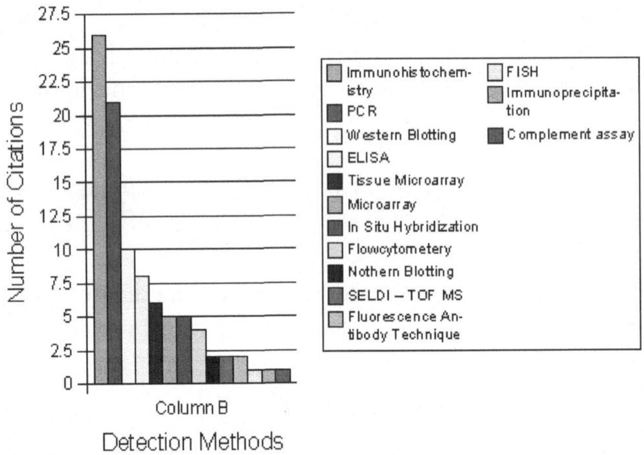

Fig. 2: Relation between Detection Method and Number of Citations

Table 1: The Result of Curation

S. No.	*Queries*	*Number of Citations*
1.	Prostate Cancer AND Biomarker AND free full text[sb]	2051
2.	Prostate Cancer AND Biomarker AND Diagnosis AND free full text[sb]	1656
3.	"Prostatic Neoplasms"[MeSH] AND "Biological Markers"[MeSH] AND "loattrfree full text"[sb]	1864
4.	"Prostatic Neoplasms"[MeSH] AND "Biological Markers"[MeSH] OR "Prostatic Neoplasms/diagnosis"[MeSH] AND "loattrfree full text"[sb]	4155

CONCLUSION

Some databases related to prostate cancer biomarkers already exist. A variety of computational tools are also available to replace the tedious mining and manual curation. This dataset is a gateway to provide or link prostate cancer information and useful for the cancer researchers. The study therefore concluded that out of 90 different candidate biomarkers, 60 were up regulated candidate biomarkers, 8 were down regulated and 22 were differentially expressed. Besides summarising candidate prostate cancer biomarkers' information, an advanced user-friendly database can be developed using computational program in current dataset. The existing dataset comprises ten parameters to allocate the information, some more parameters can be added in this dataset to make it more informative. This dataset provides a more comprehensive understanding of differentially regulated genes with different stages of prostate cancer. A number of innovative researches are going on in related field, therefore, dataset has to be updated with recent information to make it more advance, more utilisable and more informative database.

REFERENCES

[1] Bensalah, K., Lotan, Y., Karam, J.A. and Shariat, S.F. (2007). "New circulating biomarkers for prostate cancer", *Prostrate Cancer Prostatic Dis.*

[2] Etim, A., Zhou, G., Wen, X., Liu, H., Ruotti, V., Twigger, S., Jin, W., Matysiak, B., Mathis, J., Tonellato, P.J., Datta, M.W. and ChromSorter, P.C. (2004). "A database of chromosomal regions associated with human prostate cancer", *BMC Genomics*, 5, p. 27.

[3] Gonzalez, G., Uribe, J.C., Tari, L., Brophy, C. and Baral, C. (2007). "Mining gene-disease relationships from biomedical literature: weighting protein-protein interactions and connectivity measures", *Pac Symp Biocomput*, pp. 28–39.

[4] Holena, M., Sochorová, A. and Zvárová, J. 1999. "Increasing the diversity of medical data mining through distributed object technology", *Stud Health Technol Inform*, 68, pp. 442–7.

[5] Jemal, A., Murray, T., Ward, E., Samuels, A., Tiwari, R.C., Ghafoor, A., Feuer, E.J. and Thun, M.J. (2005). "Cancer statistics, 2005", *CA Cancer J. Clin*, 55, pp. 10–30.

[6] Karlsson, J., Eklund, P., Hallgren, C.G. and Sjödin, J.G. 1999. "Data warehousing as a basis for web-based documentation of data mining and analysis", *Stud Health Technol Inform*, 68, pp. 423–7.

[7] Lee, F.W. (1996). "How to use entity-relationship diagrams in developing personal database systems", *J AHIMA*, 2, pp. 42–5.

[8] McDonald, R., Scott Winters, R., Ankuda, C.K., Murphy, J.A., Rogers, A.E., Pereira, F., Greenblatt, M.S. and White, P.S. (2006). "An automated procedure to identify biomedical articles that contain cancer-associated gene variants", *Hum Mutat.*, 27(9), pp. 957–64.

[9] Parkin, D.M., Bray, F., Ferlay, J. and Pisani, P. (2002). "Global cancer statistics 2002." *CA Cancer J. Clin,* 55, pp. 74–108 .

[10] Roberts, P.M., 2006. "Mining literature for systems biology", *Brief Bioinform,* 7(4), pp. 399–406.

[11] Zucht, H.D., Lamerz, J., Khamenia, V., Schiller, C., Appel, A., Tammen, H., Crameri, R. and Selle, H. (2005). "Datamining methodology for LC-MALDI-MS based peptide profiling", *Comb Chem High Throughput Screen*, 8(8), pp. 717–23.

Section–V

Ergonomics in Agriculture

Measurement of Driver's Fatigue Using Dynamic Body Pressure Distribution

K. Adalarasu and Venkatesh Balasubramanian

Rehabilitation Bioengineering Group, Department of Biotechnology, IIT Madras
Chennai–600 036, India
E-mail: chanakya@iitm.ac.in; chanakyaaa@yahoo.com

ABSTRACT

A major problem faced in road safety systems is the impact of driver fatigue and the difficulty in its detection, which plays an important role in road accidents. Generally heavy truck drivers are required to drive for long duration at odd irregular hours. A driver who is driving for an extended period of time is prone to have discomfort and fatigue due to seated pressure on tissue. The main objectives of this study were: (1) to measure driver's fatigue during driving using body pressure distribution and (2) to analyse the relationship between comfort index and drivers' fatigue.

Five healthy male volunteers aged between 24–32 years, participated in this study after giving an informed consent. Each subject played Need for Speed TM (NFS) in a simulated driving condition for one hour. While driving, seat pressure distribution was continuously measured using a pressure mat (TekscanTM). Six physiological criteria monitored were: (1) Dynamic area, (2) contact area for pressures 40–60 mm Hg, (3) contact area for pressures 60–80 mm Hg, (4) contact area for pressures 80–100 mm Hg, (5) contact area for pressures over 100 mm Hg, (6) comfort index (CI). The parameters were calculated during 5th and 55th minutes of driving.

In all subjects, contact areas for pressure over 100 mm Hg and comfort index values are significantly low ($p < 0.05$) in 55th minute as compared to 5th minute. Similarly dynamic values are significantly high ($p < 0.05$) in 55th minute as compared to 5th minute.

The study implies that a decreasing contact area, comfort index and increasing dynamic area lead to increased fatigue in an hour long driving. Corollary to this finding is that body pressure distribution can be used as a tool to detect driver fatigue.

INTRODUCTION

The lack of any system for detecting driver fatigue is a major concern in developing a road safety system. According to a report by National Transportation Safety Board, USA driver fatigue may be a contributing factor in about 30 to 40% of all heavy truck (commercial vehicles) accidents (Transafety Reporter, 2003). Truck drivers spend long-hours at their wheel than an average passenger car driver. These long hours of seated activity can lead to discomfort and fatigue, which can cause decreased awareness, observation, decision-making, alertness, and reaction time. Many factors can cause drowsiness or fatigue while driving including driving with sleep debt, night-time driving, extended driving time, physical work, in addition to monotonous driving conditions (Ahmadian and Seigler, 2002).

Simulators are popularly used to quantify a driver's ability. It has been reported that driving simulators provide an adequate representation of the real world. Main advantages of a simulator are its safety, controlled exposure to high-risk environment, low likelihood of adverse events and cost-effectiveness (Schiff *et al.*, 1994). This has lead to the extensive use of simulators to study cockpit ergonomics in heavy vehicles.

Changes in drivers' vigilance can be quantified using three methods. The first approach is to quantify physical changes during fatigue such as the inclination of the driver's head, his sagging posture, and decline in the gripping-force on steering wheel. The second approach is to measure physiological changes of drivers such as eye activity, heart beat rate, Electroencephalographic (EEG) activities, Electromyography (sEMG) and seat pressure variation (Balasubramanian *et al.*, 2007). The third approach is to study certain behavioral aspects of drivers such as sleepiness, personality behavior and reaction times. A large number of contemporary literatures have used interface pressures (seat pressure) between the driver's body and his seat to evaluate riding comfort and fatigue. In driving vehicles the seat creates a crucial interface between the vehicle and its driver, high local pressure at the human-seat interface might cause soft tissue deformation leading to controlled blood and nutrient flows, discomfort and fatigue (Wu *et al.,* 1998). Comfort is the short-term effect that a seat has on the human body and fatigue defines the physical impairment that results from exposure to the seat dynamics for a long time. Several recent papers have shown a good correlation between distribution of interface pressures and pressure change rate with riding comfort and fatigue (Ahmadian and Seigler, 2002).

Pressure maps at the body-seat interface give details about how forces are acting between man and seat (Guenaelle, 1995), which can be used for evaluating and estimating fatigue. Ahmadian and Seigler (2002) used pressure map at seat to study the effects of two different vehicle seats (foam and an air-inflated seat cushion) using area pressure change root-mean-square (aPcrms) and Seat Pressure Distribution (SPD) to quantify driver comfort. In order to analyse seat pressure, two evaluation parameters of pressure were considered: a) contact area as a function of pressure range, which is a time-independent measure of pressure, and b) area pressure change root-mean-square (aPcrms), which analyses the dynamic behaviour of pressure (Dhingra *et al.,* 2003). Both these parameters were used in the present study to quantify fatigue and CI during one hour of simulated driving.

The aim of this study was to quantify driver fatigue from the change of interface body pressure distribution between the driver and the seat during one hour of driving by fixing drivers' posture.

METHODS

Driving Simulator

A realistic, simulator environment (with a steering wheel, accelerator and brake) was used in conjunction with a computerised driving task ('Need for Speed', Electronic Arts

1998). Players get to drive high speed luxury cars (Ferrari f50, Ford Indigo) at a range of speeds while playing this driving game, whose race circuit can be described as a highway. Video screen presented an 'in-car' view to the driver and the projected video screen was two meters away from the driver. All experiments were performed at the Rehabilitation Bioengineering Group (RBG), IIT Madras after due consent from all the participants.

Subject Detail and Protocol

Five healthy college students participated in this study. All had driving experience and none had history of musculoskeletal diseases. Mean age and weight of the subjects was 28 ± 2.4 years and 66.8 ± 10.3 kg respectively. Subjects were asked to drive for about 10 minute (training session) before starting the experiment to familiarise themselves with the various control devices and thereby allowing them to achieve a certain degree of proficiency with the driving setup. After training session subjects were asked to take a short break (20 minute). During experimentation subjects were asked to race continuously along a chosen track for 60 minute.

In this study, we used a body pressure measurement system (Tekscan's Inc, MA, USA) over the seat to capture interface seat pressure. BPMS helps to capture dynamic pressure distributions at the seat interface and is related to driver comfort and fatigue. The seat pressure distribution was continuously measured for 60 minute during simulated driving.

Data Collection and Processing

Body pressure measurement system (BPMS) is a thin flexible-resistive based sensor pad featuring a 42×48 array of 1.03×10^{-4} m^2 pressure-sensing elements, was used for obtaining pressure maps. Application of force to the sensor pad changes the resistance of the active sensing element in inverse proportion to the applied pressure. The pressure map, which is an instantaneous distribution of pressure in the matrix of sensing elements, was captured onto a frame to be viewed in the computer. Dynamic maps were captured at sampling frequency of 9.009 Hz. Pressure distribution as viewed on a movie frame is shown in Figure 1.

Raw data from the sensor pad was converted to engineering units, namely kPa, through the calibrated signal. The movie files were exported into ASCII format for post-processing in MATLAB 6p5 software (MathWorks Inc, MA, USA). Recorded movie frames were filtered to remove cells loaded with less than 0.001 kg.

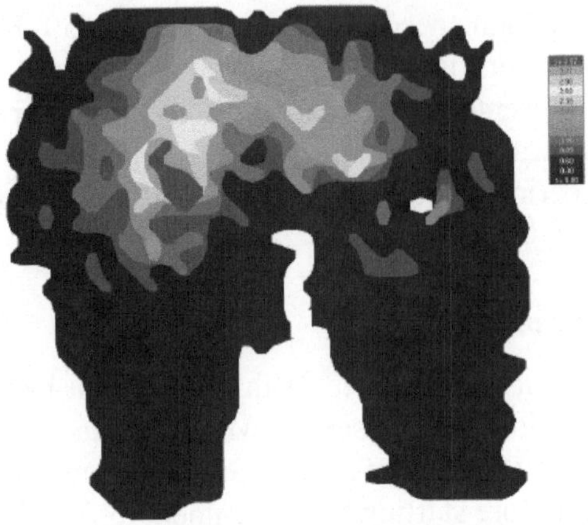

Fig. 1: Pressure Distributions at the Seat during Simulated Driving as Viewed on a Movie Frame. Numerical Values on the Scale Indicate the Pressure in PSI Corresponding to the Color

Seat Pressure Distribution

Pressure change root-mean-square (Pcrms) was calculated at 5th, 25th, 35th and 55th minutes Pcrms is calculated by the following expression,

$$Pcrms = \left(\frac{1}{T} \int_0^T \left(\frac{dP(t)}{dt} \right)^2 dt \right)$$

In the above expression, T is the total time period, which in this case is 60 minutes and P is the dynamic pressure.

The parameter, aPcrms, is derived from Pcrms taking into account the pressure distribution. It fine-tunes Pcrms by using a discrete weighting function. It was calculated by assigning different weighting factors to four different pressure ranges: 40 to 60 mm Hg, 60 to 80 mm Hg, 80 to 100 mm Hg and over 100 mm Hg. Specifically the numerical values of the four different weighting factors are 1, 2, 3 and 4. The selections of the weighting factors are based on the physiological effect of pressure on human cells in different pressure ranges. A proper justification of these ranges of pressures and corresponding weighting factors is available elsewhere (Balasubramanian and Adalarasu, 2007). After calculating Pcrms for each pressure map cell and segregating these cells based on the weighting function, aPcrms was calculated by the expression given below,

$$aPcrms = \sum_{i=1}^{4} A(r_i) Pcrms(r_i) W(r_i)$$

where,

$A(r_i)$ is the total area of cells within pressure range r_i

$W(r_i)$ ε {1, 2, 3, 4} is the weight allotted to pressure range, r_i

Contact area, $A(r_i)$ was another variable which was considered in this study. $A(r_i)$ was calculated separately for the four different pressure ranges ($r_i \ni i$ {1, 2, 3, 4}) at 5th, 25th, 35th and 55th minutes $A(r_i)$ is simply the total area of cells in the average pressure map which fall within the pressure range r_i.

Statistical Analysis

Seat pressure distribution features—(a) dynamic area, (b) contact area for pressures 40–60 mm Hg, (c) contact area for pressures 60–80 mm Hg, (d) contact area for pressures 80–100 mm Hg, (e) contact area for pressures over 100 mm Hg (f) comfort index (CI) of 5th and 55th minutes were statistically compared during one hour of simulated driving. Wilcoxon signed rank test (a non-parametric test) was performed on the features to evaluate the level of significance.

Slope was calculated using linear regression for each respective feature parameters extracted from each individual subject. SPSS 12.0 for Windows (SPSS Inc, Chicago, IL) was used for the statistical calculations.

RESULTS AND DISCUSSION

Wilcoxon signed rank test, performed on the pressure evaluation parameters—contact area for pressures over 100 mm Hg and a Pcrms, revealed a significant difference ($p < 0.05$) between 5th and 55th minute of simulated driving. Inferences on CI were made from the values of aPcrms. CI and contact area were significantly low in the 55th minute when compared to the 5th minute (Figure 2a). Similarly dynamic area was significantly ($p < 0.005$) high in 55th minute when compared to the 5th minutes of simulated driving; it reveals a higher fatigue and lower comfort during the one hour simulated driving. Also, the contact area for pressures over 100 mm Hg and CI have negative slope, dynamic area has positive slope during one hour driving (Figure 2b). Comfort index decreased with increasing time, it reveals that the subjects have higher fatigue and lower comfort. Remaining seat pressure distribution parameter did not show any significant difference ($p = 0.13$) between the two-time intervals.

Main objective of this study was to evaluate the driver fatigue during simulated driving for one hour, using features extracted from dynamic seat pressure distribution images. The dynamic body pressure distribution has been used to determine the truck driver comfort and fatigue in simulated condition (Ahmadian and Seigler, 2002). Although the objectives of the study of Ahmadian and Seigler (2002) were markedly different from ours, it clearly manifests the potential of investigating pressure distribution for quantifying driver comfort and fatigue in two different vehicle seat design conditions. We have analysed pressure distribution in single seat to investigate driver fatigue during an hour of simulated driving.

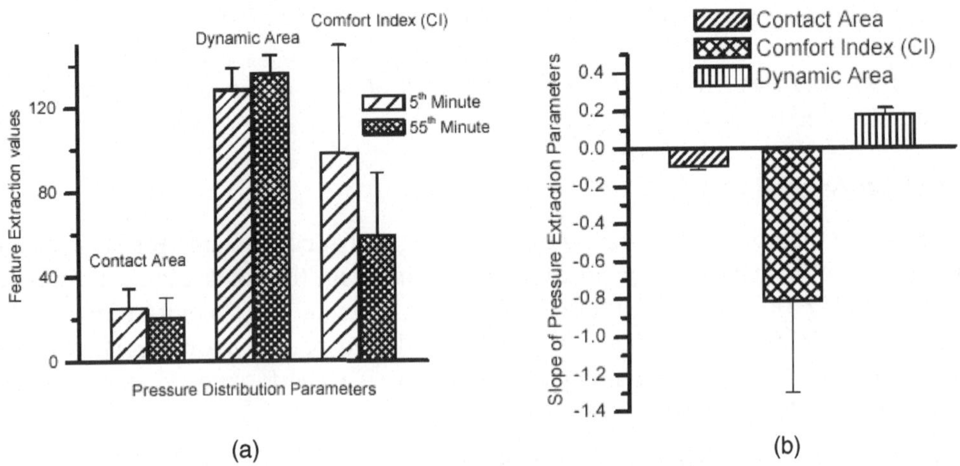

Fig. 2: Dynamic area and comfort index for 5th and 55th minutes in simulated driving. (s) Shows the mean and one-standard error for the area for pressures over 100 mmHg, (b) shows the slope of pressure extraction parameters.

The result of negative slope for comfort index shows that at long hour of driving, seat comfort index levels decreases as the driving period increases. The mean comfort index for initial driving period was 97.85, and that final driving was 59.1. As the driving period increased, the body pressure variables such as contact area and comfort index decreased (Figure 2a). In other words the body pressure of the seat span region near the seat back decreased as driving period increased. Due to fatigue, the subject's body seat interface contact area could be decreased. Feature extraction of dynamic area was affected by the driving period, and increased as driving period increased. Increasing dynamics area means that subject moved more frequently, due to fatigue over one hour driving.

Studies performed by Andreoni *et al.* (2002) involve body pressure measurement under static conditions, whereas the present study involves pressure measurement under dynamic condition. Our study shows the possibility of using the body pressure distribution dynamically and using it as a tool to evaluate driver fatigue and discomfort.

CONCLUSION

Our study proves that, the body pressure distribution gives the information on fatigue that can be used as a rehabilitation tool for drivers. Body pressure variable *i.e.,* contact area and comfort index were found to decrease as driving period increased. The drivers tend to move more frequently when they feel fatigued. Our study supports the usefulness of body pressure distribution for quantifying driver's fatigue and analyses the changes in total body pressure and seat pan. In future, correlation between the pressure variables with some physiological signal (EMG, EEG) can be studied to quantify driver fatigue.

ACKNOWLEDGMENT

The authors gratefully acknowledge RBG members, IIT Madras and others for their valuable contribution. Authors also thank all the volunteers for their time and effort to make this study possible.

REFERENCES

[1] Ahmadian, M. and Seigler, M. "A Comparative Analysis of Air-Inflated and Foam Cushions for Truck Seats", *SAE Technical Paper* 2002–01–3082.

[2] Andreoni, G., Santambrogio, C.G., Rabuffetti, M. and Pedotti A. (2002) "Method for the analysis of posture and interface pressure of car drivers", *Applied Ergonomics* 33, pp. 511–522.

[3] Dhingra, H., Tewari, V. and Singh, S. (2003). "Discomfort, Pressure Distribution and Safety in Operator's Seat—A Critical Review". *Agricultural Engineering International: the CIGR Journal of Scientific Research and Development.* Invited Overview Paper, Vol. V.

[4] Guenaellc, P. (1995). "One methodology to evaluate Automotive Seat Comfort." In: *Proceedings of the Third International Conference on Vehicle Comfort and Ergonomics,* Bologna, Italy, 29–31 March, pp. 231–240.

[5] Schiff, W., Arnone, W. and Cross, S. (1994). "Driving assessment with computer-video scenarios: more is sometimes better", *Behav. Res. Methods Instrum. Comp.* 26 2, pp. 192–194.

[6] Transafety Reporter (2003). "Truck Survey Highlights Causes of Drowsy Driving and Suggests Preventative Measures", Transafety Incorporated.

[7] Venkatesh, Balasubramanian and Adalarasu, K. "EMG-based analysis of change in muscle activity during simulated driving", *Journal of Bodywork and Movement Therapies,* Vol. 11, Issue 2, April 2007, pp. 151–158.

[8] Venkatesh, Balasubramanian, Adalarasu, K. and Regulapati, Rahul (2007). "Comparing Stationary Standing with an Intermittant Walking Posture during Assembly Operations", *Human Factors and Ergonomics in Manufacturing* (Accepted for publication).

[9] Wu, X., Rakheja, S., Boileau, P.-E., 1998 "Study of human-seat interface pressure distribution under vertical vibration", *International Journal of Industrial Ergonomics,* Vol. 21, No. 6, pp. 433–449.

Seat Design for Vehicle: Status and Recent Development

S.K. Mandal, P.K. Maji and A. Maity

Scientist, Central Mechanical Engineering Research Institute
M.G. Avenue, Durgapur–713 209, India
E-mail: sub_mandal@rediffmail.com

ABSTRACT

Overall seating comfort is influenced by both static and dynamic characteristics of seat system. Seats are one of the most important components of vehicles and they are the place where professional driver spend most of their time. For example, according to Occupational Outlook Handbook by United State Department of Labor, the truck drivers frequently work 50 or more hours a week. The truck drivers sit while they drive their 50 hours per week. Assuming four weeks vacation and one more for holidays, which is about 2350 hours driving time per year. Automotive seats, which are in contact with vehicle occupants, play an important role in improving the comfort and work environment of a driver and passengers. The improvement of automotive seating systems, particularly for the driver, has been the subject of intense interest for many years since a driver feels more fatigue than passengers. The paper describes some case studies and up-to-date techniques developed for vehicle seats that are used by different types of transportation system such as cars, trucks, tractors, trains, etc. The objective of this paper is to review the state of the art of vehicle seat design.

Keywords: Vehicle, Seat, Operator, Discomfort and Safety.

INTRODUCTION

Comfort is an attribute that today's consumers demand more and more. The seat has an important role to play in fulfilling these comfort expectations. Seating comfort is a major concern for drivers and other members of the work force who are exposed to extended periods of sitting and its associated side effects. Research literature has shown that some of the factors that affect seating comfort are seat interface pressure distribution, whole-body vibration and pressure change rate. Drivers of commercial vehicles, particularly heavy trucks, are required to drive long and sometimes irregular hours. In America, the driving limit for truck drivers, as defined by the Federal Highway Administration Hours of Service (HOS) regulations, is 10 hours. Almost 20% of drivers, however, reported that they "always or often" exceed that limit (Transafety Reporter, 1998). Commercial trucks are unique in the sense that, they are specifically designed to transport heavy loads over long distances, where for the trucks, high priority has been given to durability and functional efficiency. On the contrary, automobiles are made to comfortably accommodate passengers over relatively short distances (Gillespie, 1985). The personal vehicles are emphasising the factors such as ride comfort, handling, technology and appearance are of high importance in vehicle market. The different requirements of commercial trucks and personal automobiles have led to separate directions in design. Considering the long hours of hauling, it can be argued that one of the most important parts of the truck driver's working environment is the truck seat. Similarly the tractor operator comfort is dependent on vehicle design, the way the vehicle is operated and the expectations and sensations of drivers. Seat is one of the most important components, which influence the comfort of a tractor operator (Stikeleather, 1991). Because it provides the interface between a mechanical system, the tractor, and the delicate and sensitive human biological system, the operator. The seat design process should consider the relative and combined effects of different causes of discomfort and consider whether customer's opinion of good comfort can be achieved without unnecessary reductions in stimuli commonly associated with discomfort (Mehta and Tewari, 2000). Recently the design of the truck seat experienced major improvements. Improvements have emphasised bolster design to increase stability, and adjustments for backrest angle, contouring, and seat height to promote good posture. Also, the development of air suspension system has made the seat better capable of absorbing vibration transferred from the road surface to the driver (Ahmadian *et al.,* 2001). This paper describes the research and development for vehicle seats based on literature review of journals, technical reports and thesis. The focus groups for the vehicle seat are cars, buses, trucks, tractors, trains, airplanes and other vehicles such as motorcycles. Current research and development efforts as well as areas for future development and the projected impact on better performance of these seats are described.

Seating Discomfort

The term "seat comfort" is typically used to define the short-term effect of a seat on a human body. Seating discomfort has been examined from a number of different perspectives. The problem with evaluating comfort in regard to pressure or any other factor is that, comfort is very subjective and not easily quantified. Seating discomfort varies from subject to subject and depends on the task at hand. Comfort, however, is a vague concept and subjective in nature. It is generally defined as lack of discomfort (Shen and Vertiz, 1997). For example, truck drivers require sitting for long periods of time approximately eight hours. The extended period of sitting includes higher risk of back problems, numbness and discomfort in the buttocks due to surface pressure under the thighs (Floyd and Roberts, 1958). The sources of such discomfort are listed in Table 1.

The short-term comfort offered by a seat is relatively easy to determine by many measures [5, (Viano and Andrzejak, 1992), 16], the most effective of which is to survey potential users of the seat as they compare the "feel" of a seat for a short period of time against other seats in the same class. This practice is often adopted for different vehicles, ranging from passenger vehicles to commercial vehicles such as trucks, buses, and off road vehicles. The problem, however, with subjective evaluations is that they are costly and time-consuming. In response, a great deal of research has been performed in recent years to find objective measures for predicting seat comfort perception. Some of the proposed objective measures include vibration, interface pressure, and muscle activity. These objective measures are correlated with subjective data to determine the relative effects of each measure related to comfort (Boggs, 2004). Research has shown that some of the factors that affect seating comfort are seat-interface pressure distribution, whole-body vibration and pressure change rate (Ofori-Boetang, 2003).

Table 1: Causes of Seating Discomfort

Human Experience Mode	*Biomechanical*		*Seat/Environment*
	Physiology Causes	*Engineering Causes*	*Source*
Pain	Circulation occlusion	Pressure	Cushion stiffness
Pain	Ischemia	Pressure	Cushion stiffness
Pain	Nerve occlusion	Pressure	Seat contour
Discomfort	–	Vibration	Vehicle ride
Perspiration	Heat	Material breathability	Vinyl upholstery
Perception	Visual/auditory/tactile	Design/vibration	Vehicle cost

Vibration Approach

A major portion of the vibrations experienced by the occupants of an automobile enters the body through seat (Van Niekerka *et al.,* 2003). Whole-body vibrations, which are vertical vibrations, tend to affect human body the most. These vibrations are transmitted to the buttocks and back of the occupant along vertebral axis via base and back of the seat (Falou *et al.,* 2003). Since the natural frequency for the human trunk falls in the range of 4–8 Hz, it is expected that the whole-body vibrations that will most largely affect passengers will occur in this frequency range [10]. Many researchers have performed vibration studies of seats. Van Niekerk *et al.* (Van Niekerk *et al.,* 2003) conducted research to compare experimental data to the Seat Effective Amplitude Transmissibility (SEAT) values. The experiment used 16 different automobile seats ranging from sedans to SUVs and pickups. The SEAT value is the ratio of the vibration experienced on top of the seat and the vibration that one would be exposed to when sitting directly on the vibrating floor. SEAT values have been widely used to determine the vibration isolation efficiency of a seat. Frechin *et al.* (Frechin *et al.,* 2004) studied on an active seat to isolate the equipment and passengers in vehicle from vibration and to compensate, to certain extent, for acceleration in all directions. The main objective is to reduce the negative effects of vibration and acceleration on embarked equipment and passengers. Also, vibrations to the whole body were investigated and different driving conditions have been measured (Qiu and Griffin, 2005)-(Nawayseh and Griffin, 2005).

Pressure Approach

Drivers' comfort is as important as the functional and aesthetic design of automobiles since consumers are more and more concerned about safety and comfortable driving.

Enhancement of comfort performance of automotive seats necessitates considerations of the human-seat interface pressure distribution under dynamic vibration environment. The characteristics of pressure distribution on a rigid seat under whole-body vehicular vibrations have recently been reported (Boileau and Rakheja, 1990). Gyi *et al.* (1998) evaluated the seat pressure measurement technologies that used in the prediction of driver discomfort for various car seat designs, and provided designers and manufacturers with rapid information early in the design process. Seigler and Ahmadian (Seigler and Ahmadian, 2003) formulated two techniques, namely, Seat Pressure Distribution (SPD%) and Area Pressure Change (aPcrms) for the purpose to highlight the relative dynamics between different types of seat cushion, and their effect on driver comfort. The results show that the air-inflated seat cushion can provide significant improvements in pressure distribution between the seat cushion and the driver, therefore providing a more comfortable ride (Hinz *et al.*, 2006; Seigler *et al.,* 2003). The dynamics of air-inflated seat cushions is very different from that of foam cushions in terms of their interface with the human body.

ERGONOMIC APPROACH

The automotive industry strongly encourages research in the field of objective comfort assessment, especially dedicated to the seat and related postures (Gyi *et al.*, 1998, Guenaelle, 1995). Driver posture is one of the most important issues to be considered in the vehicle design process (Porter and Gyi, 1998) regarding not only the car and the user (Zhang *et al.*, 1996, Ebe and Griffin, 2001) but also the experimental conditions. Alem and Strawn (Alem and Strawn, 1996) designed and evaluated an energy absorbing truck seat for a 5 tonne military truck for increased protection from landmine blasts. Chang *et al.* (1996) developed a practical method for measuring seat pan and seatback contours and a graphical presentation for visual evaluation. Seat designers can use the methods for evaluating seat comfort such as support, fitness and accommodation. Cho and Yoon (Cho and Yoon, 2001) developed a biomechanical model of humans on a seat with a backrest for evaluating the vehicular ride quality. Rakheja *et al.* (2006) developed a model to study the seated occupant interactions with seat backrest and pan, and biodynamic response under vertical vibration. Wang *et al.* (2006) studied the role of seat geometry and posture on the mechanical energy absorption characteristics of seated occupants under vertical vibration. The results show that the absorbed power quantity increases approximately quadratically with the exposure level by the person. The results also reveal that the absorbed power is strongly dependent upon the individual anthropometry variables such as body mass, fat and mass index. Coelho and Dahlman (Coelho and Dahlman, 1999) conducted a pilot evaluation and experimental study on car seat side support. The introduction of side supports is annoying the problem of designing a seat to fit every human being in the range of accommodation targeted in the seat's design. The side support is built-in in both sides of the seat and people with varying body widths must be accommodated back. Some research used the seated man model (Boileau and Rakheja, 1998; Kim *et al.,* 2005) and anthropometrics (Broughton, 2004; Latchford *et al.,* 2005) to study the driver's ergonomics. Besides ergonomics, low back pain research is reported in (Bovenzi *et al.*, 2006; Okunribido *et al.,* 2007). Ample research has been performed on building up specific biodynamic models based on certain experimental data under prescribed testing conditions; a thorough investigation of mathematical human models in seated posture has not yet received the same level of attention. In future studies, research should be concentrated on knowledge improvement of the exposure–response relationship between whole-body vibration and the occurrence of low back disorders, and to advance understanding of the other physical and psychosocial factors that combine to result in the progression of low back symptoms.

Simulation and Modelling

A simulation is the execution of a model, represented by a computer program that gives information about the system being investigated. The simulation approach of analyzing a model is opposed by the analytical approach, where the method of analysing the system is purely theoretical. A simulation approach may be more reliable, depending on the quality of the model. Mathematical models are used to simulate and study the seat comfort virtually (Bouazara and Richard, 2001; Fatollahzadeh, 2006). Bouzara and Richard (Bouazara and Richard, 2001) developed a mathematical model to simulate the dynamic behaviour of a 3-D vehicle. With this model, various types of non-linear suspensions, such as active and semi-active suspensions, can be investigated. Fatollahzadeh [43] constructed a mathematical model, which clarifies and predicts the drivers' comfortable sitting posture and position. Computer-Aided Engineering (CAE) methods such as finite element analysis and simulation techniques have also been used to study and to develop vehicle seats (Verver *et al.*, 2004; Tewari and Prasad, 1999). Verver *et al.* (Verver *et al.,* 2004) used a Finite Element (FE) model of the human buttocks to predict the static pressure distribution between human and seating surface by its detailed and realistic geometric description. For modelling purposes and for evaluation of driver's seat performance in the vertical direction various computer-aided design models of the seated human body have been developed and standardised by the ISO.

MECHANICAL DESIGN

Mechanical research on vehicle seats concentrated on the actuators, the suspensions and dampers. Liu and Wagner (Liu and Wagner, 2002) designed the controllers and evaluated semi-active hydraulic and active electromechanical hydraulic actuators. A series of four model-free and model based control strategies are presented for each actuator to attenuate road vibrations transmitted to the occupant's lower and upper torso. The controller designs include relative, proportional-integral, variable structure system, and optimal control. Perisse and Jezequel [51] overviewed a theoretical study on active seat suspension. The principal objective of this study is to improve ride passenger comfort by reducing transmitted seat forces. In addition, Perisse and Jezequel (Perisse and Jezequel, 2000) also investigated the experimental feasibility of the active seat suspension to improve ride comfort. The prototype of a reversible electromechanical actuator used in an active seat suspension is presented. A reversible electromechanical actuator was used to provide a continuous variable damper/an active force generator to realise a control policy.

SUMMARY

This paper surveys a large number of studies and up-to-date techniques developed for vehicle seats used by different types of transportation systems. The objective of this paper is to describe the state of the art and recent development of vehicle seat design, which are available in current literature, and to give a general idea about unsolved problems that arise in practice. From the literature review, it can be seen that the majority of vehicle seats studies are concentrated on vibration, pressure and ergonomics. Besides these topics, driver "fatigue" also needs to be considered in vehicle seat design. The term "fatigue" is used to define the physical impairment that results from exposure to the seat vibrations for a long period of time. Fatigue may lead to decreased attention, perception, decision-making, vigilance, and reaction time. Future research on seat thermal and humidity comfort can be concentrated on intelligent thermal and humidity control system and evaluation of the physiological seat comfort for drivers by using on-the-road experiment testing.

REFERENCES

[1] Ahmadian, M., Seigler, M., Clapper, D. and Sprouse, A. (2001). "A comparative analysis of air-inflated and foam seat cushions for truck seats." Society of Automotive Engineers Technical Paper, Document no. 2002–01–3108.

[2] Alem, N.M. and Strawn, G.D. (1996). "Evaluation of an energy absorbing truck seat for increased protection from landmine blasts." US Army Aeromedical Research Laboratory report no. 96–06.

[3] Boggs, C.M. (2004). "Field study to evaluate driver fatigue on air-inflated truck curshions." Master of Science thesis, Faculty of the Virginia Polytechnic Institute and State University, Blacksburg, Virginia, USA.

[4] Boileau, P.-E. and Rakheja, S. (1990). "Vibration attenuation performance of suspension seats for off-road forestry vehicles." *International Journal of Industrial Ergonomics,* 5, pp. 275–291.

[5] Boileau, P.E. and Rakheja, S. (1998). "Whole body vertical biodynamic response characteristics of the seated vehicle driver: measurement and model development." *International Journal of Industrial Ergonomics*, 22, pp. 449–472.

[6] Bouazara, M. and Richard, M.J. (2001). "An optimization method designed to improve 3-D vehicle comfort and road holding capability through the use of active and semi-active suspensions." *European Journal of Mechanics. A. Solids*, 20, pp. 509–520.

[7] Bovenzi, M., Rui, F., Negro, C., D'Agostin, F., Angotzi, G., Bianchi, S., Bramanti, L., Festa, G., Gatti, S., Pinto, I., Rondina, L. and Stacchini, N. (2006). "An epidemiological

study of low back pain in professional drivers." *Journal of Sound and Vibration*, 298, pp. 514–539.

[8] Broughton, J. (2004). "The actual threat posed by unrestrained rear seat car passengers." *Accident Analysis and Prevention*, 36, pp. 627–629.

[9] Chang, S.R., Son, K. and Choi, Y.S. (1996). "Measurement and three-dimensional graphic representations of Korean seatpan and seatback contours." *International Journal of Industrial Ergonomics*, 18, pp. 147–152.

[10] Cho, Y. and Yoon, Y.S. (2001). "Biomechanical model of human on seat with backrest for evaluating ride quality." *International Journal of Industrial Ergonomics*, 27, pp. 331–345.

[11] Chou, J.R. and Hsiao, S.W. (2005). "An anthropometric measurement for developing an electric scooter." *International Journal of Industrial Ergonomics*, 35, pp. 1047–1063.

[12] Coelho, D.A. and Dahlman, S. (1999). "A pilot evaluation of car seat side support: leading to a redefinition of the problem." *International Journal of Industrial Ergonomics*, 24, pp. 201–210.

[13] Ebe, K. and Griffin, M.J. (2001). "Factors affecting static seat cushion comfort." *Ergonomics*, 44(10), pp. 901–921.

[14] Falou, E.W., Duchêne, J., Grabisch, M., David, Hewson, Yves Langeronand Lino, F. (2003). "Evaluation of Driver Discomfort during Long-Duration Car Driving." *Journal of Applied Ergonomics*, 34(3), pp. 249–255.

[15] Fatollahzadeh, K. (2006). "A laboratory mock-up research work on truck driver's seat position and posture." *Doctoral thesis,* Royal Institute of Technology Division of Industrial Ergonomics and Scania CV AB Vehicle Ergonomics, Stockholm, Sweden.

[16] Floyd, W.F. and Roberts, D.F. (1958). "Anatomical and Physiological Principles in Chair and Table Design." *Ergonomics* 2, 1–16, Kharagpur, India.

[17] Frechin, M.M., Arino, S.B. and Fontaine, J. (2004). "ACTISEAT: Active Vehicle Seat for Acceleration Compensation", *Proceeding Institution of Mechanical Engineers, Part D: Journal of Automobile Engineering*, 218: 925–933.

[18] Gillespie, Thomas D. (1985). Heavy Truck Ride. SAE Paper 850001.

[19] Guenaelle, P. (1995). "One methodology to evaluate Automotive Seat Comfort." In: *Proceedings of the Third International Conference on Vehicle Comfort and Ergonomics, Bologna , Italy,* 29–31 March, pp. 231–240.

[20] Gyi, D.E., Porter, J.M. and Robertson, K.B. (1998). "Seat pressure measurement technologies: considerations for their evaluation." *Applied Ergonomics*, 27(2), pp. 85–91.

[21] Gyi, D.E., Porter, J.M. and Robertson, N.K.B. (1998). "Seat pressure measurement technologies: consideration for their evaluation." *Appllied Ergonomic*, 27(2), pp. 85–91.

[22] Hanowski, R.J., Perez, M.A. and Dingus, T.A. (2005). "Driver distraction in long-haul truck drivers." *Transportation Research*, Part F8, pp. 441–458.

[23] Hinz, B., Bluthner, R., Menzel, G., Schust, M. and Seidel, H. (2006). "On the significance of body mass and vibration magnitude for acceleration transmission of vibration through Seats with horizontal suspensions." *Journal of Sound and Vibration*, 298, pp. 627–637.

[24] Hix, K., Ziemba, S. and Shoof, L. (2000). "Truck seat modelling—a methods development approach." *International ADAMS User Conference.*

[25] Kim, T., Kim, Y. and Yoon, Y. (2005). "Development of a biomechanical model of the human body in a sitting posture with vibration transmissibility in the vertical direction." *International Journal of Industrial Ergonomics*, 35, pp. 817–829.

[26] Latchford, J., Chirwa, E.C., Chen, T. and Mao, M. (2005). "The relationship of seat backrest angle and neck injury in low-velocity rear impacts." *Proceeding of Institution of Mechanical Engineers, Part D: Journal of Automobile Engineering,* 219, pp. 1293–1302.

[27] Liang, C. and Chiang, C. (2006). "A study on biodynamic models of seated humanb subjects exposed to vertical vibration." *International Journal of Industrial Ergonomics,* 36, pp. 869–890.

[28] Liu, X. and Wagner, J. (2002). "Design of a vibration isolation actuator for automotive seating systems." Part II: controller design and actuator performance. *International Journal of Vehicle Design*, 29(4), pp. 357–375.

[29] Mehta, C.R. and Tewari, V.K. (2000). "Seating discomfort for tractor operators: a critical review." *International Journal of Industrial Ergonomics* 25 (2000) pp. 661–674.

[30] Nawayseh, N. and Griffin, M.J. (2005). "Effect of seat surface angle on forces at the seat surface during whole-body vertical vibration." *Journal of Sound and vibration*, 284, pp. 613–634.

[31] Ofori-Boetang, A.B. (2003). "A study of the effect of varying air-inflated seat cushion parameters on seating comfort." *Faculty of the Virginia Polytechnic Institute and State University, Blacksburg, Virginia, USA*.

[32] Okunribido, O.O., Shimbles, S.J., Magnusson, M. and Pope, M. (2007). "City bus driving and low back pain: a study of the exposures to posture demands, manual materials handling and who-body vibration." *Applied Ergonomics*, 38, pp. 29–38.

[33] Perisse, J. and Jezequel, L. (2000). "An original feedback control with a reversible electromechanical actuator used as an active isolation system for a seat suspension." *Part II: experimental study. Vehicle Systems Dynamics*, 34, pp. 381–399.

[34] Porter, J.M. and Gyi, D.E. (1998). "Interface pressure and the prediction of car seat discomfort." *International Journal of Vehicle Design,* 19(3), pp. 255–266.

[35] Qiu, Y. and Griffin, M.J. (2005). "Transmission of roll, pitch and yaw vibration to the backrest of a seat supported on a non-rigid car floor." *Journal of Sound and Vibration*, 288, pp. 1197–1222.

[36] Rakheja, S., Stiharu, I., Zhang, H. and Boileau, P.E. (2006). "Seated occupant interactions with seat backrest and pan, and biodynamic response under vertical vibration." *Journal of Sound and Vibration*, 298, pp. 651–671.

[37] Rebelle, J. (2000). "Development of a numerical model of seat suspension to optimize the end-stop buffers." *35th United Kingdom Group Meeting on Human Responses to Vibration, ISVR*, University of Southampton, Southampton, England, 13–15 September.

[38] Seigler, M. and Ahmadian, M. (2003). "Evaluation of an alternative seating technology for truck seats." *Heavy Vehicle Systems,* 10(3), pp. 188–208.

[39] Seigler, M., Ahmadian, M. and Boggs, C. (2003). "Validation of an indentor system for evaluating truck seat cushions." *Proceeding of Institution of Mechanical Engineers, Part D: Journal of Automobile Engineering,* 217:343–348.

[40] Shen, W. and Vertiz, A. (1997). "Redefining Seat Comfort." *SAE Paper* 970597.

[41] Stikeleather, L.F. (1991). "Seat vibration and ride comfort In Human Factors, module 6 of a series of quality instructional materials." American Society of Agricultural Engineers, St. Joseph, MI 49085.

[42] Tewari, V.K. and Prasad, N. (1999). "Three DOF modelling of tractor seat-operator system." *Journal of Terramechanics,* 36, pp. 207–219.

[43] Transafety Reporter (1998). "Truck Survey Highlights Causes of Drowsy Driving and Suggests Preventative Measures." Transafety Incorporated.

[44] Van Niekerk, J.L., Pielemeier, W.J. and Greenberg, J.A. (2003). "The use of Seat Effective Amplitude Transmissibility (SEAT) values to predict dynamic seat comfort." *Journal of Sound and Vibration*, 260, pp. 867–888.

[45] Van Niekerka, J.L., Pielemeierb, W.J. and Greenberg, J.A. (2003). "The use of Seat Effective Amplitude Transmissibility (SEAT) Values to Predict Dynamic Seat comfort." *Journal of Sound and Vibration*, 260, pp. 867–888.

[46] Verver, M.M., de Lange, R., van Hoof, J. and Wismans, J.S.H.M. (2005). "Aspects of seat modelling for seating comfort analysis." *Applied Ergonomics*, 36, pp. 33–42.

[47] Verver, M.M., van Hoof, J., Oomens, C.W.J., Wismans, J.S.H.M. and Baaijens, F.P.T. (2004). "A finite element model of the human buttocks for prediction of seat pressure distributions." *Computer Methods in Biomechanics and Biomedical Engineering*, 7(4), pp. 193–203.

[48] Verver, M.M., van Hoof, J., Oomens, C.W.J., Wismans, J.S.H.M. and Baaijens, F.P.T. (2004). "A finite element model of the human buttocks for prediction of seat pressure distributions." *Computer Methods in Biomechanics and Biomedical Engineering*, 7(4), pp. 193–203.

[49] Viano, D.C. and Andrzejak, D.V. (1992). Research Issues on the Biomechanics of Seating Discomfort: an Overview with Focus on Issues of the Elderly and Low-Back Pain. *SAE Paper* 920130.

[50] Wang, W., Rakheja, S. and Boileau, P.E. (2006). "The role of seat geometry and posture on the mechanical energy absorption characteristics of seated occupants under vertical vibration." *International Journal of Industrial Ergonomics*, 36, pp. 171–184.

[51] Zhang, L., Helander, M.G., Drury, C.G. (1996). "Indentifying factors of comfort and discomfort in sitting." *Human Factors*, 38(3), pp. 377–389.

Leg Strength of Agricultural Workers in Central India

K.N. Agrawal[1], P.S. Tiwari[1], L.P. Gite[2], Joydeep Majumder[2], Sachin Pharade[2] and V. Bhusan Babu[1]

[1]Agricultural Mechanization Division
[2]AICRP on Ergonomics and Safety in Agriculture
CIAE, Nabibagh, Bhopal–462038 (MP), India

ABSTRACT

A study was therefore, carried out to collect isometric leg/foot strength of Indian male as well as female agricultural workers. A strength measurement set-up developed at CIAE, Bhopal was used for the purpose. Data were collected on 1701 subjects from twenty different locations in Madhya Pradesh in central India of which 944 were male and 757 were female agricultural workers. The mean age, stature and weight of the male subjects were 29.8 ± 9.5 years, 1649 ± 59 mm and 51.2 ± 6.4 kg whereas for female subjects the values were 33.7 ± 8.2 years, 1519 ± 54 mm and 45.0 ± 7.3 kg, respectively. The isometric leg and foot strength of male subjects was higher than those of female subjects. The mean values for isometric left and right leg strength in sitting posture were 378.5 ± 87.2 N and 414.1 ± 92.2 N, respectively for male workers and 243.6 ± 62.1 N and 262.7 ± 66.8 N, respectively for female workers. Similarly, the mean values of left and right foot strength in sitting posture were 286.7 ± 76.9 N and 310.4 ± 79.6 N, respectively for male workers and 184.0 ± 52.6 N and 197.4 ± 59.4 N, respectively for female workers. The mean left and right leg/foot strength values were statistically non-significant at 5% level. Weight of the subjects indicated a positive correlation with isometric leg strength values. The 5th percentile left and right leg strength values were 235.1 N and 262.5 N, respectively for male workers and 141.4 N and 152.8 N, respectively for female workers. These values can be used to set limits in the design of leg operated controls of tractor and other self propelled farm machinery and farm equipment requiring leg/foot pedal power depending on the frequency of movement.

Keywords: Leg Strength, Foot Strength, Agricultural Workers, Central India, Farm Machinery, Tractor Control.

INTRODUCTION

Operation of manual controls such as brake and clutch on all self-propelled agricultural machinery including tractors, require application of leg forces while the operation of accelerator pedal requires application of foot force. Apart from these, many other agricultural activities such as the operation of pedal operated pumps, threshers etc., and transporting loads on pedal operated carts require leg or foot forces to supply energy required for performing the task. Researchers have pointed out that operation of leg/foot-operated controls is responsible for high physical workload and, moreover, for musculoskeletal complaints affecting the low back and lower extremities. Jafry and Haslegrave (1995) studied the effect of tractor pedal operation on spinal loading. They indicated that the operation of foot pedals required at least 143 N force, which was an important risk factor, associated with tractor driving. The flattening of the lumber spine could explain the increasing disc pressure while operating foot pedals. An increase in the activity of the illiospinous and the abdominal muscles each time the foot is lifted onto a pedal can also contribute to the increase in disc pressure. The lumber disc pressure increased when the gears were shifted and increased even more when pedal was depressed. The results showed a significant relationship between leg operated control operation and low back pain. Dupuis (1959) investigated the strain on the tractor operators during operation of different controls. He reported that during an hour the driver operated the clutch pedal, brake pedals and hydraulic control lever of a front mounted loader 230, 100 and 250 times, respectively. In his study, the maximum pedal force, which could be applied by the operators, was measured by placing the pedal within 89–165 mm below the seat surface. It was also observed that human energy consumption could be reduced by 13–29% by making improvement in tractor controls and seat. It was concluded that a properly designed tractor workplace could improve the efficiency and comfort of the operator. Yadav (1995) observed the mean frequencies of 72, 148, 230 and 436 during an hour for operation of clutch, brakes, draft control lever and viewing backward, respectively by the Indian tractor operators. He concluded that maximum energy was spent in operation of different essential controls in tractor workplace and hence deciding suitable location of these controls was essential.

Snook (1978) pointed out that designing the job to fit the worker could reduce up to one-third of industrial back injuries. Further, the job design was found to be significantly more effective in controlling low back injuries than selecting the worker for the job or training the worker to fit the job. Many factors need to be considered for the design of the operator's workplace. If a comparison of the workplace of an automobile and a tractor is made, a large braking force (about 600 N) is required to stop a tractor, which is really very high for an Indian operator of medium stature and weighing 50 kg or less. Without getting the back support as in case of a short stature operator, it is not possible to apply a force higher than his/her weight. To achieve enhanced performance and efficiency of man-machine system along with better comfort and safety of the operators, it is necessary to design various controls and work place on tractors with due consideration to the anthropometric and strength data of Indian agricultural workers. If the operator controls are not properly adapted to his/her strength capabilities, the performance demanded of him/her may quickly reach and even exceed the limits of tolerance. As a result of excessive stress, premature fatigue, and impaired health, the possibility of accidents might increase. A tractor operator's survey in Central India indicated that there was a need to reduce force requirement for brake pedal operation as about 33% operators reported that the force required in brake operation was more (Tiwari *et al.*, 2004). Also about 36% operator's felt that the tractor controls, *viz.*, brake, clutch, and hydraulic lever were not conveniently placed on tractors.

Tractor operator's comfort and safety have received considerable attention abroad. The design of tractors in India has not changed much in the past two decades, especially from ergonomics point of view (Patel *et al.*, 2000). In India, most of the tractor designs have been adopted from the foreign designs, which are based on the anthropometric dimensions and strength capabilities of western population. There has not been any consideration for anthropometric dimensions and strength limitations of Indian agricultural workers, which has led to many occupational health problems and accidents with the tractors under field and road conditions. Tiwari (2001) ergonomically evaluated the tractor operator workplace and activities. He reported that the location of tractor controls met the requirements of the Indian Standard (IS 12343, 1998). But the range of dimensions given in Indian Standard is very large. He observed that braking force increased with increase in forward speed during field operation. It was 746 N, which was above the recommended limits as per IS 10703 (1992)/ISO TR 3778 (1987).

Few anthropometric studies have been carried out on Indian agricultural workers (Sen *et al.*, 1977; Gupta *et al.*, 1983; Gite and Yadav, 1989). However, these studies were mainly centred on male agricultural workers and that too for limited body dimensions. On the other hand the strength data of Indian agricultural workers are not available. The All India Coordinated Research Project (AICRP) on Ergonomics and Safety in Agriculture (ESA) collected such data for male as well as female agricultural workers from Madhya Pradesh (Central India) through one of its centres located at Central Institute of Agricultural Engineering, Bhopal. Data on seventy five body dimensions, four skin-fold thickness and sixteen strength parameters useful for the design of agricultural machines and equipment were collected, compiled and analysed. This paper presents the data on leg and foot strength (in sitting posture) of agricultural workers of Madhya Pradesh and outlines the significance of using these data for the design of leg operated controls on self-propelled machinery and other leg operated agricultural equipment.

MATERIAL AND METHODS

Subjects

The study was carried out in twenty districts from nine agro-climatic zones of Madhya Pradesh (Central India). The districts selected for the survey were Rewa, Panna, Shahdol, Jabalpur and Balaghat from Kemore Plateau and Satpura Hills Zone; Bhopal, Guna, Sagar and Raisen from Vindhya Plateau Zone; Hosangabad from Central Narmada Valley Zone; Gwalior from Gird Zone, Tikamgarh from Bundelkhand Zone; Chindwara and Betul from Satpura Plateau Zone; Mandsaur, Rajgarh, Ujjain and Dewas from Malwa Plateau Zone; West Nimar from Nimar Valley Zone and Jhabua from Jhabua Hills Zone. Data were collected for 1701 subjects (944 male and 757 female) from different communities including tribal population. The subjects were randomly selected from the healthy agricultural workers in the age group of 18 to 65 years. All the subjects were free from physical abnormalities and were in good health.

Tasks

The workers were required to apply leg force on a pre-designed stationary pedal while sitting in erect posture. In another attempt the workers were required to apply force on the pedal through their foot using dorso-flexion while heel of the foot was kept on the ground. They were instructed to apply their maximum leg and foot force evenly without jerks. As per the protocol for strength data collection, the subjects were required to reach their maximum strength within first two seconds and then maintain the maximum strength for next three seconds (Kumar *et al.*, 1995). During a preliminary trial it was observed that some stimulus in the form of light/sound is required to guide the subjects for applying the leg/foot force for the desired time duration. Therefore, a five seconds timer with a red light signal and beeping sound (developed at CIAE, Bhopal) was used during force application. The subjects were asked to release the applied force on the pedal smoothly as the red light went off and the beep stopped after 5 seconds.

Equipment and Procedure

A survey team of four well qualified staff (2 male and 2 female) experienced in measurement of anthropometric dimensions and human strength parameters, collected the complete data of 1701 subjects. The complete survey work was carried out in continuous supervision of one of the scientists of the Institute involved in the project.

The anthropometric dimensions and the skin-fold thickness were measured using Harpendens Anthropometer and Holtain Skinfold Calliper, respectively adopting the procedure formulated by AICRP on ESA (Gite and Chatterjee, 1999). Standard terminologies given in the Anthropometric Source Book (NASA, 1978) were used. The four skin-fold thicknesses were used to calculate Body Density (BD) using the equation:

$$\text{Body Density (BD)} = 1.1599 - (0.0717 \times \log \Sigma \text{ four skin-fold thicknesses}) \quad \ldots (1)$$

The four skin folds for which the data were collected were bicep, tricep, sub-scapular and supra iliac. The per cent body fat was calculated from the body density using the following equation as proposed by Siri (1959):

$$\text{Per cent Body Fat (\%BF)} = (495/\text{BD}) - 450 \quad \ldots (2)$$

The absolute body fat was calculated using the equation:

$$\text{Absolute Body Fat} = (\text{Body weight} \times \text{\%BF})/100 \quad \ldots (3)$$

The lean body mass of the subject was calculated by subtracting the absolute body fat from total body weight as:

$$\text{Lean Body Mass} = \text{Total Body Weight} - \text{Absolute Body Fat} \quad \ldots (4)$$

A strength measurement setup developed at CIAE, Bhopal for measuring 14 human strength parameters useful for the design of agricultural machinery was used in the study. The setup mainly consisted of the following:

- A wooden platform of 2300 × 810 mm size.
- Two vertical posts (1500 mm height) made up of 48.5 mm diameter mild steel pipes erected at a spacing of 385 mm. These posts were bolted to the wooden platform. Another vertical post made up of mild steel box section of 40 × 40 × 4 mm size and having the same height as the circular posts was erected in between the posts. This vertical post was provided with 5.5 mm holes at a spacing of 25 mm on the rear side and had a pulley at its top.
- Two braces made of mild steel angle iron for supporting the vertical posts.
- A height adjustable horizontal bar made up of 40 × 40 × 4 mm size box section, was provided to slide over the circular posts with the help of two collars welded on the bar.
- A wire rope was connected to the middle of the horizontal bar, which passed over the pulley mounted at the top of the middle vertical post. The other end of the wire was provided with a hook to anchor it into the desired hole of the middle post to adjust the height of the horizontal bar. The horizontal bar has sliding space to fit the pedal and load cell-mounting assembly.
- A stationary pedal of 200 × 75 mm size was welded on a hollow pipe of 63 mm diameter such that central axis remains orthogonal to axis passing through the centre line of pedal plate. Another hollow pipe was fitted to slide in the fixed hollow pipe such that load cell can be bolted inside the two hollow pipes using two nuts and bolts. This complete pedal assembly was mounted on a specially made quadrant type pedal mounting assembly, which can slide in the horizontal bar and can be fixed at a desired distance either on left or right side of the operator. The quadrant type pedal mounting assembly was provided with a series of equidistant holes on a pre-designed arc so that height and orientation of the pedal can be adjusted according to the need of the worker.
- The worker was asked to sit on a seat designed for the purpose. The height and the distance of the seat from pedal assembly were made adjustable to incorporate workers of different body dimensions.

A Novatech load cell (1 kN) of tension and compression type with digital load indicator was used for measuring the leg strength of the subjects. Readings obtained during the 3s force application were noted continuously and the most stable value of force, which appeared for maximum duration in the load indicator, was noted for the trial. The complete human strength measurement set up along with anthropometer and other accessories was carried to each survey site for survey work.

Experimental Protocol

During the measurement of leg and foot strength the seat height was kept as 19% of the stature and horizontal distance of foot pedal from SIP was kept as 35% of the stature for optimal location of foot pedal (Mehta *et al.*, 2007). The orientation of the pedal was selected such that ankle angle remains 900 while applying the force. The height adjustable pedal with load cell assembly was adjusted so that ankle angle remains 900 and the posture attained was as defined in Figure 1. During the measurement of foot strength, the pedal assembly was reoriented such that the knee angle was kept at 90° and operator was asked to apply force on pedal using dorso flexion without lifting his ankles from ground as shown in Figure 2.

Fig. 1: Measurement of Leg Force with Right Leg in Sitting Posture

Fig. 2: Measurement of Foot Force with Left Leg in Sitting Posture

The subject looked straight forward during the application of leg/foot force. With the start of electronic timer, the subject applied the force, to attain the maximum in first 2s and hold it until the light/sound signal stopped finally after 5s. Throughout the 5s duration, the subject was strictly prohibited to change the prescribed posture or dislodge the other legs or use hands to take support from seat or its back rest. The subject was bare-footed and advised to hold the steering wheel while applying the force. The exertions were replicated thrice for leg strength as well as for foot strength and the mean value of these replications was taken as the value of leg/foot strength. A rest of 2 minutes was given in between two successive trials for each subject (Kumar, 1991).

Data Analysis

The anthropometric as well as the leg/foot strength data for male and female subjects were analysed to get mean standard deviation, coefficient of variation, standard error of mean, 5th and 95th percentile values and minimum and maximum values, using the "Statistica" statistical tool. The calculated percentile values were true percentile values for which the following standard equations given in Anthropometric Source Book (NASA, 1978) were used:

$$5^{th} \text{ percentile value} = \text{Mean} - 1.645 \times \text{SD} \quad \ldots (5)$$

$$95^{th} \text{ percentile value} = \text{Mean} + 1.645 \times \text{SD} \quad \ldots (6)$$

The leg/foot force data were statistically analysed to know the effect of mode of force application and also the gender effects. The t-test was used for comparisons. Multiple regressions were developed between age, weight, stature, lean body mass and functional leg length of the subjects with leg/foot strengths for male and female subjects.

RESULTS AND DISCUSSION

Anthropometric Parameters of Agricultural Workers

Tables 1 and 2 present the mean, standard deviation, coefficient of variation, standard error of mean, 5th and 95th percentile values and minimum and maximum values for relevant anthropometric parameters of male and female agricultural workers, respectively. The mean age, stature and weight of male subjects were 29.8 ± 9.5 years, 1649 ± 59 mm and 51.2 ± 6.4 kg, respectively while the corresponding parameters for female subjects were 33.7 ± 8.2 years, 1519 ± 54 mm and 45.0 ± 7.3 kg. In general the male subjects were heavier and taller than female subjects. The mean lean body mass of male subjects was also higher than female subjects.

Table 1: Anthropometric Parameters of Male (N = 944) Agricultural Workers

Parameters	*Mean*	*SD**	*CV†*	*SE of Mean‡*	*Percentile*		*Range*	
					5th	*95th*	*Min*	*Max*
Age, years	29.8	9.5	31.88	0.098	14.1	45.5	18.0	65.0
Weight, kg	51.2	6.4	12.5	0.066	40.6	61.8	35.0	77.0
Lean body mass, kg	44.7	1.7	3.8	0.049	41.8	47.3	38.5	49.9
Stature, mm	1649	59	3.58	0.610	1552	1747	1424	1854
Trochanteric height, mm	859	43	5.01	0.442	788	930	701	1025
Functional leg length, mm	1005	45	4.48	0.462	931	1079	860	1180
Thigh circumference, mm	436	39	8.94	0.401	373	500	310	575
Calf circumference, mm	301	23	7.64	0.244	263	339	210	440

Table 2: Anthropometric Parameters of Female (N = 757) Agricultural Workers

Parameters	*Mean*	*SD**	*CV†*	*SE of Mean‡*	*Percentile*		*Range*	
					5th	*95th*	*Min*	*Max*
Age, years	33.7	8.2	24.40	0.299	20.2	47.2	18.0	60.0
Weight, kg	45.0	7.3	16.28	0.266	32.9	57.0	28.0	77.0
Lean body mass, kg	38.5	5.7	14.84	0.208	29.1	47.9	24.5	64.4
Stature, mm	1519	54	3.47	1.920	1430	1607	1237	1687
Trochanteric height, mm	797	41	5.19	1.500	729	865	667	961
Functional leg length, mm	929	45	4.86	1.640	855	1003	805	1060
Thigh circumference, mm	431	51	11.86	1.860	347	515	280	665
Calf circumference, mm	289	27	9.41	0.991	244	334	200	400

*Standard deviation; †Coefficient of variation; ‡Standard error of mean.

Leg/Foot Strength of Agricultural Workers

Table 3 presents the mean, standard deviation, coefficient of variation, standard error of mean, 5th and 95th percentile values and minimum and maximum values of leg and foot strength of male and female agricultural workers. The mean values for leg strengths for left and right legs were 378.5 ± 87.2 N and 414.1 ± 92.2 N for male and 243.6 ± 62.1 N and 262.7 ± 66.8 N, respectively for female workers. The t-test conducted between left and right leg strength values of male and female workers revealed that strength values observed for right and left legs were not significantly different at 5% level of significance. Similarly, the mean values of left and right feet strength in sitting posture were 286.7 ± 76.9 N and 310.4 ± 79.6 N, respectively for male and 184.0 ± 52.6 N and 197.4 ± 59.4 N, respectively for female workers. These values indicated that leg strength was more than foot strength ($p < 0.01$). This is in agreement with the findings reported by Grandjean (1980) and van der Beek *et al.* (2000) both for male and female workers.

Table 3: Leg and Foot Strength of Male and Female Agricultural Workers of Central India

Parameters	*Mean*	*SD**	*CV†*	*SE of Mean‡*	*Percentile*		*Range*	
					5th	*95th*	*Min*	*Max*
Male (n = 944)								
Right leg, N	414.1	92.2	22.27	0.897	262.5	565.7	128.5	498.3
Left leg, N	378.5	87.2	23.04	0.948	235.1	521.9	150.1	557.2
Right foot, N	310.4	79.6	25.64	0.791	179.4	441.3	110.9	586.6
Left foot, N	286.7	76.9	26.82	0.891	160.2	413.1	147.2	693.6
Female (n = 757)								
Right leg, N	262.7	66.8	25.43	1.209	152.8	372.6	114.8	478.7
Left leg, N	243.6	62.1	25.49	2.255	141.5	345.7	101.0	502.3
Right foot, N	197.4	59.4	30.09	1.458	99.7	295.0	80.4	383.6
Left foot, N	184.0	52.6	28.59	1.913	97.4	270.5	61.8	377.7

*Standard deviation; †Coefficient of variation; ‡Standard error of mean.

Leg and Foot Strength of Male and Female Agricultural Workers

Different studies on leg strength have shown that muscular strength plays an important role in most leg operated tasks. Age and some anthropometric dimensions, *viz.*, weight, stature, functional leg length, thigh and leg circumference as well as the posture adopted during force application also affect the maximum leg strength without any musculoskeletal injury. In the present study, the leg strength was higher than foot strength for male as well as female workers. The higher value for leg strength was due to more active participation of muscles in the thighs and lower leg in force generation. During foot force application as the subject kept his ankles on the ground a part of the force applied by the muscles got restricted by the posture adopted while using dorso flexion. In this posture the weight of the leg was not that effective and force generated by leg and thigh muscles also got reduced. The weight of the leg of the subject got partially supported by the ankle on the ground rather than being used for active force generation.

According to NASA (1978) and Grandjean (1980) women can generally exert forces about two-third of that exerted by men. A close perusal of the mean values of leg and foot strengths of agricultural workers in the present study showed that female workers could exert 63-64% of leg and foot forces of those exerted by male workers. This was mainly because male and female subjects differed in anthropometric characteristics: men were heavier and taller than women.

Increased leg strengths were observed with an increase in body weight in case of both male and female workers. The body weight of the subjects had a positive correlation with leg strength with correlation coefficients (R^2) of 0.244 and 0.197 for male and female workers, respectively. The correlation coefficients for foot strengths were 0.198 and 0.102, for male and female workers, respectively. Thus, body weight of the subjects predicted the maximum leg/foot strengths. Lean body mass of the subjects was also a better predictor of maximum strength when compared to stature and functional leg length of the subjects. On the other hand age of the subjects (male as well as female) was observed to be the worst predictor of maximum leg/foot strengths. Furthermore, taller subjects might exert higher maximum leg forces. However, the relationship between stature and maximum leg/foot forces is less straight forward because of the influence of the handling posture including body segment angles and position of the feet. In the present study, the correlations coefficients (R^2) of regression of stature over leg and foot strength were 0.08 and 0.086, respectively for male workers and 0.076 and 0.076, respectively for female workers (Table 4).

Design Considerations

The maximum work tolerance on a working day can be indirectly obtained from the maximum isometric strength for a single exertion (Waters *et al.*, 1993). The determination of maximum acceptable forces depends on the assumption that an individual can estimate his/her maximum work tolerance without experiencing health complaints during a certain work period. In terms of external exposure, frequency of a certain activity and working hours are controlled during the experiments, while the subjects are given control of the level of the exposure. The subject adjusts his/her maximum acceptable force depending on his/her own feelings of exertion or fatigue. In general the risk of developing musculoskeletal disorders increases when exerted forces on a working day approximate the maximum strength and maximum acceptable forces are exceeded.

In terms of work-related factors, the exposure to force application can be expressed with three factors: Intensity (magnitude and direction), frequency and duration. The risk of musculoskeletal disorders increases, if any of these dimensions deviates from its optimum value. On the other hand, a combination of sub-maximal values of all three dimensions may together also increase the risk of health complaints. Therefore, these dimensions must be examined in view of maximal as well as the combinations of their sub-maximal values.

Table 4: Multiple Regressions of the Leg and Foot Forces

Parameters		*MR*	R^2	*DF*	*F*	*p<*
Male (n = 944)						
Leg	Age	0.060	0.003	1	3.46	0.0628
	Weight	0.495	0.244	1	304.95	0.000
	Stature	0.287	0.081	1	89.25	0.000
	Lean body mass	0.474	0.2248	1	262.84	0.000
	Functional leg length	0.262	0.068	1	68.10	0.000
Foot	Age	0.08	0.006	1	1.96	0.169
	Weight	0.446	0.198	1	222.97	0.000
	Stature	0.297	0.088	1	86.52	0.000
	Lean body mass	0.474	0.198	1	213.63	0.000
	Functional leg length	0.308	0.095	1	92.03	0.001
Female (n = 757)						
Leg	Age	0.012	0.001	1	0.108	0.743
	Weight	0.449	0.197	1	180.17	0.000
	Stature	0.279	0.076	1	61.72	0.000
	Lean body mass	0.498	0.248	1	274.6	0.000
	Functional leg length	0.301	0.090	1	71.44	0.000
Foot	Age	0.002	0.000	1	0.004	0.9484
	Weight	0.346	0.102	1	75.98	0.000
	Stature	0.271	0.072	1	53.15	0.000
	Lean body mass	0.351	0.121	1	86.30	0.000
	Functional leg length	0.297	0.088	1	63.46	0.000

One of the problems encountered by a designer is that in most cases the posture of the user during force exertion cannot be adequately anticipated. The force that can be exerted is influenced to a high degree by the subject's posture. Standardized postures are generally used, though the methods of description tend to vary considerably. Leg strength capability depends on a complex interaction of posture, shoe/floor friction, pedal position and subject anthropometry (Mehta *et al.*, 2007; Snook, 1978). Generally, it is recognized that persons with better muscle built-up, higher leg reach and high body weight can achieve higher leg/foot force capability if proper lumbar support is provided.

For design purposes the 5th percentile strength values should be used to make the design suitable for 95% of the workers population. The 5th percentile left and right leg strength values for male workers were 235.1 N and 262.5 N, respectively. Brake is a right leg operated control, which is operated for short duration; therefore, its design may be limited to 85% of 5th percentile value of right leg strength (van Wely, 1970). Thus the suggested limits for the operation of brake pedal shall be 223 N. On the other hand clutch is a left leg operated control but as it is an occasionally operated control, its design should be limited to 60% of the 5th percentile value of the left leg strength. Therefore, the suggested limits for the operation of clutch pedal shall be 130 N.

The foot strength of male workers in sitting posture was 310.4 and 286.7 N for right and left legs, respectively and the corresponding 5th percentile values were 179.4 and 160.2 N. Similarly the foot strengths of female workers in sitting posture were 197.4 and 184.0 N for right and left legs, respectively and the corresponding 5th percentile values were 99.7 and 97.4 N. The accelerator pedal is a frequently operated control. In India tractors are mostly operated by male agricultural workers using their right foot; therefore 30% of 5th percentile foot strength value of right foot of male agricultural workers should be taken as the limiting value. This value thus comes to about 54 N. However, the minimum force requirement for accelerator pedal should not be less than the segmental weight of the foot so that operator's foot can rest on the pedal without pressing it when it is not so desired.

CONCLUSION

The study indicated that the leg and foot strength of male agricultural workers is higher than those of female workers. The mean values for isometric leg strength in sitting posture with left and right legs are 378.5 ± 87.2 N and 414.1 ± 92.2 N, respectively for male workers and 243.6 ± 62.1 N and 262.7 ± 66.8 N, respectively for female workers. The mean values of foot strength in sitting posture with left and right foot were 286.7 ± 76.9 N and 310.4 ± 79.6 N, respectively for male workers and 184.0 ± 52.6 N and 197.4 ± 59.4 N, respectively for female workers. Weight of the subjects indicates a positive correlation with leg strength. The 5th percentile left and right leg strength values were 235.1 N and 262.5 N, respectively for male workers and 141.4 N and 152.8 N, respectively for female workers. Leg operated activities of frequent loading in nature should be designed such that the force requirement does not exceed 60% of the 5th percentile strength value, although it may rise to 85% as long as the effort is not occasional in nature (like braking). Agricultural activities requiring continuous force application either by foot or leg should be designed such that the force requirement is below 30% of the 5th percentile strength value to have a margin for static loading of the muscles.

ACKNOWLEDGEMENT

The authors are grateful to Dr. M.M. Pandey, Director and Dr. V.V. Singh, Head, Agricultural Mechanization Division, Central Institute of Agricultural Engineering, Bhopal, for their help and encouragement during the study. Assistance received from Miss. Vineeta Singh, Miss Sweta Prajapati and Miss Anita Mungale, Research Fellows,

AICRP on ESA and Mr. R.J. Raina, Laboratory Technician is duly acknowledged. Thanks are also due to the subjects for their wholehearted cooperation in data collection.

REFERENCES

[1] Alvi, S.A.A. (1971). "Physiological studies on human performance during various modes of work applicable to agriculture." *Unpublished Ph.D. Thesis*. Agricultural Engineering Deptt., Indian Institute of Technology, Kharagpur.

[2] Ayoub, M.M. and McDaniel, J.W. (1974). "Effects of operator stance on pushing and pulling tasks." *Transactions of American Institute of Industrial Engineering*, 6(3), pp. 185–195.

[3] Bureau of Indian Standards (IS) 1992. "Agricultural tractors—maximum actuating forces required to operate controls." IS 10703–1992.

[4] Bureau of Indian Standards (IS) 1998. "Agricultural tractors—operator's seat—technical requirements." IS 12343–1998.

[5] Daams, B.J. (1993). "Static force exertion in postures with different degrees of freedom." *Ergonomics,* 36, pp. 397–406.

[6] Damkot, D.K., Pope, M.H., Lord, J. and Frymoyer, J.W. (1984). "The relationship between work history, work environment and low back pain in men." *Spine*, 9, pp. 395–399.

[7] Davis, P.R. and Stubbs, D.A. (1977). Safe levels of manual forces for young males (2). *Applied Ergonomics,* 8(4), pp. 219–228.

[8] Dupuis, H. (1959). "Effect of tractor operation on human stress." *Agric. Eng.* 40: 510–519, 525.

[9] Frymoyer, J.W., Pope, M.H. and Constanza, M.C. (1980). "Epidemiologic studies of low back pain." *Spine*. 5, pp. 419–423.

[10] Gite, L.P. and Chatterjee, D. (1999). "All India anthropometric survey of agricultural workers-Proposed action plan." *All India Coordinated Research Project on Human Engineering and Safety in Agriculture,* Central Institute of Agricultural Engineering, Bhopal.

[11] Gite, L.P. and Yadav, B.G. (1989). "Anthropometric survey for agricultural machinery design." *Appl. Ergon*. 20, pp. 191–196.

[12] Grandjean, E. (1980). "Fitting the task to the man—An ergonomic approach." Taylor and Francis Ltd. London.

[13] Gupta, P.K., Gupta, M.L. and Sharma, A.P. (1983). "Anthropometric survey of Indian farm workers." *Agricultural Mechanisation in Asia, America and Latin America*, 14:27–30.

[14] International Organisation for Standardisation (ISO) (1987). "Agricultural tractors—maximum actuating forces required to operate controls." *ISO/TR* 3778.

[15] International Organisation for Standardisation (ISO) (1997). "Tractors and self-propelled machinery for agriculture and forestry—operator controls—actuating forces, displacement, location, and method of operation." *ISO/TC 23/SC3/N525*.

[16] Jafry, T. and Haslegrave, C.M. (1995). "The effects of tractor foot pedal operation on spinal loading." *Paper Presented at the United Kingdom Informal Group Meeting on Human Response to Vibration held at the Silsoe Research Institute, Silsoe.*

[17] Keyserling, W.M., Herrin, G.D., Chaffin, D.B., Armstrong, T.J. and Foss, M.L. (1980). "Establishing an industrial strength testing program." *American Industrial Hygiene Association Journal*, 41, pp. 730–736.

[18] Kuiper, J.I., Burdore, A., Verbeek, J.H.A.M., Frings-Dresen, M.H.W., van der Beak, A.J. and Vhkari-Juntura, E.R.A. (1999). "Epidemiologic evidence on manual materials handling as risk factors for back disorders: A systemic review." *International Journal of Industrial Ergonomics*. 24, pp. 389–404.

[19] Kumar, S. (1991). "Arm lift strength in workspace." *Applied Ergonomics*, 22(5), pp. 317–328.

[20] Kumar, S., Narayan, Y. and Bacchus, C. (1995). "Symmetric and asymmetric two-handed pull-push strength of young adults." *Human Factors*, 37(4), pp. 854–865.

[21] Mehta, C.R., Tiwari, P.S., Rokade, S., Pandey, M.M., Pharade, S.C., Gite, L.P. and Yadav,, S.B. (2007). "Leg strength of Indian operators in the operation of tractor pedals." *International Journal of Industrial Ergonomics*. 37, pp. 283–289.

[22] NASA (1978). "Anthropometric Source Book. Vol-I." Reference Publication 1024, Scientific and Technical Office, Washington D.C., U.S.A.

[23] Patel, R., Kumar, A. and Mohan, D. (2000). "Development of an ergonomical facility for Indian tractors." *Appl. Ergon*. 31, pp. 311–316.

[24] Snook, S.H. (1978). "The design of manual handling tasks." *Ergonomics*. 21(132), pp. 963–985.

[25] Tiwari, K. (2001). "Ergonomical evaluation of tractor operator's workplace and activity." *Unpublished M. Tech. Thesis*, Jawaharlal Nehru Krishi Vishwa Vidyalaya, Jabalpur, India.

[26] Tiwari, P.S., Gite, L.P., Kot, L.S., Dubey, A.K. and Singh, V.V. (2004). "Users survey on ergonomical problems during tractor operation." *Paper presented during National Conference on Humanizing Work and Work Environment held at NITIE, Mumbai* during April 22–24, 2004.

[27] van Wely, P. (1970). "Design and disease." *Applied Ergonomics*. 1, pp. 262–269.

Design of Tractor Controls Based on Ergonomical Considerations

C.R. Mehta, M.M. Pandey, P.S. Tiwari, L.P. Gite and Joydeep Majumder

Central Institute of Agricultural Engineering
Nabibagh, Bhopal–462038, India
E-mail: crmehta@ciae.res.in

ABSTRACT

The locations of tractor controls were optimised based on anthropometric data of 5434 male agricultural workers. The longitudinal distance of brake and clutch pedals from seat index point (SIP) based on optimum leg reach envelope of 5th percentile agricultural workers should be 510 mm. Based on 5th percentile all India data, the vertical distance of brake and clutch pedals should not be more than 370 mm from SIP taking 90 mm downward movement of brake pedal. The brake pedals should not be placed farther than 75–125 mm from the centre line. The vertical distance of accelerator pedal based on 5th percentile all India data should not be more than 430 mm below SIP with 30 mm downward movement of accelerator pedal. The accelerator pedal should be placed to the right side of brake pedal but should not be placed at a distance of less than 75 mm from centre line. The centre of steering wheels on a tractor must be 425–525 mm forward and 175–385 mm above the SIP.

The control dimensions of some of frequently operated controls like brake and clutch pedals, accelerator pedal etc. were designed based on anthropometric data of 5434 and strength data of 3423 Indian male agricultural workers. Brake pedal is a right leg operated control. The 5th percentile value of right leg strength sitting of 250 N was taken as the limiting force. Clutch pedal is a left leg operated control. The 5th percentile value of left leg strength sitting is to be considered as the limiting force. The length of brake and clutch pedals should be minimum of 60 mm and optimum of 75 mm (based on ergonomical guidelines). The breadth of these pedals should be decided based on 95th percentile value of foot breadth. The accelerator pedal is a frequently toe operated (foot operated) control. The limiting force should be 75 N considering 30% of the 5th percentile value of right foot strength. The minimum length of accelerator pedal should be equal to 95th percentile values of toe and width of accelerator pedal should be wider than 95th percentile value of foot breadth.

Keywords: Ergonomics, Tractor Control, Indian Workers.

INTRODUCTION

Operating a tractor imposes a lot of physical and mental stress upon the operator. If the operator, seat and controls are not properly located, his work performance may be poor and there is also a possibility of accident also in tractors. Proper design of controls is also important for comfortable and safe operation of the equipment. The design involves location of control as well as strength limits for operating these controls. At present, the data available for Indian operators on these aspects are very limited. Some case studies for workplace layout for operators of tractors/self-propelled equipment have been carried out in this country (Yadav, 1995; Tiwari, 2001). However, the Bureau of Indian Standards at present has adopted ISO standards. These standards are primarily based on anthropometric data of Western workers.

Matthew and Knight (1971) recommended the dimensions for optimum pedal area for British driver, when the controls are operated either by heel or toe. They suggested that the vertical distance of pedal should vary with type of task and should not be more than 400 mm. The pedal should be displaced no farther than 75 to 125 mm from the midline.

Tiwari (2001) conducted a study on ergonomical evaluation of tractor operator workplace and activities. He reported that nearly all locations of tractor controls met the requirements given in the Indian Standard (IS: 12343, 1988). But the range of dimensions presented in the standard is very large.

To achieve enhanced performance and efficiency of man-machine system along with better comfort and safety of the operators, it is necessary to locate and design various controls on tractors with due consideration to the anthropometric and strength data of Indian male agricultural workers.

MATERIAL AND METHODS

Indian tractors are mainly operated by male agricultural workers of the country. Therefore, the anthropometric data of 5434 (mean age of 33.4 years) and strength data of 3423 male agricultural workers were used to suggest control dimensions and minimum actuating force for operating tractor controls on Indian tractor. The methodology given by Matthew and Knight (1971) was followed for developing optimum hand and leg reach envelops. Some of the important anthropometric data of agricultural workers compiled for development of hand and leg reach envelopes were age, weight, stature, elbow grip length, shoulder grip length, elbow rest height, sitting acromial height, knee height sitting, popliteal height sitting, buttock popliteal length, buttock knee length, foot length etc. The mean, 5th percentile and 95th percentile values of data were calculated.

RESULTS AND DISCUSSION

The locations of frequently operated tractor controls were optimised based on anthropometric data of 5434 male agricultural workers. It was recommended that longitudinal distance of brake and clutch pedals from SIP based on optimum leg reach envelope of 5th percentile agricultural workers should be 510 mm. Based on 5th percentile all India data, the vertical distance of brake and clutch pedals should not be more than 370 mm from SIP including downward movement of brake pedal. The brake pedals should be placed as close to the centre line as possible but should not be placed at a distance of less than 75 mm from centre line. The vertical distance of accelerator pedal based on 5th percentile all India data should not be more than 430 mm below SIP with 30 mm downward movement of accelerator pedal. The accelerator pedal should be placed to the right side of brake pedal but should not be placed at a distance of less than 75 mm from centre line. The position of steering wheel must be 425–525 mm forward and 175–385 mm above the SIP.

The approach for optimising control dimensions and limits for maximum actuating force is as follows:

Accelerator Pedal

The accelerator pedal is of great significance for accurate regulation of engine speed. If the pedal pressure is too high, the energy required to continuously operate the pedal will be excessive. The 5th percentile value of right leg strength of Indian male agricultural workers is 248 N. The accelerator pedal is a frequently operated control. Therefore, we take 30% of the 10th percentile value of right foot strength for design purpose, which is equal to 58 N (~60 N). The maximum actuating force for operation of accelerator pedal should not exceed 60 N.

The accelerator pedal is a toe operated (foot operated) control. The 95th percentile values of foot length and instep length of Indian male agricultural workers are 268 and 208 mm, respectively. Therefore, the minimum length of accelerator pedal should be 60 mm. The width of accelerator pedal should be wider than 95th percentile value of foot breadth (110 mm) of Indian tractor operators. Therefore, minimum width of accelerator pedal should be 110 mm. As force required to operate accelerator pedal is less, the foot breadth of 5th percentile operator (79 mm) may also be considered. Therefore, the minimum width of accelerator pedal may be kept as 79 mm.

Brake Pedal

Operating a brake pedal of a tractor requires the driver to exert a force so as to move the pedal at prescribed distance along its axis of travel. In accordance with the generally accepted practice, tractor brake pedals are usually operated with the right foot. It is a right leg operated control and force required in actuating it is quite high. The 5th percentile value of right leg strength of Indian male agricultural workers is 248 N. We take 5th percentile value of right leg strength for design purpose which is equal to 250 N. Therefore, the maximum actuating force for operation of brake pedals should be less than 250 N. The 95th percentile values of foot length and instep length of Indian male agricultural workers are 268 and 208 mm, respectively. Therefore, the minimum length of brake pedals should be 60 mm and optimum should be 75 mm based on ergonomical guidelines. The width of brake pedals should be wider than 95th percentile value of foot breadth (110 mm) of Indian tractor operators. Therefore, minimum width of brake pedals should be 110 mm.

Clutch Pedal

The clutch pedal is used very frequently, even more frequently, than the brake pedals. This is a left leg operated control. The force necessary for the operation of clutch pedal is usually smaller than that required for brake pedals. The 5th percentile value of left leg strength sitting of Indian male agricultural workers is 258 N. We take 50% of the 5th percentile value of left leg strength for design purpose which is equal to 129 N. Therefore, the maximum actuating force for operation of clutch pedals should be less than 130 N.

The minimum length of clutch pedal should be 60 mm and optimum should be 75 mm based on ergonomical guidelines. The width of clutch pedal should be wider than 95th percentile value of foot breadth (110 mm) of Indian tractor operators. Therefore, minimum width of clutch pedal should be 110 mm.

Steering Wheel

The steering wheel in a tractor is used to transmit steering forces and to support the body of the operator. The 5th percentile value of torque strength both hands (sitting) is 175 N. It is a frequently hand-operated control. We take 30% of 5th percentile value of torque strength of both hands (sitting) as the limit of maximum actuating force which is equal to 52 N. Therefore, the force required in steering wheel operation should be less than 52 N.

As per ergonomics guidelines, the minimum and maximum diameter of steering wheel should be 200 and 500 mm, respectively. The 95th percentile value of 'middle finger palm grip diameter' and 5th percentile value of 'inside grip diameter' are 39 and 38 mm, respectively. For a tractor, the steering wheel diameter may be kept between 35–38 mm and preferable is 35 mm based on anthropometric data.

Gear Selection Lever and Speed Selection Lever

According to Dupuis (1959), gearshift lever and hydraulic controls are used most frequently. Therefore, these controls should be located close to steering wheel in order to minimise the time for hand movements and prevent excessive movements of the body, as in all such functions energy is consumed.

The power grip with round knob or knob with flat head is preferable for gear shift lever/ speed selection lever (high/low). The 5th percentile value of 'grip span' of Indian male agricultural workers is taken for design purpose, which is equal to 48 mm. Therefore, the diameter of gear shift knob/speed selection lever should be 50 mm. The optimum and maximum values of gear shift knob diameter as per ergonomical guidelines should be 40 and 75 mm, respectively.

Hand Throttle Lever

It is usually fitted on right hand side of the steering column. The shape of knob of hand throttle lever may be cylindrical or as that of gear selection lever. The 5th percentile value of 'grip span' of Indian male agricultural workers is taken for design purpose, which is equal to 48 mm. Therefore, the diameter of hand throttle lever should be 50 mm. The optimum and maximum values of hand throttle lever knob diameter as per ergonomical guidelines should be 40 and 75 mm, respectively. The 60% of 10th percentile value of right hand push strength sitting should be taken as the limiting force for operation of hand throttle lever on a tractor which is equal to 30 N. Therefore, the force required for operation of hand throttle lever on a tractor should not exceed 30 N.

Foot Step

The 95th percentile values of foot length and foot breadth of male Indian agricultural workers are 268 mm and 111 mm, respectively. Therefore, the minimum length and width of foot step on a tractor should be 268 mm and 110 mm, respectively. The 10th percentile value of standing knee height of Indian agricultural workers is 426 mm. Therefore, the foot step on a tractor should be provided at a height of 426 mm above the ground level.

CONCLUSIONS

The following conclusions can be drawn from the study.

1. On the basis of optimum hand and leg reach envelopes of Indian male agricultural workers, it is recommended that the frequently hand operated controls on a tractor should be placed 390–546 mm forward and 270–450 mm above the SRP and frequently leg operated controls should be placed 430–870 mm forward and 130–370 mm below the SRP.
2. Based on anthropometric data of Indian male agricultural workers, the minimum length of clutch and brake pedals should be 60 mm and optimum should be 75 mm based on ergonomical guidelines. The minimum width of the pedals should be 110 mm.
3. The maximum actuating force limits of 600 N and 350 N for brake and clutch pedals, respectively as recommended by Bureau of Indian Standards (IS: 10703, 1992) need to be revised. The limit for maximum actuating force for brakes and clutch for existing work place layout of Indian tractors should be 250 N and 130 N, respectively.

REFERENCES

[1] Bureau of Indian Standards (IS), 1992. "Agricultural tractor—maximum actuating forces required to operate controls." IS 10703–1992.

[2] Dupuis, H. (1959). "Effect of tractor operation on human stress." *Agricultural Engineering*, 40, pp. 510–519, 525.

[3] Matthews, J. and Knight, A.A. (1971). "Ergonomics in agricultural equipment design." National Institute of Agricultural Engineering, Silsoe, UK.

[4] Tiwari, K. (2001). "Ergonomical evaluation of tractor operator's workplace and activity." *Unpublished M. Tech Thesis,* Jawaharlal Nehru Krishi Vishwa Vidyalaya, Jabalpur, India.

[5] Yadav, R. (1995). "Some ergonomical investigation on tractor operator workplace design." *Unpublished Ph.D. Thesis*, Indian Institute of Technology, Kharagpur.

Assessment of Controls Layout of Indian Tractors

Adarsh Kumar, Bhaskar Gaikwad and J.K. Singh

Division of Agricultural Engineering
Indian Agricultural Research Institute, New Delhi–110 012, India
E-mail: adarsh_iari@rediffmail.com

ABSTRACT

Tractors in low-income countries are used both for farm and non-farm activities. Most of the tractors being manufactured in India are product of collaboration with other countries. The design of tractors manufactured in India has not changed much in the past five decades especially from ergonomic point of view because of economic considerations. This paper describes the control layout assessment with respect to the Indian population and compares the location of controls with workspace envelopes and IS 12343 standard for commonly used tractors on Indian farms. Controls like steering and foot clutch, foot brake, foot accelerator are located in areas defined by IS 12343 standard in some tractors but these are not placed in the workspace envelopes of Indian population. It results in mismatch between the workspace envelope and location of controls as defined by the standard. The controls need a complete change in their layout to be in the workspace envelopes, as this cannot be achieved by providing seat movement in horizontal and vertical direction in the present tractor design.

Keywords: Tractor, Controls Layout, Workspace Envelopes.

INTRODUCTION

The use of tractors on Indian farms has increased from a population of 52,000 in the sixties to 3.2 million in the nineties (Motor Transport Statistics of India, 2003). In Low Income Countries (LICs) ergonomic studies on tractors have been very few. In India the previous studies are confined to anthropometric surveys of agricultural workers (Gite and Yadav 1989) and there are a few qualitative reports (Bhatia and Rawal, 1976; Rawal and Bhatia, 1978; Singh, 1979; Tandon and Sirohi, 1987), which highlight the problems related to agricultural equipment in general.

Studies from High-Income Countries (HICs) underline the importance of ergonomic considerations for improving safety and comfort of tractor operators. In a recent study, Hsiao *et al.* (2005) used the anthropometric dimensions for tractor cab design and concluded that the SAE J 2194 standard for ROPS is improper for the surveyed population. Thirteen anthropometric dimensions were identified for establishing body models to improve tractor cab design. Mismatch between tractor design and ergonomic criteria like operator capability in terms of body dimensions was studied. Dupuis (1959) claimed that injuries could result from poor ergonomics. It was stated, "If the operator's controls are not properly adapted to his harmony, the performance demanded of him may quickly reach and even exceed the lines of tolerance. As a result of excessive stress, premature fatigue and impaired health, the possibility of accident will result". Access and egress is also one of the common causes of injuries among tractor drivers (Suutarinum, 1992). Pheasant and Harris (1982) investigated the pedal position with respect to Seat Reference Point (SRP) and appropriate driver posture and concluded that ideally pedal location should be 12.5% stature below SRP and 47.5% in front of SRP to have a better driving posture and optimum force application. Bottoms and Barber (1978) reported that most agricultural equipment workers behind the tractor, and driver spend a large proportion of his time looking backwards and adopting a poor posture. They studied a mechanism that allowed a tractor seat to swivel up to 20 degrees from the normal forward facing position. The results showed a decrease in muscle activity in the shoulder and neck regions with swivelled seat. Zander (1972) conducted studies on workspace envelopes for combined harvester controls and suggested seat movement for comfortable control operations.

The present study was aimed to carry out comparative assessment of different tractor models with respect to workspace envelope and ISO standard for tractor controls. Anthropometric dimensions for Indian population were used for development of workspace envelopes (Patel *et al.*, 2000). A control layout-measuring device was used for carrying out comparative evaluation of different tractor models displays and controls.

METHODS

Work Place Envelope Template

A two dimensional templates in 1:5 scale were used for both in plan and elevation for 50th percentile Indian male based on methodology used by Zander (1972). This envelope shows optimum and maximum areas for control operated by upper and lower limbs (Patel *et al.*, 2000). The template was superimposed on control layouts prepared using layout measuring device to detect the location of the controls, which were outside the optimum and maximum areas.

Measurement of Controls Using Layout Measuring Device

Layout-measuring device was mounted on the tractor seat to find out different control positions and movements in three dimensions with respect to the seat reference point. The basic design was adopted from Zander (1972) and modified further to incorporate the seat reference-measuring device of ISO 3462, 1980 (Patel *et al.*, 2000). The combination of layout measuring device and ISO device standardised the procedure for measuring the control locations with respect to the seat reference point irrespective of different seat designs. Observations of horizontal, vertical and radial coordinates of different controls were taken. Correction for seat compression due to operator's weight was incorporated in vertical coordinate observations.

The tractors selected for the experimentation were commonly used tractors both in terms of horsepower as well as models. The rating and year of manufacturing of the tractors are given in Table 1.

Table 1: Rating and Year of Manufacturing of Tractors Used for Experimentation

Tractor	*A*	*B*	*C*	*D*	*E*	*F*	*G*	*H*	*I*	*J*
Year of manufacturing	1972	1990	1995	1982	1993	1997	2000	2003	2002	2003
HP	35	35	35	40	45	45	45	48	50	60

Comparison of Control Location on Different Tractors

The spatial arrangement of controls was compared with the workspace envelopes and standards (ISO 4253, 1993; IS 12343, 1998). The standard defines the dimension of control (Figure 1) with respect to Seat Index Point (SIP) for the tractors with track width greater than 1150 mm. As per standard, SIP is 140 mm ahead and 90 mm above the Seat Reference Point (SRP). It specifies that steering wheel should lie 425 to 525 mm forward and 175 to 385 mm above SIP and the angle of steering column should be in the range of 0–40°. The details of prescribed location of other controls are shown in Figure 1.

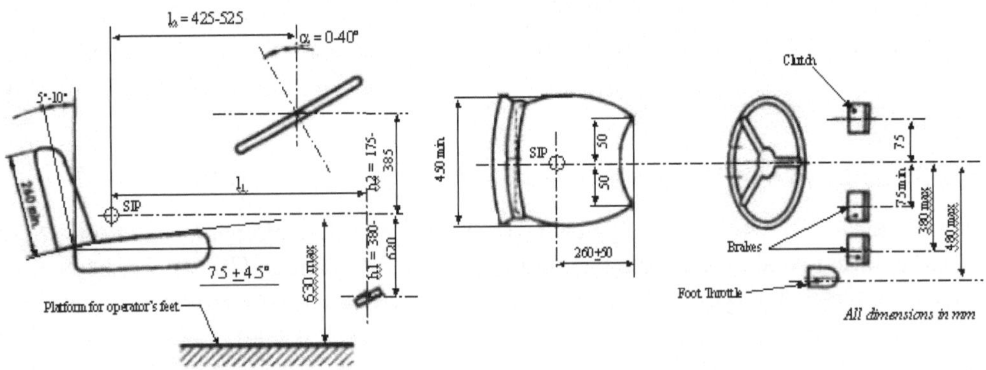

Side View Top View

Fig. 1: Location of Controls (IS 12343, 1998)

RESULTS AND DISCUSSION

Different tractor controls were superimposed on workspace. Table 2 gives the location of control with respect to workspace envelope in terms of optimum area, maximum area and located outside the workspace envelope both in vertical and horizontal planes. Combining the location in both planes of workspace envelope the location is defined in terms of control located inside or outside the workspace envelope. Table 3 indicates the placement of control in the defined area of IS 12343 standard with respect to horizontal and vertical placement, considering both planes; if the control lies inside the area then it is defined rightly placed.

Table 2: Location of Tractor Controls with Respect to Workspace Envelope

Tractors		*Controls*												
	Plane	*Steer-ing*	*Hand Accelerator*	*Gear Lever*	*High/ Low Gear*	*Horn*	*Light Switch*	*Engine Stop Lever*	*PTO*	*Depth Control lever*	*Foot Brake*	*Foot Clutch*	*Foot Accelerator*	*Controls in Workspace Envelope*
A	Horizontal	OA	OA	O	MA	OA	OA	OA	MA	MA	O	O	O	
	Vertical	OA	OA	O	O	O	O	O	MA	O	O	O	O	
	Location	2	2	X	X	X	X	X	2	X	X	X	X	3
B	Horizontal	O	O	O	MA	MA	MA	OA	OA	MA	MA	MA	O	
	Vertical	O	O	O	O	O	O	O	O	O	OA	OA	O	
	Location	X	X	X	X	X	X	X	X	X	2	2	X	2
C	Horizontal	OA	O	MA	OA	OA	OA	OA	MA	MA	O	O	O	
	Vertical	OA	O	MA	O	O	O	O	O	O	O	O	O	
	Location	2	X	2	X	X	X	X	X	X	X	X	X	2
D	Horizontal	OA	OA	OA	OA	OA	O	MA	MA	MA	O	O	O	
	Vertical	OA	MA	O	O	O	O	O	O	O	MA	O	O	
	Location	2	2	X	X	X	X	X	X	X	X	X	X	2
E	Horizontal	OA	O	MA	MA	OA	OA	OA	MA	MA	O	O	O	
	Vertical	OA	O	O	O	O	O	O	O	O	MA	O	O	
	Location	2	X	X	X	X	X	X	X	X	X	X	X	1
F	Horizontal	OA	O	MA	MA	OA	OA	OA	MA	MA	O	O	O	
	Vertical	OA	O	MA	O	O	O	O	O	MA	O	O	O	
	Location	2	X	2	X	X	X	X	X	2	X	X	X	3
G	Horizontal	OA	MA	MA	MA	OA	OA	OA	MA	MA	O	O	O	
	Vertical	OA	O	MA	O	O	O	O	O	O	O	O	O	
	Location	2	X	2	X	X	X	X	X	X	X	X	X	2
H	Horizontal	OA	O	MA	MA	OA	OA	OA	O	MA	O	O	O	
	Vertical	OA	O	MA	O	O	O	O	O	O	O	O	O	
	Location	2	X	2	X	X	X	X	X	X	X	X	X	2
I	Horizontal	OA	O	OA	OA	OA	O	OA	OA	MA	O	O	O	
	Vertical	OA	O	O	O	O	O	O	O	O	O	O	O	
	Location	2	X	X	X	X	X	X	X	X	X	X	X	1
J	Horizontal	OA	O	MA	MA	OA	OA	OA	MA	MA	O	O	O	
	Vertical	OA	O	MA	O	O	O	O	O	O	O	O	O	
	Location	2	X	2	X	X	X	X	X	X	X	X	X	2

Note: OA = Optimum area; MA = Maximum area; O = Outside workspace envelope; 2 = Located inside workspace envelope in both horizontal and vertical plane; X=Located outside workspace envelope in either or both planes

It is evident from the result of the study that many controls are not located within the optimum work place envelope and some even exceed the limit of maximum area of workplace envelope. Therefore, the operator has to adopt uncomfortable postures and stretch his limits of normal reach to operate the controls. Frequently used controls such as foot brake, clutch, accelerator, depth and draft control lever need proper placement so that they can be operated conveniently and do not cause excessive stress to the operator. Foot accelerator pedal is located much away from the longitudinal axis of tractor as compared to the other foot-operated controls. Location of foot brake and foot clutch pedals lies horizontally below the knee and vertically at average foot height of 50th percentile of Indian population, in the tractors of 45 hp (34 kW) and above power range, whereas it lies horizontally ahead of knee in low range tractors (less than 45 hp (34 kW)). This is contradictory to the fact that more pedal force is required for the operation of these controls in higher HP tractors. Higher HP tractors have higher momentum and need more pedal force to disengage clutch or apply brakes, if these controls are not hydraulically operated. None except tractor E and H had hydraulic operated pedal brakes. The pelvic muscles and body weight are used for operation of these pedals instead of thigh muscles. So the location of foot brakes and clutch needs to be shifted ahead of the knee position for higher horsepower range tractors. The foot operated controls, also suggested by Pheasant and Harris (1982), should be located at 47.5 and 12.5 per cent of stature below and in front of SRP, which falls in workspace envelope of the present study.

Table 3: Location of Tractor Controls with Respect to IS 12343

Controls	*Plane*	*Tractors*									
		A	*B*	*C*	*D*	*E*	*F*	*G*	*H*	*I*	*J*
Steering	Horizontal	N	I	I	N	N	I	N	I	I	I
	Vertical	N	I	I	N	N	I	N	I	I	I
	Location	X	2	2	X	X	2	X	2	2	2
Foot brake	Horizontal	I	I	I	I	I	I	I	I	I	I
	Vertical	N	I	I	I	I	I	I	I	I	N
	Location	X	2	2	2	2	2	2	2	2	X
Foot Clutch	Horizontal	I	I	I	I	I	I	I	N	I	I
	Vertical	N	I	I	N	N	I	N	I	I	I
	Location	X	2	2	X	X	2	X	X	2	2
Foot accelerator	Horizontal	N	N	N	I	N	N	I	I	N	I
	Vertical	I	I	I	I	I	I	I	I	I	I
	Location	X	X	X	2	X	X	2	2	X	2
Controls in designated area		0	3	3	2	1	3	2	3	3	3

Note: I = Inside designated area; N = Outside designated area; 2 = Located inside designated area in both the planes; X = Located outside designated area in either or both the planes

Table 4: Steering Angle of Different Tractors

Tractor	*A*	*B*	*C*	*D*	*E*	*F*	*G*	*H*	*I*	*J*
Steering angle (°)	40.8	33.48	24.05	37.8	37.07	29.42	36.8	32.8	38.66	42.87

Locations of gear selection levers which were provided on dash board lie in optimum area of workspace envelope but those provided on floor do not fall in optimum work space and lie as low as knee level between legs. These floor gear selection levers overlap with the operational space of other controls like foot clutch and brake which otherwise should have been located near the longitudinal axis of the tractor for easy operation.

Controls like horn button, light switch and engine stop lever are located at lower positions and require bending posture for operating these controls, and need to be placed on dashboard at a higher elevation. Parking brakes fall completely outside optimum workspace area. To overcome the problem, seat position should be made adjustable in both vertical and horizontal planes. The seat adjustment should be made at the beginning of the operation by the tractor operator, so that the tractor control lies within optimum workspace for different population groups.

Foot brake, in the workspace envelope is only for one tractor but this control is in the designated area of the IS standard for all tractors except one. Similarly, foot clutch falls in the workspace envelope in case of one tractor and foot accelerator in none of the tractors. But these two controls in five and four tractors are as per IS 12343. This clearly signifies that the tractor controls are located as per IS standard but not matching with the workplace envelopes of Indian population, which are based on anthropometric dimensions. The mismatch between the control location and workspace envelopes can be because of discrepancy between anthropometric dimension of Indian population and tractor layout design. Most of the tractors manufactured in India are in collaboration from developed countries and their layouts were designed according to anthropometric dimensions of the population of collaborating countries. The same were used in India, which might be one possible reason of improper location of controls. Moreover, the controls in contemporary machines in HICs are sophisticated and require less force to operate. Therefore, location may not be very critical as compared to mechanical controls where force requirement is high.

ISO standard designated control locations are varying widely considering the workspace envelopes of Indian population; this deviation may be because of environmentally controlled cabins, altering the location of controls which have high degree of automation. In an attempt to bring controls within workspace envelopes by adjusting the relative position of controls with respect to seat, it was found that by raising the location of all controls by 12 cm, resulted in hand controls falling in workspace envelopes, in both vertical and horizontal directions but foot controls were still outside the work space envelope (Figure 2a). The further horizontal forward movement of 18 cm was required to bring the foot controls in the workspace envelopes (Figure 2b). But this relocation places the hand controls outside the workspace envelope. It means that only adjustment in seat location will not give the desired results of locating the controls in the workspace envelopes. Therefore, it is necessary to relocate controls on tractor so that they are placed in workplace envelope of Indian population and seat adjustment can take care of anthropometric variability.

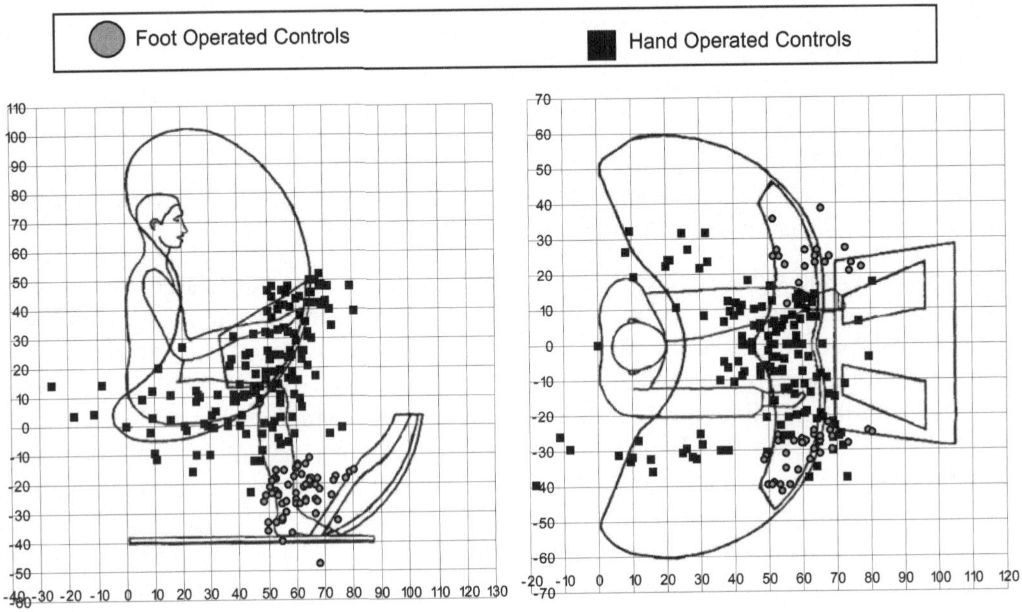

(a) Controls Raised by 12 cm

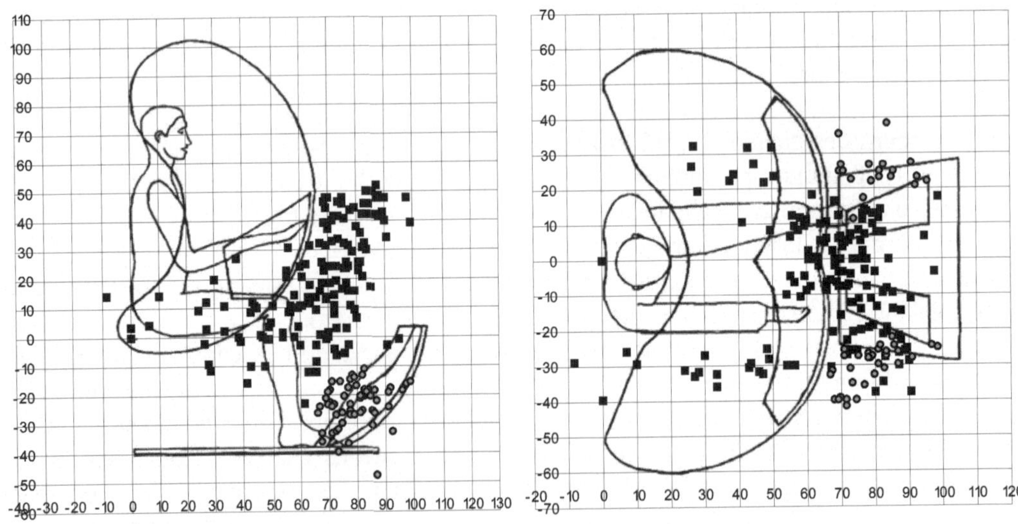

(b) Controls Shifted 12 cm up and 18 cm Horizontally Ahead

Fig. 2: Relocation of Controls of Different Tractors

CONCLUSION

The location of controls on Indian tractors falls in optimum, maximum and outside the workspace envelope. The controls like PTO, depth control lever, horn, light switch, engine stop lever high and low lever need adjustment in vertical plane to make them located in workspace envelopes. However, the controls like steering, foot clutch, foot brake foot accelerator are located in areas defined by IS 12343 standard in some tractors whereas these are not placed in workspace envelopes resulting in mismatch between workspace envelope of Indian population and location of controls as defined by standard.

The controls need a complete change in their layout to be in the workspace envelopes, as this cannot be achieved by providing seat movement in horizontal and vertical directions in the present tractor design. The seat should have provision for horizontal and vertical movements to accommodate the maximum population for comfortable operation of tractor controls once they are located within the workspace envelopes.

REFERENCES

[1] Bhatia, B.S. and Rawal, G.S. (1976). "Human factors related to tractor accidents." *Food Farming and Agriculture,* 7, pp. 46–47.

[2] Bottoms, D.J. and Barber, T.S. (1978). "A swiveling seat to improve tractor drivers posture." *Applied Ergonomics*, 9, pp. 77–84.

[3] Duppis, H. (1959). "Effects of tractor operation on human stresses." *Agricultural Engineering*, 40, pp. 510–525.

[4] Gite, L.P. and Yadav, B.G. (1989). "Anthropometric Survey for Agricultural Machinery Design: An Indian case study." *Applied Ergonomics,* 20, pp. 191–196.

[5] Hsiao, H., Whitestone, J., Bradtmiller, B., Whisler, R., Zwiener, J., Lafferty, C., Kau, T.Y. and Gross, M. (2005). "Anthropometric criteria for the design of tractor cabs and protection frames." *Ergonomics,* 48, pp. 323–53.

[6] IS 12343. "Agricultural tractors—Operator's seat—Technical requirements", Bureau of Indian Standards, New Delhi, 1998.

[7] ISO 3462 (1980). "Methods for determination of seat reference point of agricultural tractors", International Organization of Standardization, Switzerland.

[8] ISO 4253 (1993). "Agricultural tractors—Operator's seating accommodation—Dimensions." International Organization for Standardization, Switzerland.

[9] Motor Transport of India (2003). Transport Research Wing, Ministry of Surface Transport, Government of India.

[10] Patel, R., Kumar, A. and Mohan, D. (2000). "Development of ergonomic evaluation facility for Indian tractors." *Applied Ergonomics*. 31, pp. 311–316.

[11] Pheasant, S.T. and Harris, C.M. (1982). "Human strength in the operation of tractor pedals." *Ergonomics,* 25, pp. 53–63.

[12] Rawal, G.S. and Bhatia, B.S. (1978). "Farm Machinery and Safety." *Progressive Farming*, 4, pp. 13–15.

[13] Singh, S. (1979). "Tractor Safety, in Report of Faculty Programme in Human Engineering", College of Agricultural Engineering, Punjab Agricultural University, Ludhiana, India.

[14] Suutarinum, J. (1992). "Tractor accidents and their prevention." *International Journal of Industrial Ergonomics*.10, pp. 321–329.

[15] Tandon, S.K. and Sirohi, N.P.S. (1987). "Safety in agricultural machines and tractor operations" in *Proceedings of VIII National Conference on Critical Care Medicine*, New Delhi.

[16] Zander, J. (1972). "Ergonomic in machine design: a case study of the self propelled combine harvester." Netherlands, Wageningen: H. Veenman and Zonen.

Physiological Cost and Discomfort during Operation of an Agricultural Foot Sprayer

Hemant Shrimali[1], Ravi Mathur[1] and L.P. Gite[2]

[1]College of Technology and Engineering, Udaipur–313 001, India

[2]Central Institute of Agricultural Engineering, Bhopal–462 038, India

E-mail: [1]hemantshrimali@rediffmail.com

ABSTRACT

Physiological and discomfort parameters were evaluated for male users in terms of HR, OCR, EER, BPDS and ODR during the operation of an agricultural foot sprayer. During pedalling operation of the sprayer mean values of HR, •HR, OCR, •OCR, EER, ODR, and BPDS ranged from 98.17 to 100.67 beats/min, 23.6 to 27.3 beats/min, 0.48 to 0.88 l/min, 0.32 to 0.48 l/min, 3.18 to 3.48 kcal/min, 6.4 to 7.4 and 57.3 to 59.3, respectively. For spraying operation with lance mean HR, •HR, OCR, •OCR, EER, ODR, and BPDS ranged from 85.7 to 92.0 beats/min, 12.3 to 14.6 beats/min, 0.29 to 0.42 l/min, 0.11 to 0.24 l/min, 1.95 to 2.49 kcal/min, 3.74 to 5.56 and 33.24 to 46.62, respectively.

The mean HR, •HR, mean OCR, •OCR, EER, BPDS and ODR of the subjects were found to be less for lance operation as compared to the corresponding values during pedalling operation for spraying. Spraying with an agricultural foot sprayer was rated light, considering pedalling, which is a primary operation during spraying.

Keywords: Physiological Cost, Foot Sprayer, Discomfort.

INTRODUCTION

The increased use of hybrid seeds, fertilizer and irrigation water in Indian agriculture has resulted in higher incidence of pest attacks and thus necessitates the use of pesticides. Pesticides are indispensable for getting high crop yields, protecting forests and horticultural crops, ensuring public and animal health, safeguarding agricultural commodities in stores and house hold articles, etc., by controlling and preventing the various pests involved. Their use has been progressively increasing with the growing awareness. The use of the plant protection equipment is becoming a necessity for the mankind for protection of their crops and also to kill the pests damaging the crop. In recent years there have been many development in the equipments for pesticides' application and various kinds of modern pesticide-application machines are now available. These machines have better efficiency as well as reduced labour requirement and production costs.

In India current population of foot sprayers is about 0.6 million and about 10 firms are manufacturing this sprayer (Shrimali and Mathur, 2004).

An ergonomic study was carried out on agricultural foot sprayer and found that the performance of sprayer depends to a great extent upon the operator's ease in operating its different controls and operating the machine for a larger period without fatigue (Shrimali, 2005). Heart Rate (HR), Oxygen Consumption Rate (OCR), Energy Expenditure Rate (EER), Body Part Discomfort Score (BPDS) and Overall Discomfort Rating (ODR) are the major parameters to rate the physiological cost and discomfort of the operator to operate the machine. These parameters have also been considered by various scientists for ergonomical evaluation of an agricultural machine.

MATERIAL AND METHODS

For the working of an agricultural foot sprayer, two operators are required, i.e., one for pedalling and the other for the spraying operation. Physiological cost and discomfort were found for the operation of the sprayer by male subjects. The independent parameters included operating pressure (of 600 kPa specified by Anon, 1994) and three male subjects, and dependent parameters were HR, OCR, EER, ODR, BPDS. Three male subjects representing 5th, 50th and 95th percentile of the user's population were selected to operate the sprayer. The HR and OCR were measured with the help of an ambulatory metabolic measurement system Metamax II. The heart rate and the oxygen consumption rate were measured for 2 hrs duration having 15 minutes rest and 5 minutes spraying. From the data so obtained ΔHR, and ΔOCR were then calculated. For the measurement of discomfort parameters, the operator was asked about its fatigue level after 2 h of continuous operation. The ODR was measured with the help of a 100 cm scale. Corlett and Bishop (1976) technique was used for the measurement of BPDS.

RESULTS AND DISCUSSION

Working Heart Rate (HR) and Increase in Heart Rate (•HR)

Figure 1(a) shows mean HR values of the subjects during the performance of pedalling and spraying operation. The mean HR of the subjects ranged from 98.2 to 100.7 beats/min for pedalling operation and 85.7 to 92.0 beats/min for spraying operation. The working heart rate for the pedalling operation was more than spraying operation for all the subjects. Thus pedalling operation was more strenuous than spraying operation. Further, considering pedalling operation as a primary operation, spraying with an agricultural foot sprayer may be classified under light category (Christensen, 1953).

Working Oxygen Consumption Rate and Increase in Oxygen Consumption Rate

Figures 2a and 2b show mean OCR and ΔOCR values of the subjects during the performance of pedalling and spraying operation. The mean OCR of the subjects ranged from 0.48 to 0.88 litres/min for pedalling operation and 0.30 to 0.42 litres/min for spraying operation. From the figure it was found that working oxygen consumption rate for the pedalling operation was more as compared to spraying operation for all the subjects as pedalling operation was more strenuous.

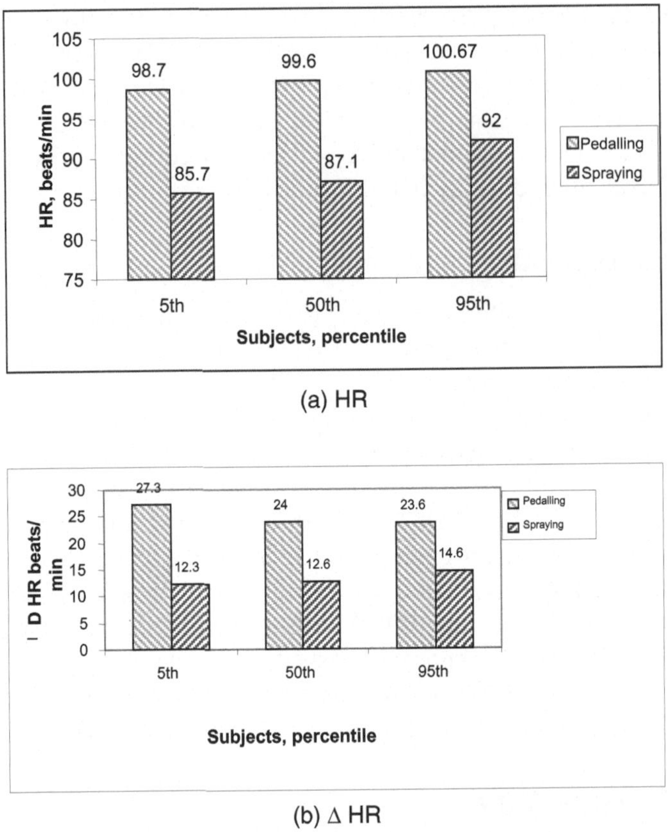

Fig. 1: Mean HR and •HR for the Operation of an Agricultural Foot Sprayer

Overall Discomfort Rating (ODR)

Figure 3 shows the mean ODR for the subjects during pedalling operation. The mean ODR for 5th, 50th and 95th percentile subjects were 7.40, 6.76 and 6.40 respectively for pedalling operation and 5.56, 4.66 and 3.74 respectively for spraying operation.

Body Part Discomfort Score (BPDS)

Figure 4 shows the mean BPDS for the selected subjects during pedalling and spraying operation. The BPDS 5th, 50th and 95th percentile subjects were 59.30, 58.60, and 57.30 respectively for pedalling and 46.62, 37.50 and 33.24 for spraying operation. The subjects felt more fatigue during pedalling operation as compared to spraying operation.

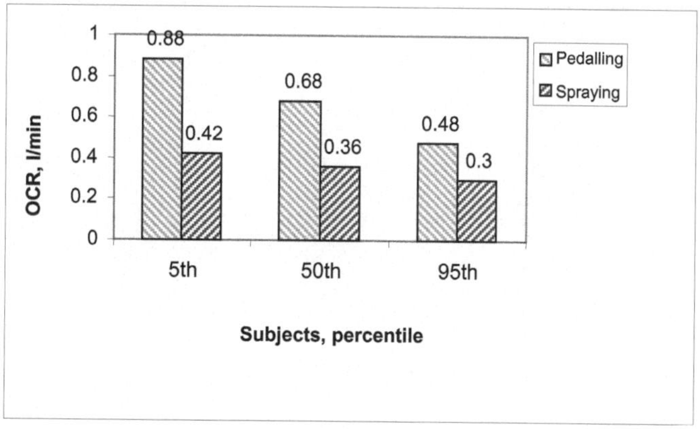

(a) OCR

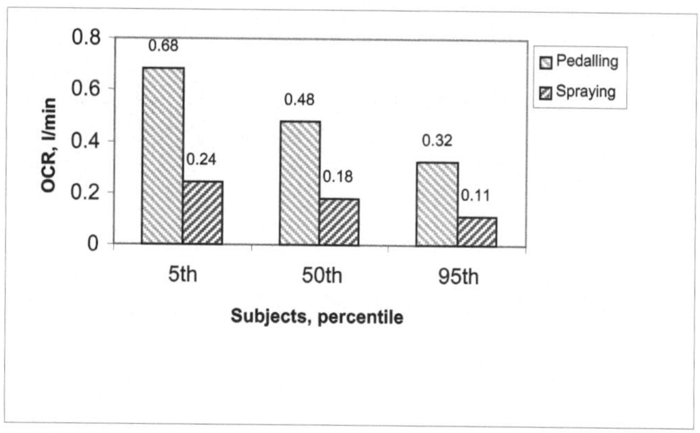

(b) Δ OCR

Fig. 2: Mean OCR and •OCR for the Operation of an Agricultural Foot Sprayer

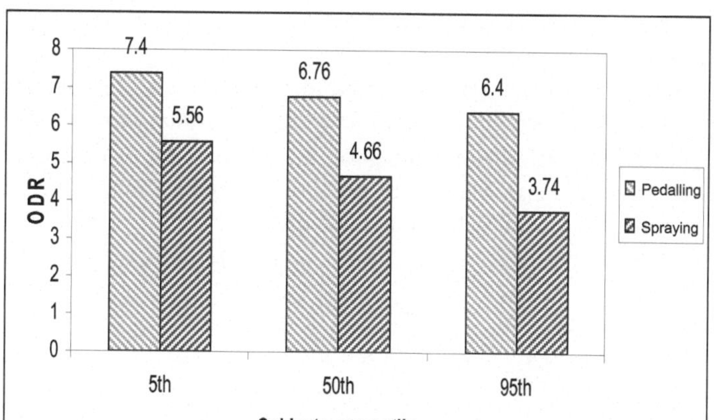

Fig. 3: Mean ODR for the Operation of an Agricultural Foot Sprayer

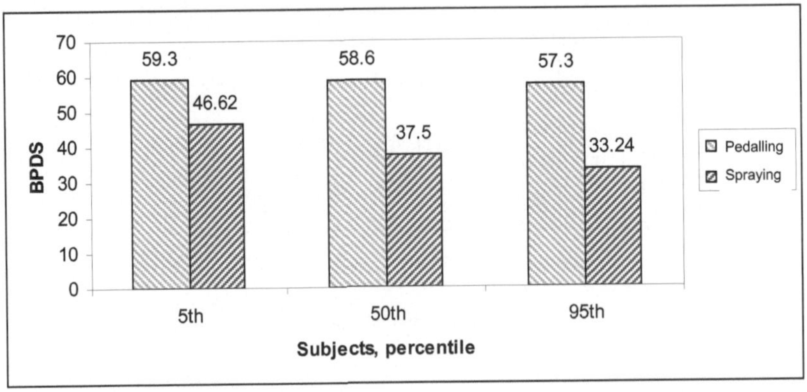

Fig. 4: Mean BPDS for the Operation of an Agricultural Foot Sprayer

CONCLUSION

The mean HR, ΔHR, OCR, ΔOCR, BPDS and ODR of subjects during spraying operation was lesser than the corresponding values during pedalling operation.

ACKNOWLEDGEMENT

The assistance received from Aspee Research and Development Foundation, Mumbai and facilities extended by CIAE, Bhopal and MPUAT, Udaipur are sincerely acknowledged.

REFERENCES

[1] Anonymous (1994). BIS 10134:1994 "Method of Testing for Manually Operated Sprayers."

[2] Christensen, E.H. (1953). "Physiological valuation of work in Nykroppa Iron works in symposium on fatigue." (eds. Floyd, W.F. and Welford, A.T.) London. H.K. Lewis.

[3] Corlett, E.N. and Bishop, R.P. (1976). "A technique for assessing postural discomfort." *Ergonomics*, 19:175–182.

[4] Shrimali, H. and Mathur, R. (2004). "Plant protection equipments for small farm mechanization." *Proceedings of National seminar on small farm mechanization held at MPUAT*, Udaipur.

[5] Shrimali, H. (2005). "Ergonomic Studies on an Agricultural Foot Sprayer", *Unpublished Ph.D. Thesis.* Deptt. of Farm Machinery and Power Engineering, CTAE, Maharana Pratap University of Agriculture and Technology, Udaipur.

Development and Ergonomical Evaluation of Wheeled Fertilizer Broadcaster

Prafull Mange, C.J.S. Pannu, S.C. Gandhi and Santokh Singh

Department of Farm Power and Machinery
Punjab Agricultural University, Ludhiana–141 004, India

ABSTRACT

The wheeled fertilizer broadcaster was developed using the wheel hand hoe and manual fertilizer broadcaster. The power transmission system was designed to provide power from ground wheel to the fertilizer broadcaster. The ground wheel of wheel hand hoe was modified to generate sufficient traction to run double disc spinners of fertilizer broadcaster. The ergonomical evaluation was carried on newly developed wheeled fertilizer broadcaster. The independent variable was subjects (S) to find their effect on the physiological responses like heat rate, postural configuration, Overall Discomfort Rating (ODR) and Body Part Discomfort Score (BPDS) in light and heavy soils. Three subjects of different heights (S1, S2 and S3) were selected for the study. At forward speed of 2.0 km/h, all physiological responses were found to be within permissible limits in both light and heavy soils. The average heart rate and corresponding work load of three subjects were 108.67, 107.67, 107.67 beats/min and 41.59, 38.82, 53.00 W in light soil and 113.00, 111.33, 111.33 beats/min and 46.51, 44.68, 58.68 W in heavy soil at forward speed of 2 km/h. The bending angles of three subjects were 3.22, 3.39 and 2.83 degree in light soil and 3.39, 3.72 and 3.17 degree in heavy soil at forward speed of 2 km/h. The overall discomfort ratings of three subjects were 3.00, 2.83 and 2.67 in light soil and 3.33, 3.17 and 3.00 in heavy soil at forward speed of 2 km/h. The body part discomfort scores of three subjects were 32.33, 31.00 and 27.33 in light soil and 34.33, 31.67 and 30.67 in heavy soil at forward speed 2 km/h. Heart rate and work load were comparable and net bending angle, overall discomfort rating and body part discomfort scores were less with wheeled fertilizer broadcaster than the manually operated fertilizer broadcaster.

Keywords: Fertilizer Broadcaster, ODR, BPDS.

INTRODUCTION

In modern farming, fertilizers are often applied in the field with broadcast spreaders. The objective is to secure a uniform lateral or transverse distribution of the fertilizer. Hand broadcasting of fertilizer is still a common practice followed by majority of the farmers. Weight of manually operated fertilizer broadcaster (12.9 kg) filled with fertilizer causes stress on neck. It is also difficult for a person to carry weight of fertilizer and to operate machine for 8 to 10 hours. The principle disadvantage of hand broadcasting is the non-uniformity of distribution causing uneven growth of crop, which ultimately results in poor yield. Griffin *et al.* (1983) developed mathematical model to simulate the discharge of granules on the rotating impeller, the motion of particles on impeller, the trajectory of air borne particles and ground impact point along line parallel to the direction of travel.

Parish (2003) studied the effect of impeller angle on uniformity pattern found and the importance of maintaining the impeller level while operating a rotary spreader. The magnitude of angular error that would cause pattern problems with a professional walk-behind turf spreader has not previously been determined. With a walk behind spreader, impeller angle is determined by handle height, thus impeller angle is dependent on how the operator holds the handle. Yildirim and Kara (2003) studied the effect of vane height on uniformity of distribution pattern in rotary fertilizer spreaders with different flow rates through an orifice onto the disc. It was studied for triple superphosphate and calcium ammonium nitrate. The most uniform distribution pattern was obtained for a vane with 35 mm height and an orifice with 35 mm for both fertilizers. Hence, there was a need to develop a mechanical device, which could broadcast fertilizer uniformly. Keeping this in view, an effort was made to develop a wheeled fertilizer broadcaster and its evaluation from ergonomical point of view.

METHODS

The development work of wheeled fertilizer broadcaster (Figure 1) was accomplished in the research hall and the study was conducted in the research farm of the department of Farm Power and Machinery at PAU Ludhiana in the year 2007. The detailed specifications of wheeled fertilizer broadcaster are given in Table 1.

Table 1: Specification of Wheeled Fertilizer Broadcaster

S. No.	*Description*	*Specifications*
1.	Overall dimension Height, mm Width, mm Length, mm	 1100 540 1500
2.	Wheel (Traction) Diameter, mm Width, mm No. of Lugs	 400 75 13
3.	Power transmission Gear (Teeth) Chains Sprocket (Teeth)	 Two (45), Two (10), One (24) & One (12) Two One (32) & Three (16)
4.	Spinners Spinner Diameter, mm Fins	 Two 200 Six
5.	Hopper Height, mm Width, mm Length, mm	 200 250 540
6.	Total weight, kg	22.4

Fig. 1: Double Spinner Wheeled Fertilizer Broadcaster

The traction wheel was designed to provide sufficient traction to operate fertilizer broadcaster (Figure 2). The traction wheel is made of two outer rings and lugs of angular shape. The thirteen lugs were welded on the outer rings. Transmission was designed to provide sufficient rotation of spinner to get a uniform distribution. Power is transmitted from ground wheel to spinners with speed ration of 1:18. Physiological responses, *viz.*, heart rate, postural configuration, overall discomfort rating and body part discomfort score were measured for ergonomic studies on wheeled fertilizer broadcaster. The detailed anthropometric data and other physiological characteristics of three subjects (S1, S2 and S3) were measured. Subjects were calibrated in the ergonomics laboratory during the working hours on treadmill and the heart rates were recorded at different speeds up to 6.0 km/h with an increment of 0.5 km/h. The relationship between heart rate and work load was obtained for each subject. The work loads of the subjects in the field experiments were calculated from the characteristics curve obtained during the calibration of subjects.

For measurement of heart rate, a polar heart rate monitor was used. Body discomfort is quantified in two ways. One is Overall Discomfort Rating (ODR) and the other is Body Part Discomfort Score (BPDS). For the assessment of ODR a 7-point scale (0 = no discomfort, 7 = extreme discomfort) was used. Body part discomfort score is the measure of localized discomfort and the technique suggested by Corlett and Bishop (1976) was used to assess the body discomfort. Postural configuration was measured in the form of net bending angle by using the flexicurve between the working and normal posture in the lumber region of the spinal cord. For this study four forward speed (1.5, 2.0, 2.5 and 3.0 km/h) and three subjects were taken and heart rate, postural configuration, overall discomfort rating and body part discomfort score were measured. The field experiments were conducted for the combination of the subjects (S) and four forward speeds (F). The suitable values of forward speeds were decided on the basis of trial run on the wheeled fertilizer broadcaster and it was used throughout the experiment in both light and heavy soils. During the trial runs, the dependent parameters like heart rate, net bending angle, overall discomfort rating and body part discomfort score were recorded at different forward speeds of wheeled fertilizer broadcaster in both light and heavy soils.

Fig. 2: Traction Wheel of Wheeled Fertilizer Broadcaster

RESULTS AND DISCUSSION

During the field experiment in light soil, when wheeled fertilizer broadcaster operated at forward speed of 2 km/h, then minimum heart rates for subjects S1, S2 and S3 were recorded as 108.67, 107.67 and 107.67 beats/min (Table 2), respectively and maximum heart rates for subjects S1, S2 and S3 were recorded as 128.67, 128.33 and 126.33 beats/min respectively at forward speed of 1 km/h. When wheeled fertilizer broadcaster operated in heavy soil at forward speed of 2 km/h, the minimum heart rates for subjects S1, S2 and S3 were recorded as 103.00, 102.33 and 99.00 beats/min, respectively and maximum heart rate for subjects S1, S2 and S3 were recorded as 132.67, 134.33 and 131.33 beats/min, respectively at forward speed of 3 km/h. It is obvious that the heart rate increases with the increase in speed of operation. The corresponding work load calculated from calibration curves also has the same trend. At forward speed of 2.0 km/h the heart rate or workload is in the moderate category.

Table 2: Average Heart Rate and Work Load of Different Subjects in Both Light and Heavy Soil

Subjects	*Forward Speed (km/h)*	*Light Soil*		*Heavy Soil*	
		Heart Rate (beats/min)	*Work Load (W)*	*Heart Rate (beats/min)*	*Work Load (W)*
S1	1.5	98.67	30.22	103.00	35.15
	2.0	108.67	41.59	113.00	46.51
	2.5	121.33	55.99	125.67	60.91
	3.0	128.67	64.32	132.67	68.86
S2	1.5	98.67	24.43	102.33	30.29
	2.0	107.67	38.82	111.33	44.68
	2.5	120.67	59.61	124.67	66.01
	3.0	128.33	71.87	134.33	81.46
S3	1.5	94.67	32.87	99.00	39.58
	2.0	107.67	53.00	111.33	58.68
	2.5	117.33	67.97	122.67	76.22
	3.0	126.33	81.90	131.33	89.64

When wheeled fertilizer broadcaster operated in light soil, the net bending angle of spine varied from 2.50° to 5.67°, 2.78° to 5.89° and 2.11° to 4.89° for subjects S1, S2 and S3 at different forward speeds (Table 3). When wheeled fertilizer broadcaster operated in heavy soil, the net bending angle of spine varied from 2.67° to 6.11°, 3.05° to 6.11° and 2.50° to 5.89° in case of subjects S1, S2 and S3 at different forward speeds. Net bending angle of spine increased as forward speed of wheeled fertilizer broadcaster increased in both light and heavy soils. At forward speed of 2 km/h, the net bending angle was comfortable for operator.

When wheeled fertilizer broadcaster operated in light soil, the overall discomfort rating varied from 2.50 to 4.50, 2.33 to 4.33 and 2.17 to 4.17 for subjects S1, S2 and S3 at different forward speeds (Table 4). When wheeled fertilizer broadcaster operated in heavy soil, the overall discomfort rating varied from 2.83 to 5.00, 2.67 to 4.83 and 2.50 to 4.67 in case of subjects S1, S2 and S3 at different forward speeds. Overall discomfort rating increased as forward speed of wheeled fertilizer broadcaster increased in both light and heavy soils. When wheeled fertilizer broadcaster operated in light soil, then body part discomfort score varied from 27.33 to 47.67, 26.00 to 45.67 and 21.33 to 41.00 for subjects S1, S2 and S3 respectively at different forward speeds (Table 5). When wheeled fertilizer broadcaster operated in heavy soil, the body part discomfort score varied from 27.67 to 54.67, 26.00 to 52.67 and 25.00 to 51.00 in case of subjects S1, S2 and S3 respectively at different forward speeds. Body part discomfort score increased as forward speed of wheeled fertilizer broadcaster increased in both light and heavy soils.

Table 3: Average Net Bending Angles of Different Subjects in Both Light and Heavy Soil

Subjects	*Forward Speed (km/h)*	*Net Bending Angle (Degree)*	
		Light Soil	*Heavy Soil*
S1	1.5	2.50	2.67
	2.0	3.22	3.39
	2.5	3.94	4.45
	3.0	5.67	6.11
S2	1.5	2.78	3.05
	2.0	3.39	3.72
	2.5	4.17	4.50
	3.0	5.89	6.11
S3	1.5	2.11	2.50
	2.0	2.83	3.17
	2.5	3.61	4.11
	3.0	4.89	5.89

Table 4: Average Overall Discomfort Ratings of Different Subjects in Both Light and Heavy Soils

Subjects	*Forward Speed (km/h)*	*ODR*	
		Light Soil	*Heavy Soil*
S1	1.5	2.50	2.83
	2.0	3.00	3.33
	2.5	3.83	4.17
	3.0	4.50	5.00
S2	1.5	2.33	2.67
	2.0	2.83	3.17
	2.5	3.67	4.00
	3.0	4.33	4.83
S3	1.5	2.17	2.50
	2.0	2.67	3.00
	2.5	3.33	3.83
	3.0	4.17	4.67

CONCLUSION

In the field experiment, the average heart rate and corresponding work load of three subjects were 108.67, 107.67, 107.67 beats/min and 41.59, 38.82, 53.00 W in light soil and 113.00, 111.33, 111.33 beats/min and 46.51, 44.68, 58.68 W in heavy soil at 2 km/h forward speed. The net bending angle of three subjects were 3.22°, 3.39° and 2.83° in light soil and 3.39°, 3.72° and 3.17° in heavy soil at forward speed of 2 km/h. The overall discomfort ratings of three subjects were 3.00, 2.83 and 2.67 in light soil and 3.33, 3.17

and 3.00 in heavy soil at forward speed of 2 km/h. The body part discomfort score of three subjects were 32.33, 31.00 and 27.33 in light soil and 34.33, 31.67 and 30.67 in heavy soil at 2 km/h forward speed. The variations caused in the heart rate, postural configuration, Overall Discomfort Rating (ODR) and Body Part Discomfort Score (BPDS) due to subjects were found to be less with wheeled fertilizer broadcaster as compared with manual fertilizer broadcaster.

Table 5: Average Body Part Discomfort Score of Different Subjects in Light and Heavy Soil

Subjects	*Forward Speed (km/h)*	*BPDS*	
		Light Soil	*Heavy Soil*
S1	1.5	27.33	27.67
	2.0	32.33	34.33
	2.5	39.33	42.33
	3.0	47.67	54.67
S2	1.5	26.00	26.00
	2.0	31.00	31.67
	2.5	38.67	40.33
	3.0	45.67	52.67
S3	1.5	21.33	25.00
	2.0	27.33	30.67
	2.5	34.00	38.33
	3.0	41.00	51.00

ACKNOWLEDGEMENT

The financial support received from the Indian Council of Agriculture Research, New Delhi for "All India Coordinated Research Project on Ergonomics and Safety in Agriculture", is thankfully acknowledged.

REFERENCES

[1] Corlett, E.N. and Bishop, R.P. (1976). "A Technique for Assessing Postural Discomfort." *Ergonomics,* 19(2): pp. 175–82.

[2] Griffin, C.L., Rifter, D.W. and Mathens, E.S. (1983). "Simulation of Rotary Spreader Distribution Pattern." *Trans of ASAE*, 26 (1), pp. 33–37.

[3] Parish, R.L. (2003). "Effect of Impeller Angle on Pattern Uniformity." *Appl Engg. in Agric,* 19(5): pp. 531–33.

[4] Yildirim, Y. and Kara, M. (2003). "Effect of Vane Height on Distribution Uniformity in Rotary Fertilizer Spreaders with Different Flow Rates." *Trans ASAE*, 19(1), pp. 19–23.

Agricultural Accidents Scenario in Rajasthan

Abhay Kumar Mehta, Hemant Shrimali and S.S. Meena

All India Coordinated Research Project on Ergonomics and Safety in Agriculture
MPUAT, Udaipur, India

ABSTRACT

So far, the accidental data in Rajasthan has not been documented and compiled statistically. The agricultural accident survey work was carried out in some selected districts of Rajasthan from the cluster of thirty two districts. The overall accident incidence rate per 1000 agricultural workers in the state was 3.73. The overall accident incidence rate per 1000 worker was 2.36.There were 52 major accidents reported in the state due to agricultural hand tools and the overall accident incidence rate was 0.16 per 1000 hand tools. Total 235 accidents were reported due to other reasons like snake bite, falling in well, hitting by animal, slip, fall, etc. The overall accident incidence rate due to other reasons was found is 1.93 per 1000 agricultural workers. Majority of accident victims were farmers (47%) followed by agricultural labourers (26.93%) and family members (26.07%). Major cause of accidents with thresher in the area concerned was due to dangling of loose clothes which was enhanced by high velocity winds blowing mainly in major portion of threshing season and lack of information about safe operational procedures of threshers.

Keywords*:* Accidents, Agriculture, Rajasthan.

INTRODUCTION

The increasing demand of food grain and oilseed is being met out by increased production not only by virtue of use of high yielding varieties, fertilizers, pesticides, insecticides and irrigation facility but mechanization also played an important role in it. Accidents are very common in various agricultural operations due to ignorance, lack of training, lack of knowledge about the operation and improper design of tools and implements. However, these accidents, at least their chances and severity can be controlled or minimized by adequate consideration and strategic applications of human factors. The database on accidents would help in having a clear picture of the situation in the state and decide the priorities, and the work plan for preventing/minimising these accidents. Continuous monitoring and analysis of accident data would help in assessing the impact of various preventive measures (Anonymous, 2002).

So far the accidental data in Rajasthan have not been documented and compiled. These accidental data will be useful in deciding the area of attention for preventing these accidents. After getting the reasons of accidents, the researchers/scientists may give due consideration of the impacts in design/development of agricultural machinery. These accident data are also useful for government policy makers to modify the safety law to reduce the accidents and save the human life. Recommendations can be made by researcher to use particular system, sign or colour of symbol to reduce the accidents. If any accidents are made due to drudgery, proper modification in agricultural machinery may also be suggested.

METHODS

The agricultural accident survey work was carried out in selected 280 villages of seven districts of Rajasthan from the cluster of thirty two districts, which were already stratified, based on different agro-climatic zones. Selection of districts and villages was based on the basis of tractor density in different districts of the state (Anonymous, 2003).

A set of questionnaire of two parts was developed for collection of accident data. The questionnaire was especially designed to meet the requirements of the investigation. The first part included collection of basic information of the village, *viz.*, population, households, farming families, net cultivated area, irrigated area, major crops grown, etc. It also included data on farm power sources and agricultural machinery available in the village.

The second part included information of accident victims including identification particulars, state of perceptual senses/abnormality before the occurrence of accident, details of the machinery involved in the accident, the type of feed chute used for feedings in case thresher/crusher involved in accident; details of the accident, *i.e.*, date, place, nature of injury sustained, period of hospitalization, absence of work due to injury, income loss, machine repair cost, etc. The type of accident and causes for the accident were also recorded.

In this survey the direct interview method and filling up the questionnaires by the interviewers were adopted as the method of data collection. The data were collected for the agricultural year 2005–06 starting from July, 2005 to June, 2006. Field investigators were given training for collection of data and filling up of questionnaire.

RESULT AND DISCUSSION

The total number of agricultural accidents reported and the total number of accident victims in selected villages during the year 2005–06 are furnished in Table 1. The table shows that the total number of accident victims in the selected villages was same as the total number of accidents *i.e.*, 453. Region wise distribution of agricultural accidents is presented in Figure 1 which shows higher occurrence rate in north eastern part of the state. The major reasons for accidents in this area were snake bite, falling in well, slip, fall etc. in addition to machinery accidents.

Highest number of agricultural accidents was reported in Alwar district followed by Dausa, Udaipur, Bikaner, Jalore, Sri Ganganagar and Jhalawar; while the highest incidence rate per 100,000 workers was reported in Alwar (1169), followed by Dausa (547), Udaipur (537), Bikaner (200), Sri Ganganagar (82), Jhalawar (44) and Jalore (25).

The overall accident incidence rate per 100,000 workers in the state of Rajasthan was 373 (Table 2). The accident incident rate for farm machinery accidents was 236 per 100,000 machinery. Major cause of accidents with thresher in the area concerned was due to dangling of loose clothes which was enhanced by high velocity winds blowing mainly in major period of threshing season and lack of information about safe operational procedures of threshers and other machinery.

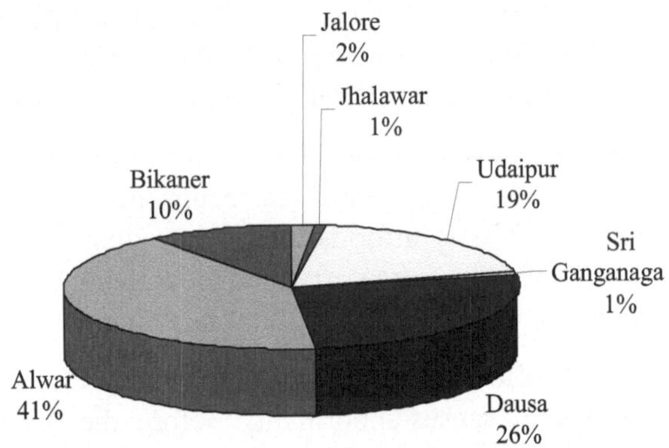

Fig. 1: Agricultural Accidents Reported in Selected Districts

Table 1: Total Numbers of Accidents in Selected 280 Villages of Seven Districts

S. No.	*District*	*No. of Accidents*	*No. of Accident Victims*
1.	Jalore	8	8
2.	Jhalawar	3	3
3.	Udaipur	86	86
4.	Sri Ganganagar	5	5
5.	Dausa	117	117
6.	Alwar	188	188
7.	Bikaner	46	46
Total		453	453

The classification of accident victims in the state according to their occupational status is presented in Table 3. The majority of accident victims were farmers (47%) followed by agricultural labourers (26.93%) and family members (26.07%) except Alwar district where maximum victims were agricultural labourers.

The age wise classification of accident victims in the selected villages is presented in Table 4. The maximum accident victims were of the age group of 30–45 years followed by 16 to 29 years and more than 45 years. Minimum victims were in the age group of below 15 years as this age group is not involved in agricultural activities.

Table 2: Agricultural Accidents Reported in Selected 280 Villages of Seven Districts

S. No.	*District*	*Population Engaged in Agricultural and Allied Activities*	*No. of Accidents*	*Accident Incident Rate/ 100,000 Workers*
1.	Jalore	32255	8	25
2.	Jhalawar	6762	3	44
3.	Udaipur	16024	86	537
4.	Sri Ganganagar	6118	5	82
5.	Dausa	21391	117	547
6.	Alwar	16079	188	1169
7.	Bikaner	22942	46	200
Total		121571	453	373

Table 3: Occupational Status Wise Distributions of Accident Victims

Status	*Farmer*	*Family Member*	*Labourer*	*Total*
Jalore	7	0	1	8
Jhalawar	2	0	1	3
Udaipur	50	16	20	86
Sriganganagar	4	0	1	5
Dausa	62	38	17	117
Alwar	60	54	74	188
Bikaner	28	10	8	46
Total	213	118	122	453

Table 4: Age Wise Classifications of Victims

Age (year)	*< 15*	*16–29*	*30–45*	*> 45*	*Total*
Jalore	0	2	3	3	8
Jhalawar	0	1	2	0	3
Udaipur	4	24	43	15	86
Sri Ganganagar	0	2	3	0	5
Dausa	24	30	44	19	117
Alwar	19	55	76	38	188
Bikaner	7	9	20	10	46
Total	54	123	191	85	453
Percent of victims	11.92	27.15	42.17	18.76	100

Source wise classification of agricultural accidents in selected villages of different districts is furnished in Table 5. It is interesting to note that more than 50 per cent accidents were not due to use of agricultural machinery and the main reason for these accidents was unawareness of farm workers, climatic condition, and absence of implement-use training. Tractor was mainly responsible for the machinery accidents followed by chaff cutter and threshers. Out of the seventy tractor accidents more than half of the accidents were reported during transportation, therefore the on road accidents of tractor are of major concern.

Table 5: Fatal and Non-fatal Accidents

S. No.	*Source of Agricultural Accidents*	*Fatal Accidents*	*Non-fatal Accidents*	*Total*
Accident Prone Agricultural Machinery				
1.	Plough	1	25	26
2.	Tractor	7	27	34
3.	Power Tiller	0	0	0
4.	Cane Crusher	0	2	2
5.	Chaff Cutter	0	35	35
6.	Thresher	0	35	35
7.	Sprayer	5	3	8
8.	Seed drill	0	10	10
Hand Tools				
	Spade & Axe	0	18	18
1.	Sickle	0	34	34
Other Accidents				
1.	Snake bite	16	58	74
2.	Electric shock	8	15	23
3.	Fall from tree	2	14	16
4.	Fall in well	7	19	26
5.	Others	6	106	112
Total Accidents		52	401	453

CONCLUSION

The agricultural accidents occurrence rate in Rajasthan was 373 accident victims per 100,000 agricultural workers and the accident rate for farm machinery related accidents was 236 per 100,000 machinery. The north-eastern part of the state experienced higher accident incident rate and the major concerns were snake bite, fall in well, slip, fall, etc.

The majority of accident victims were farmers and were of the age group of 30–45 years. Tractor accidents were mainly on roads when the tractor was used for transportation.

ACKNOWLEDGEMENT

Authors are thankful to AICRP on ESA and ICAR New Delhi for providing the help and financial support for the project.

REFERENCES

[1] Anonymous (2002). *Report of the All India Coordinated Research Project on Human Engineering and Safety in Agriculture*, CIAE, Bhopal.

[2] Anonymous (2007). *Annual Report of AICRP on Ergonomics and Safety in Agriculture.* Udaipur Centre.

Pattern of Use of Personal Protective Equipment and Measures during Application of Pesticides by Agricultural Workers

Bhoopendra Singh[1] and Mudit Kumar Gupta

Dept. of Forensic Medicine and Toxicology, Rural Medical College
Loni, Ahmednagar, India
E-mail: [1]drsinghb@gmail.com; fmt@pmtpims.org

ABSTRACT

Pesticides, despite their known toxicity, are widely used in developing countries for agricultural purposes. The objectives of this study are to find the most important defects in agricultural/farm-workers' protection, use of Personal Protective Equipment (PPE), to detect dangerous practices, and to find the habits of agricultural/farm-workers when they apply pesticides or these products.

The selected 100 agricultural workers aged between 21 and 60 years were included in a case-control survey study carried out in the Rural Medical College, Loni. The workers who had been involved in pesticide application for agricultural purpose were personally interviewed to gain information on types, treatment equipment, use of personal protection and safety measures during the application and knowledge of the risks of pesticide exposure.

Pesticides were mostly applied manually using Knapsack sprayer (70%), only 5% farmers were using tractor-mounted sprayer. Workers frequently performed tasks involving additional exposure to pesticides during mixing chemicals (66%) or washing equipment (65%). Majority of the workers/applicators used no personal protection or used it defectively/partially. Most of the workers/respondents (77%) did not bother for safety and health risks of pesticide exposure. Workers involved in pesticide application hardly used personal protection measures. Almost half of the applicators did not care for wind direction while spraying, thus working at risk of exposure. There is a clear need to develop specific training and prevention programmes for these workers.

Keywords: Personal Protective Equipment (PPE), Agricultural Worker, Pesticide Application, Protective Measure.

INTRODUCTION

In a country like India poisoning by pesticides represents an important problem. The harmful effects on human beings in the form of acute and chronic toxicity exposed to insecticides are an established fact. The incidence of poisoning by pesticide is increasing according to existing reports, and it is estimated that about 5 million people die every year as a result of intentional, accidental and occupational exposure worldwide[1].

Furthermore, chronic exposure has been linked to other health effects, such as poly-neuropathy, dermatitis, behavioural changes and cancer as stated in number of publications.

The accidental chronic poisoning due to failure in adopting proper preventive measures during spraying/application is known but very few studies have been carried out. Large sections of farmers exposed to the insecticides are illiterate and not aware of the proper toxic effects and preventive measures. Absorption of insecticides through the skin, respiratory passages or by oral route may result in acute toxicity. Long-term exposure to the sub-toxic doses results in harmful effects on various systems, *viz.*, Neurotoxicity, neuro-endocrinotoxicity, carcinogenicity, etc. The proposed study was planned to find out the pattern of uses of the preventive measures for the safe use of insecticides.

MATERIALS AND METHODS

Methods of application of pesticides by the agricultural workers were studied from 100 houses of the selected villages of Loni and Pravaranagar of Ahmednagar district. The agricultural workers who had been involved in pesticide application for agricultural purpose were interviewed face-to-face to gain information on the patterns of hardware used for spraying of insecticide, prevalent storage practice adopted by the user, type of equipment for handling of the chemicals, and the extent to which safety norms being followed by the users during the application/treatments and knowledge of the risks of pesticide exposure. Pre-designed Standard Proforma was used to gather information. The data was analysed by using SPSS-10 software.

RESULT AND DISCUSSION

The results of survey to study of patterns of hardware used for spraying of pesticide, prevalent storage practices adopted by users and type of pesticide application equipment used for pesticide application based on views of pesticide applicator are presented and discussed here.

Protective Gears Used

Majority of pesticide applicators were not using recommended protective gears. Only 18% of respondents were in complete clothing as recommended while handling pesticide. Forty per cent pesticides applicators were taking bath and washing their cloth every day after spraying. Majority (93%) of respondents were not using goggles and some of them (33%) were not using gloves during spraying. People of this area were conscious about inhalation exposure and 81% of respondents were using cloth on face and 5% of respondents were using dust mask during mixing/application of pesticide. Majority (35%) were doing spray barefooted, 33% applicators were using slippers and only 32% applicators were wearing shoes. Nobody was wearing gumboots (Table 1). People were not using recommended protective gears either due to discomfort faced from protective gears or ignorance. Similar findings were reported by Dogra *et al.* (1998).

Table 1: Distribution of Subject in Respect of Protective Gears

Protective Gears		*No. of Participants at Loni*
Clothing	Complete Partial	23 77
Ablution	Cloth changing Bathing & washing cloth	60 40
Gloves	User Non user	67 33
Goggles	User Non user	07 93
Nose protection	Gas mask Dust mask Cloth on face None	00 05 81 14
Foot wear	Gum boot Shoes Slipper None	00 32 33 35

Safety Measures

Out of 100 respondents only 59 had read the instruction given by pesticide safety manual and only 23% followed the safety manual. Majority of farmers did not bother for the safety. It was observed that considerable proportions of respondents (20%) were smoking, chewing tobacco or consuming other items during application, which was not acceptable in any case. Fewer respondents (18%) were working alone while handling highly toxic pesticide. The findings of this study differ with the findings of Dogra *et al.* (1998) in that respondents those were smoking, chewing tobacco or consuming other items during application were higher in this study.

Many pesticide applicators (34%) did not follow right direction with respect to wind direction, which may increase exposure with pesticide and affect application efficiency adversely. The number of applicators who had not followed right direction with respect to wind direction were more in the study of Dogra *et al.* as compared to present study[(2)]. About 50% of respondents were checking their equipment for working/leakage before use. To check the leakage of equipment most of farmers were using soap, rubber tube, cloth, etc., in exceptional cases they were changing connecting pipes and most of the respondent were avoiding regular servicing of equipment.

Storage of Pesticides

Farmers of these villages generally avoid storage of pesticide due to its easy availability and risk of health hazards if not stored properly. Fifty per cent of respondents had kept pesticide in original container at field and other respondents stored pesticides at home, which is not recommendable from safety point of view.

Method of Application

Seventeen per cent of respondents were using knapsack sprayer due to its easy availability and suitability for spraying in their crops. However, some farmers (5%) were using tractor-mounted sprayer and 34% were using manual sprayer. Majority (44%) of farmers were using some other method for spraying the pesticide. When we compared these findings with findings of Dogra *et al.* (1998), it was observed that farmers of this area were more ignorant about the health hazards due to wrong practice of pesticide application since 44% of farmers were using other methods (manual) for spraying the pesticide.

Pesticide Container Disposal Practices

The survey included questions mostly related to management and disposal of left over pesticide and empty (used) pesticide container. Proper rinsing is necessary so that it may not contaminate surrounding atmosphere and ground water. Most of pesticide applicators (45%) sold the empty containers after rinsing and 34% disposed without proper rinsing/ washing which was not acceptable. Only 12% of respondents used empty container for domestic purpose after well rinse, which was also not healthy practice. Remaining respondents threw empty container in land which is not acceptable.

Exposure to Pesticide

This part of survey included questions related to make up of participants and their involvement in pesticide application. It is difficult to estimate correct duration of pesticide exposure for each pesticide applicator because most of them are illiterate and they do not maintain record of such practices. A thumb rule was adopted to estimate relative exposure to pesticide. Number of years of pesticide exposure and number of sprays done by each pesticide applicator per year were recorded.

It is evident from analysis that majority (55%) of pesticide applicators were of 21–40 years age (Table 2). Out of these 100 respondents, 75 were spraying pesticide 10–30 times per year. Major proportion of respondents (82%) was exposed to pesticide for 6–15 years. Therefore it may be concluded that people of age group of 21–40 years were relatively highly exposed to pesticides.

During the handling of pesticides, the dermal route is considered as the major route of exposure [(3)]. Several methods are used for monitoring dermal exposure, such as washing or wiping, pseudo-skin methods, biological monitoring and video imaging. The 'whole body' technique is a method to determine the potential toxicity and actual exposure by analysing the active substance in the clothing, preferably work clothing and underclothing, as well as inhalation monitoring and biological monitoring. In this study, the effectiveness of use of PPE was estimated by using the whole body technique.

Table 2: Age Wise Distribution of Participants

Age Groups	*No. of Participants (%)*
21–30	25 (25%)
31–40	30 (30%)
41–50	20 (20%)
51–60	13 (13%)
> 60	12 (12%)

Most of studies showed that the use of personal protective equipment, changes in application procedures, packaging, mixing, and biological monitoring reduced the pesticide exposure under controlled conditions. Cholinesterase monitoring and urinary metabolites can identify workers with a higher risk of overexposure. Most studies suggested that the interventions should be examined for their ability to reduce pesticide overexposure in actual working populations.

Lander and Lings (1991) found that Cholinesterase Enzyme (ChE) activity decreased as weekly spraying time increased among greenhouse applicators. They found that gloves were helpful in preventing ChE inhibition. Putnam *et al.* (1983) reported reduced pesticide exposure with use of rubber gloves but concluded that gloves did not completely eliminate exposure while working with vegetable crops treated with nitrofen.

Some studies showed that although most of the farmers were aware of the importance of use of protective measures while applying pesticides, there was no significant positive relationship between awareness and use of protective measures. The main reason for not using protective measures was discomfort. These findings are similar to the findings of the study.

CONCLUSIONS

The following conclusions can be drawn from the study:

- The uses of organophosphate group of pesticides were very common by farmers in this area.
- Among the Organochlorine group of pesticide, the farmers in this area more commonly use the Endosulfon.
- Majorities of pesticide applicators were not using protective gears.
- Considerable/significant numbers of the respondents were not following recommended safety measures.
- Almost half of the applicators were not following right direction with respect to wind direction thus increasing the risk of exposure.

SUGGESTIONS

Since this study was a random study with the objective to examine the present status of preventive measures adopted by pesticide applicators in agricultural sector, it is difficult to arrive at specific conclusion. But it is clear from the findings of the present study that there is a critical lack of awareness regarding pesticide poisoning and its health hazard among pesticide applicators/farmers. It was also observed during study that symptoms of toxicity were observed in pesticide applicators.

Following factors were contributing to risk of pesticide toxicity:

- Lack of suitable protective clothing for tropical climate.
- Poor knowledge and understanding of safe practices in pesticides usage.
- Farmers using pesticide in concentration in excess of requirement.
- Poor maintenance facilities for spray equipment thereby giving rise to hazardous contamination.
- Use of pesticide mixture in some cases.
- Therefore, following steps are recommended to reduce pesticide poisoning up to some extent.
- Manufacturers of pesticide/pesticide application equipment must organise training programmes for all persons related with pesticides (production, formulation, storage, transportation, application, monitoring, etc.) through professionally trained people preferably Agricultural Engineers.
- Proper monitoring of pesticide consumption at state and national level so that its misuse can be restricted.
- There should be facilities of periodical health check-up for pesticide applicators at least at Taluka/district level so that toxic hazards/chronic poisoning could be detected at initial stage and precautionary measures can be adopted.
- There should be service/monitoring centres for spraying equipment at least at Taluka/District level so that faulty equipment can be repaired/replaced before it creates health hazard to human being.
- Further study should be conducted for suitable protective clothing/preventive gears subject to terms and conditions of farmers of different parts of country.
- There is need to carry out study to evaluate the risk of toxicity due to pesticides using control group.

ACKNOWLEDGEMENT

We are thankful to ICMR, New Delhi for their financial support under STA research scheme. Further we are also thankful to the management, Prof. & Head Dr. R.S. Choudhari and Principal Dr. K.S. Soodan for their co-operation at various levels.

REFERENCES

[1] Dogra, T.D., Bhardwaj, D.N., Singh, Bhoopendra and Chauhan, N.S. (1998). "Study of Deficiencies in Preventive Measures and Resulting Health Hazards Among Agricultural Workers using Pesticides." *Final Report of Research Project of Indian Council of Agricultural Research*, New Delhi.

[2] Fenske, R.A. (1993). "Dermal Exposure Assessment Techniques." *Ann Occup Hyg*, 37: pp. 687–706.

[3] Keifer, M.C. (2000). "Effectiveness of Interventions in Reducing Pesticide Overexposure and Poisonings." *Am J. Prev Med.*, 18(4 Suppl), pp. 80–9.

[4] Lander, F. and Lings, S. (1991). "Variation in Plasma Cholinesterase Activity among Greenhouse Workers, Fruit Growers, and Slaughter Men." *Brit J Indus Med,* 48, pp. 164–66.

[5] Maroni, M., Colosio, C., Ferioli, A. and Fait, A. (2000). "Biological Monitoring of Pesticide Exposure: a review." *Toxicology*, 143(1), pp. 1–118.

[6] Ohayo-Mitoko, G.J., Kromhout, H., Karumba, P.N. and Boleij, J.S. "Identification of Determinants of Pesticide Exposure among Kenyan Agricultural Workers using Empirical Modelling." *Ann Occup Hyg.,* 43(8), pp. 519–25.

[7] Putnam, A.R., Willis, M.D., Binning, L.K. and Boldt, P.F. (1983). "Exposure of Pesticide Applicators to Nitrofen: Influence of Formulation, Handling Systems and Protective Garments." *J. Agri. Food Chem*, 31, pp. 645–650.

[8] Report of the WPRO Regional Workshop on Epidemiology of Poisoning by Pesticides held in Singapore (December 1999).

[9] Sivayoganathan, C., Gnanachandran, S., Lewis, J. and Fernando, M. (1995). "Protective Measures use and Symptoms among Agro Pesticide Applicators in Sri Lanka." Soc. Sci Med., 40 (4), pp. 431–436.

[10] Vander, Jagt K., Tielemons, E., Links, I., Brouwer, D. and Van Hemmen, J. (2004). "Effectiveness of Personal Protective Equipments: Relevance of Dermal and Inhalation Exposure to Chlorpyrifos among Pest Control Operators." *J. Occup Environ Hyg*, 6, pp. 355–362.

[11] Wolfe, H.R., Durham, W.F. and Armstrong, J.F. (1967). "Exposure of Workers to Pesticides." *Arch Environs Hlth*, 14, pp. 662–633.

Vibration Characteristics of Self Propelled Power Weeders

K. Kathirvel, S. Thambidurai, D. Ramesh and D. Manohar Jesudas

Department of Farm Machinery, Agricultural Engineering College and Research Institute
Tamil Nadu Agricultural University, Coimbatore–641 003, India

ABSTRACT

The power weeder is a multi-purpose hand tractor designed primarily for intercultural operation. The operator's environment includes all the factors in the surroundings, which have an effect on man-machine system. Among these factors, mechanical vibration is most important as it significantly accelerates fatigue and affects sensitivity and reaction rates of the operator. This paper deals with Hand Transmitted Vibration (HTV) of selected power weeders (TNAU-Varun power weeder, Balram power weeder and Oleo power weeder) during weeding operation in cotton crop at 1.5, 1.8 and 2.1 km/h forward speeds of operation. The HTV was measured and analysed as per the guidelines of International Standards ISO 5349 (1986). Vibration level was measured using the portable four-channel multi-analyzer system (Bruel and Kjaer Type 3560 C). The increase in engine speed from 1600 to 3200 rpm resulted in three fold increase in peak acceleration at handle of all the three power weeders. Vibration at the top of the engine was the highest, followed by handle, gearbox, chassis and root of handle in all the three selected power weeders. The operator's exposure time to HTV was limited to 30 to 60 min with increase in speed from 1.5 and 1.8 km/h, further increase in speed to 2.1 km/h reduced the limit to less than 30 min for all the three power weeders. Provision of vibration isolators resulted in the reduction of peak values of r.m.s acceleration by 21.9 to 25.5% for TNAU-Varun power weeder, 15.5 to 21.1% for Oleo power weeder and 8.2 to 14.8% for Balram power weeder respectively with the increase in forward speed from 1.5 to 2.1 km/h. Provision of vibration isolators increased the exposure limit to 1–2 h for TNAU Varun and Oleo power weeders at 1.5 and 1.8 km/h forward speeds. With the increase in forward speed from 1.5 to 2.1 km/h, the exposure limit to noise level was reduced from 7.0 to 5.7 h for TNAU-Varun power weeder, 5.9 to 4.6 h for Oleo power weeder and 4.1 to 3.4 h for Balram power weeder.

Keywords: Power Weeder, Hand Transmitted Vibration.

INTRODUCTION

Weed control is one of the most expensive operations in crop production. Hand weeding requires more labour and consumes more time leading to higher cost of weeding. Hence inter-cultivation of weeds using mechanical power is a viable and economical solution to this problem. Though tractor has been introduced in our country for over four decades, till now, there is no visible change in the scenario of inter cultivation practices in the mechanised farm. The reasons are many, among which the initial cost of the tractor, the traditional cropping pattern and layout are some of them. Power weeders have been developed for the mechanical control of weeds. The ergonomics aspects of power weeder are of great importance as working with power weeder involves considerable physical strain to the operator. Hand-transmitted vibration of a power weeder is very high since the handle grip of a walking tractor is a cantilever beam, and the power is obtained from a single cylinder diesel engine (Ying *et al.*, 1998). Daily exposure to hand-arm vibrations over a number of years can cause permanent physical damage known as 'white finger syndrome' or it can damage the joints and muscles of the wrist and the elbow. White finger syndrome in its advanced stages is characterised by blanching of the extremities of the fingers, which is caused by damage to the arteries and nerves in the soft tissues of the hand. Measurement and evaluation of vibrations are necessary for assessing operator comfort and to suggest limits for the continuous operation of power weeders. Thus, this paper deals with the measurement of machine vibration and HTV at handle grip level at different forward speeds during weeding operations.

Handle vibration in hand operated tilling machines is mainly due to the reciprocating motion of the main moving parts. Qunying *et al.* (1989) conducted that the major excitations of the hand transmitted vibration of a walking tractor are the unbalanced inertia force of the engine and the unevenness of road surface. They also reported that the hand transmitted vibration caused by the unevenness of road makes up about 20 per cent of the total transmitted vibration of the walking tractor (Qunying *et al.* 1993).

MATERIALS AND METHODS

The types of power weeders selected for the investigations were TNAU-Varun power weeder (PW1), Balram power weeder (PW2) and Oleo Power Weeder (PW3). The TNAU-Varun power weeder (PW1) consists of 7.5 kW diesel engine, power transmission housing, ground wheels, rotary knives, handle and clutch. From the engine, the power is transmitted to the transmission house and then to the ground wheels and rotary knives. The knives when rotating enable weeding and mulching the soil. Width of the rotary is 550 mm and depth of weeding can be adjusted. The Balram power weeder consists of a 4 kW petrol start kerosene run engine, power transmission housing, rotary knives, handle and clutch. There is provision to adjust ground wheel spacing according to crop row spacing. A clutch with lever provided at the handle actuates the simple idler pulley and disengage the power transmission from the engine to rotary tynes. These knives when rotating enable weeding and mulching the soil. Width of the rotary is 350 mm and depth of weeding can be adjusted. The Oleo power weeder is compact and light weight machine (4.2 kW petrol engine) ideal for working with ease even in confined spaces. A front wheel with vertical tip-up mechanism is provided for easy transport and can be lifted to facilitate tillage work. Two-side protection discs ensure that crops remain undamaged while also maximising operator safety. The transmission guard is reinforced to withstand projected stones and accidental impact.

Instrument for Measuring Vibration and Noise

The machine vibration and human vibration of the power weeders were measured and analysed using the portable PULSE multi-analyser system (Brüel and Kjaer type 3560 C).

The transducer used for vibration measurements is the piezoelectric accelerometer. It exhibits better all round characteristics and stability than any other type of vibration transducer and its response is linear through the frequency range of interest in human vibration measurements. The PULSE multi-analyser system is a versatile, task oriented analysis system for vibration and noise analysis. It provides the platform for a range of PC based measurement solutions. Type 3650C is a portable system powered by internal batteries or an external DC supply. The base software for a PULSE system is vibration and noise analysis type 7700.

Measurement of Hand Transmitted Vibration

The power weeder was put in proper test condition before conducting the experiments, that is, in full working order with full fuel tank and radiator, without optional front weights, tire ballast and any specialised components. Tyres used for the tests were of standard size and depth of treads was not less than 70 per cent of the depth of a new tread. Pneumatic wheels with recommended tyre pressure were used during weeding operations. There were no known mechanical defects that would result in abnormal vibration in power weeders. The vibration from the handle of the power weeder is transmitted to the hand and arm of the operator through the palm of his hand. The hand transmitted vibration was measured at handle-grip level as per the guidelines issued in ISO 5349 (1986). The transducer employed was a piezoelectric accelerometer (B&K, Type 4392) mounted on a hand adapter to insert between the fingers and the grip (Ying *et al.*, 1998 and Ragni *et al.*, 1999) and fixed on the grip by tape. The transducer is inserted between the middle and index fingers of left hand of each subject since the force output from index and middle finger is larger than that from ring and little finger (Fransson and Winkel, 1991). The orientation of the measurement axes of the accelerometers is according to ISO 5349 (1986). The Z-axis is directed along the second metacarpus bone of the hand, X-axis perpendicular to the Z-axis (both these axes are normal to the longitudinal axis of the grip) and Y-axis, parallel to the longitudinal axis of the grip.

The depth of operation was maintained at constant level of about 10 cm during rototilling. The subjects were instructed to hold the handle grip with a light and constant compression force. The hand transmitted vibration was measured for power weeders during the weeding operation in cotton field. Measurements were carried out at selected levels of forward speed of power weeder for weeding operations. The speedometer was monitored continuously to ensure constant forward speed selected during each trial. Measurements were made at different forward speeds, viz., 1.5 km h^{-1}, 1.8 km h^{-1} and 2.1 km h^{-1} during field trials. The PULSE programme was activated after the power weeder was started for the operation and the measurement was recorded with an acquisition period of 60 seconds (Ying *et al.*, 1998). Each trial was repeated five times for all operating conditions. The same procedure was repeated for all the selected subjects.

RESULTS AND DISCUSSION

The vibration levels of TNAU Varun power weeder, Balram and Oleo power weeders were evaluated in terms of machine vibration in stationary mode and hand transmitted vibration in weeding operation mode. Machine vibrations at different locations of power weeder in stationary condition as well as hand transmitted vibration under operating conditions in field were measured and recorded.

Machine Vibrations in Stationary Mode

The peak acceleration values arrived from the vibration spectrum for the selected power weeders at different engine speeds are furnished in Table 1. It is observed that as the engine speed increased, the peak acceleration also increased at different locations for all the three power weeders. Since the major vibration contribution was the power stroke of the engine, as the engine speed increased more power strokes are completed per second and the different components of power weeder vibrated frequently and resulted in higher values of acceleration. The increase in engine speed from 1600 to 3200 rpm resulted in three fold increase in peak acceleration at handle of TNAU-Varun and two fold increases for Oleo and Balram power weeders. Comparing the acceleration at different locations of power weeder, it is found that the vibration at the top of the engine was the highest followed by handle, gear box, chassis and root of handle. The vibration at the top of the engine was the highest since the major excitation of the vibration of the power weeder is the unbalanced inertia force of the engine (Qunying *et al.*, 1989; Dong, 1996 and Ying *et al.*, 1998). Among the three power weeders, the vibration induced at selected locations was maximum for Balram power weeder followed by Oleo and TNAU-Varun power weeders. This might be due to the damping effect of rubber tyre in TNAU-Varun power which is absent in Balram and Oleo power weeders.

Table 1: Machine Vibrations of Power Weeders in Stationary Mode

Engine Speed, rpm	Peak Acceleration, m/s^2				
	Engine Top	Chassis	Gear Box	Root of Handle Bar	Handle
TNAU-Varun power weeder					
1600	7.45	4.30	3.80	2.10	6.21
1900	9.79	6.12	4.11	4.22	9.65
2200	11.45	7.56	5.20	5.41	11.10
2500	13.02	9.54	6.78	6.54	12.10
2700	15.41	11.78	8.12	7.89	15.25
3000	18.11	14.23	9.51	9.22	17.45
3200	20.18	16.25	11.23	10.21	19.23
Oleo power weeder					
1600	13.75	5.30	–	2.14	10.01
1900	15.21	6.89	–	3.47	13.41
2200	18.54	7.76	–	5.26	16.12
2500	19.45	8.21	–	7.81	17.45
2700	20.15	10.11	–	11.58	19.48
3000	21.32	11.48	–	13.45	21.11
3200	22.47	13.21	–	13.97	22.65
Balram power weeder					
1600	16.93	8.65	3.60	3.20	12.40
1900	18.71	9.97	5.47	5.34	16.53
2200	21.53	10.89	7.99	7.02	19.63
2500	22.84	12.05	11.75	10.36	20.61
2700	23.38	13.89	14.71	13.94	22.10
3000	24.80	15.21	17.51	15.32	23.47
3300	26.13	16.30	19.81	16.40	24.40
3600	27.10	17.40	21.50	17.50	25.10

Vibration Characteristics in Weeding Operation

The peak value of acceleration arrived from the vibration spectrum for the selected power weeders at selected levels of forward speed for first and second weeding operations are furnished in Table 2.

Table 2: Machine Vibration of Power Weeders during Weeding

Forward Speed (km/h)	*Peak Acceleration on Handle, m/s²*	
	Weeding at 30 days	*Weeding at 60 days*
TNAU -Varun power weeder		
1.5	5.48	4.10
1.8	6.11	6.50
2.1	7.59	7.98
Oleo power weeder		
1.5	6.09	6.24
1.8	7.44	6.97
2.1	9.79	7.51
Balram power weeder		
1.5	10.43	10.45
1.8	12.25	12.35
2.1	14.63	15.31

It is observed that the magnitude of acceleration on handle increased with increase in selected levels of forward speed for first and second weeding operations. Comparing the handle vibration during first and second weeding, it is evident that peak acceleration on the handle was higher in first weeding than second weeding. Since the field was relatively compact during first weeding than second weeding, the damping effect was less in weeded field. Comparing handle vibrations of selected three-power weeders during weeding, it is seen that acceleration values were higher for Balram power weeder than Oleo and TNAU-Varun power weeder for weeding on 30 and 60 days.

Hand Transmitted Vibration (HTV)

The HTV of the selected power weeders was assessed in terms of safe exposure limit and latent period or duration of exposure necessary before the onset of vascular symptoms, characterised by finger blanching for 10 percentiles of an exposed population. As per the guidelines in ISO 5349 (1986), hand transmitted vibration should be evaluated in the direction in which the most serious vibration happened. Hence in this investigation it was evaluated in x direction (Ikeda *et al.*, 1998). The mean values of rms weighted acceleration (HTV) for all the subjects at selected levels of forward speed for the three power weeders during first and weeding are furnished in Table 3.

Table 3: Hand Transmitted Vibration of Power Weeders during Weeding

Forward Speed (km/h)	*Mean Value of Root Mean Square Acceleration (m/s²)*		*Exposure Time (h)*		*Probability of White Finger Syndrome (years)*	
	Weeding at 30 days	*Weeding at 60 days*	*Weeding at 30 days*	*Weeding at 60 days*	*Weeding at 30 days*	*Weeding at 60 days*
TNAU -Varun power weeder						
1.5	3.52	3.38	1/2 – 1	1/2 – 1	6.08	6.25
1.8	3.92	3.77	1/2 – 1	1/2 – 1	5.60	5.77
2.1	4.99	4.51	< 1/2	< 1/2	4.28	4.74
Oleo power weeder						
1.5	3.65	3.52	1/2 – 1	1/2 – 1	5.92	6.08
1.8	4.25	4.15	1/2 – 1	1/2 – 1	4.94	5.06
2.1	5.07	4.63	< 1/2	< 1/2	3.94	4.48
Balram power weeder						
1.5	3.78	3.58	1/2 – 1	1/2 – 1	5.76	6.01
1.8	4.72	4.36	1/2 – 1	1/2 – 1	4.46	5.05
2.1	5.26	4.99	< 1/2	< 1/2	3.80	4.28

Vibration Isolators

From the predicted exposure limit of the subjects for the measured values of HTV, it is noticed that the safe exposure time limit for hand transmitted vibration of the selected power weeders was of serious concern restricting the safe exposure to <½ – 1–2 h in weeding. On an average the power weeders were used for more than 5 h per day. If the subjects were exposed to 5 h at this level of HTV, the prevalence of vibration induced white finger (VWF) and numbness of the hands will start very earlier. Hence, in order to increase the exposure time, the HTV of power weeder needs to be reduced. The vibration at the handle can be reduced by providing vibration isolators.

Vibration Isolators for Engine

In power weeders, the major source of vibration is the unbalanced inertia force of the engine. The vibration from the engine is transmitted to the handle through the chassis, gear box and handle bar. Four numbers of vibration isolators (isomer made of Steryl butadiene rubber) were mounted between the engine and the chassis.

The probability of occurrence of white finger syndrome for 10 percentiles of an exposed population was reduced by 14.5 to 18.7 per cent for TNAU-Varun power weeder, 9.5 to 24.2 per cent for Oleo power weeder and 3.2 to 21.0 per cent for Balram power weeder respectively with the increase in forward speed from 1.5 to 2.1 km/h after incorporating the vibration isolators.

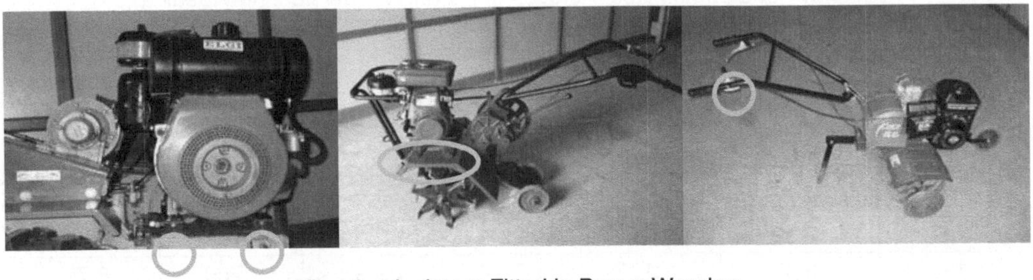

Vibration Isolators Fitted in Power Weeders

Table 4: Hand Transmitted Vibration during Weeding with Vibration Isolators

Forward Speed (km/h)	*Root Mean Square Acceleration (m/s²)*	*Exposure Time (h)*	*Probability of White Finger Syndrome (years)*
TNAU -Varun power weeder			
1.5	2.64 (3.38)	1–2 (½–1)	7.31 (6.25)
1.8	2.81 (3.77)	1–2 (½–1)	6.81 (5.77)
2.1	3.56 (4.51)	½ –1 (<½)	5.83 (4.74)
Oleo power weeder			
1.5	2.89 (3.42)	1–2 (½–1)	6.72 (6.08)
1.8	3.16 (4.15)	1–2 (½–1)	6.03 (5.06)
2.1	3.47 (4.63)	½ –1 (<½)	5.91 (4.48)
Balram power weeder			
1.5	3.23 (3.52)	½ – 1 (½–1)	6.21 (6.01)
1.8	3.74 (4.36)	½ –1 (½–1)	5.63 (5.05)
2.1	4.25 (4.99)	½ –1 (<½)	5.42 (4.28)

(The figures in parentheses are the values of peak acceleration without isolators)

CONCLUSIONS

The increase in engine speed from 1600 to 3200 rpm resulted in three-fold increase in peak acceleration at handle of TNAU-Varun and two-fold increase for Oleo and Balram power weeders. Vibration at the top of the engine was the highest and followed by handle, gear box, chassis and root of handle in all the three selected power weeders. Among the three power weeders, the vibration induced at selected locations was maximum for Balram power weeder followed by Oleo and TNAU-Varun power weeders. The exposure time to HTV was limited to ½ to 1 h with increase in speed from of 1.5 and 1.8 km/h, further increase in speed to 2.1 km/h reduced the limit to < ½ h respectively for all the three power weeders. Provision of vibration isolators increased the exposure limit to 1–2 h for TNAU-Varun and Oleo power weeders at 1.5 and 1.8 km/h of forward speeds.

ACKNOWLEDGEMENT

The financial and technical support provided under All India Coordinated Research Project on Ergonomics and Safety in Agriculture (ICAR) and TNAU is acknowledged.

REFERENCES

[1] Dong, M.D. (1996). "Testing Analysis and Evaluation of Vibration Transmitted by Handles of GN-5 Walking Tractor." *Journal of Zhejiang Agricultural University*, 22(1), pp. 68–72.

[2] Fransson, C. and Winkel, J. (1991). "Hand Strength: the Influence of Grip Span and Grip Type." *Ergonomics,* 34(7), pp. 881–882.

[3] Jiao, Qunying, Wang, Qianhua, Chen, Kuifu and Zhao, Daxing (1993). "The Excitations and Characteristics of the Vibration of Walking Tractors Handle using Isolators." *Transactions of the Chinese Society of Agricultural Machinery,* 24(4), pp. 74–79.

[4] Majumder, B. (1994). "Vibration and Noise Characteristics of a Thai-made Power Tiller." *Unpublished ME Thesis*, Asian Institute of Technology, Bangkok, Thailand.

[5] Ragni, L., Vassalini, G., Xu, F. and Zhang, L.B. (1999). "Vibration and Noise of Small Implements for Soil Tillage." *Journal of Agricultural Engineering Research,* 74, pp. 403–409.

[6] Ying, Y., Zhang, L., Xu, F. and Dong, M. (1998). "Vibratory Characteristics and Hand Transmitted Vibration Reduction of Walking Tractor." *Transactions of the ASAE,* 41(4), pp. 917–922.

Magnitude, Severity and Causes of Agricultural Accidents in Arunachal Pradesh—A Case Study

K.N. Dewangan, D.U. Gorate, C. Owary and G. Gogoi

Department of Agricultural Engineering
North Eastern Regional Institute of Science and Technology
Nirjuli–791 109, Arunachal Pradesh, India

ABSTRACT

An investigation was conducted to study the magnitude, severity and causes of agricultural accidents in one of the states of north-eastern India, Arunachal Pradesh. Survey was conducted to collect data from 160 villages of 4 districts. Survey team met the victims personally. The study revealed that the agricultural accidents incident rate in Arunachal Pradesh was 5.58/1000 workers/year and it was higher in districts, where jhoom cultivation was extensively followed. About 72.7% of the total agricultural accidents occurred when the agricultural workers were using various hand tools, while rest of 27.3% accident occurred due to other reasons. Incident rate due to hand tools was higher where hill agriculture was practised to a greater extent. The agricultural accident incident rate not involving hand tools and implements was found to be 1.55/1000 workers/year. Incident rate of other accidents was higher in areas where wetland rice cultivation was practised. Among all the activities, jungle clearing accounted for about 26.47% of the total accidents. Jungle clearing operation and fencing operation to some extent used dao and they accounted for 35.6% of the total accidents. About 95.3% of the agricultural accidents were non-fatal. About 1.18% of the accidents were minor and did not result in any injury, 56.47% of the accidents were moderate, 24.71% were serious, 3.53% were severe and 1.18% was maximum type. Most of accidents resulted in cut (45.88%) followed by snake/honey bee/spider/insect bite (24.71%).

INTRODUCTION

Accidents occur due to culmination of various factors, *viz.*, man, equipment, crop, environment etc. The leading causes of fatality in crop production is farm equipment based on the analysis of data of the All India Coordinated Research Project on Ergonomics and Safety in Agriculture (AICRP on ESA), Nag and Nag (2004) reported that most of the fatal accidents in agricultural sector resulted from the use of powered machinery, with the annual fatality rate of 22 per 100,000 farmers. Tiwari *et al.* (2002) reported that the overall incidence rate was 1.25/1000 workers/year from 1995 to 1999 in central India. About 9.2% of the incidents were fatal, and most of the fatal incidents were due to tractors and snakebites (42.9% each). About 77.6% of all incidents were due to farm machinery, 11.8% were due to hand tools, and the remaining 10.6% were due to other sources like snakes, wells, etc.

Arunachal Pradesh is distinct from the rest of India, where most of the farm operations are carried out manually using tools and equipment made locally. Use of mechanically powered equipment is limited. Since the use of human power is extensive in cultivation of crops, the accidents occur due to tiring, fatigue, falling from bunds and lack of safety aspects in the traditional equipment, undulating loose/slippery terrain, etc. This may result in severe injuries and fatalities, which due to remote location do not get reported. As sufficient and reliable data on etiology of agricultural accidents is not available, it is essential to collect the information on agricultural accidents, severity of injuries and causes for accidents so that necessary strategy can be developed to minimize the occurrence of such accidents.

METHODOLOGY

A survey was conducted in four districts *viz.*, Papum Pare, Lower Subansiri, West Siang and East Siang of Arunachal Pradesh. These districts adequately represent the entire state of Arunachal Pradesh (Dewangan *et al.*, 2004a and 2004b). Furthermore, these districts are located in four different agro-ecological sub-regions of Arunachal Pradesh and inhabited mainly by three major ethnic groups with varying nature of socio-economic conditions. As per recommendation of Indian Agricultural Statistical Research Institute, New Delhi, 40 villages were randomly selected in each district for accident survey.

The data for the period of July 2005 to June 2006 was colleted during survey. The villages were thinly populated therefore entire farming population of village was considered for the survey. The village level information was obtained from either Agricultural Development Officers (ADO) or Agricultural Assistants (AA) or Anchal Samiti Members (ASM) or village leaders (Gaon Burha). Meeting was conducted in each village at public places like Primary School, Primary Health Centre, Gram Panchayat, where all the farmers and agricultural workers were informed to attend. Preliminary information of the victim like name, age, address and brief details of the accident was collected during this meeting. Based on this information, door to door visit and personal interview of survey team with the victims of the village were done to gather the complete information of the victim. The information provided by the victims was noted in the proforma used for data collection. The information like, experience of working in farm, year of accident, type of injury, its severity, approximate cost involved in the treatment of victim, period of hospitalisation, days of absence from the work was collected directly from the victim. The personal interview of the accident victim gave the insight into the reliability of the data besides indicating the type of injury, its severity and causes of accident. The data collected during the survey was then analysed.

RESULTS AND DISCUSSION

The details of the sample size considered during the survey along with the total number of accidents and victims during 2005–2006 are presented in Table 1. A total of 85 agricultural accidents were found and 85 victims were identified in the survey. The overall agricultural accident incident rate was 5.58/1000 workers/year, which was higher than that reported in the central (1.9/1000 workers/year) and southern region of India (3.64/1000 workers/year). Furthermore, the accident incident rate was higher in the West Siang district, where, in the selected villages, jhum cultivation is extensively followed.

Table 2 indicates that 72.2% of the total agricultural accidents occurred when the agricultural workers were using various hand tools and implements. This is higher than the observation of Mohan and Patel (1992), who reported about 47% of accidents due to hand tools in the northern region of India. Furthermore, it is higher than the All India average of accidents due to hand tools which was reported to be 8% by AICRP on ESA.

Table 1: Details of the Sample Size of Agricultural Accidents Survey and Number of Victims during 2005–2006

S. No.	*District*	*Agricultural Workers Included in Survey*	*No. of Accidents*	*No. of Victims*	*Accident Incident Rate (No. / 1000 workers/year)*
1.	Papum Pare	2330	28	28	5.58
2.	Lower Subansiri	4027	18	18	4.47
3.	West Siang	4073	24	24	5.89
4.	East Siang	5015	15	15	6.44
Total		15445	85	85	5.58

Accident incident rate due to hand tools and implements was higher in Papum Pare district where hill agriculture is practised to a greater extent than other districts. Figure 1 shows that 35.6% accident was due to dao wherein a knife shaped hand tools used for cutting of shrubs, trees, etc. Plough, sickle, spade and power tiller involved respectively 14.1%, 11.4%, 8.4% and 1.2% of the total accidents. Percentage of other accidents accounted for 28.2% (Table 2). The accident incident rate of other accidents was higher in East Siang district followed by Lower Subansiri, Papum Pare, and West Siang districts. This reveals that such accidents are higher in areas where wetland rice cultivation is practised.

Table 2: Details of Accidents Related to Hand Tools and Implements and Other Accident

S. No.	*District*	*Accidents Due to Hand Tools and Implements*		*Other Accidents*	
		*Number**	*Accident Incident Rate (No. / 1000 workers/year)*	*Number**	*Accident Incident Rate (No. / 1000 workers/year)*
1.	Papum Pare	21(75.6)	4.22	6(21.4)	1.20
2.	Lower Subansiri	12(67.6)	3.02	5(27.8)	1.24
3.	West Siang	16(65.3)	3.85	4(16.7)	0.98
4.	East Siang	12(82.2)	5.29	9(60.0)	3.86
Total		61(72.7)	3.97	24(28.2)	1.55

*Number in parenthesis indicates the percentage.

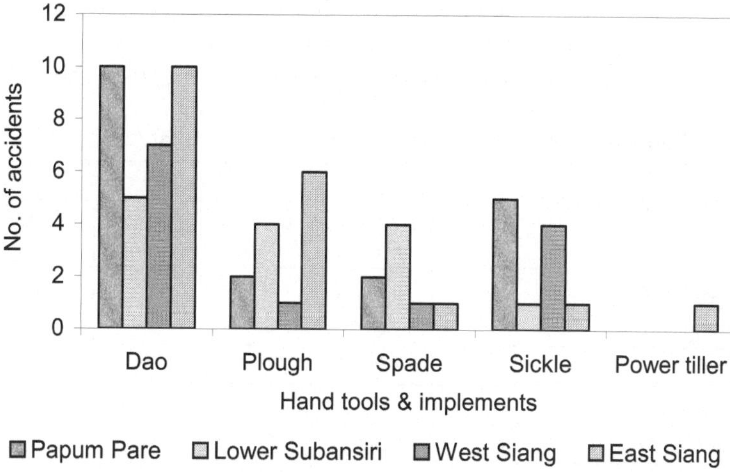

Fig. 1: Hand Tools and Implements Involved in Agricultural Accidents

Table 3 shows the number of accidents occurred during various activities. Among all the activities, jungle clearing accounted for 26.47% of accidents followed by harvesting (12.94%), ploughing (10.59%), going to field (8.24%), returning from field (7.06%), collecting fire wood, channel/bund making, fencing and weeding accounted each (5.69%), seed bed preparation for rice crop using spade (4.71%), searching animal in the jungle (3.53%) and hauling (1.18%).

Figure 2 indicates that 95.3% of the agricultural accidents were non-fatal in nature and 4.71% of the accidents were fatal in nature. About 1.18% of the accidents did not cause any injury. This accident was due to honeybee bite. About 56.47% of the accidents belong to moderate and 24.71% serious category and required professional treatment and major hospitalisation. They included accidents occurred during jungle clearing, slipping during going to and returning from field, hauling, etc. About 3.53% of the accidents were of severe type, which resulted in permanent disability in agricultural workers. These accidents mostly occurred during going to and returning from field and hauling due to slipping and falling from bunds. About 1.18% of the accidents were of maximum severity type, which resulted in the death of agricultural workers. These accidents mostly occurred due to snake bite during jungle clearing. Gite (2005) reported that among the accidents not involving powered machineries and tractors, snake bite is one of the major causes of fatal accidents. However, in Arunachal Pradesh, the farmers have local medicines and rituals purely based on the indigenous knowledge, with which they minimise the severity of injuries.

The analysis of type of accidents presented in Table 4 indicates that most of accidents resulted in cut (45.88%), snake/honey bee/spider/insect bite (24.71%), falls (8.24%), slip and hit accounted each (7.06%) and piercing of stones, thorns etc. (4.71%). The details of the causes of accidents are presented in Table 5. This table indicates that the ignorance and over speed of operation accounted each (36.5%), lack of safety devices and sloppy surface (5.9%), slippery ground and poor light (4.7%), tiredness (2.4%), imbalance of body, loose garments and overloading (1.2%) were the major causes of accidents.

Table 3: Activity Wise Classifications of Agricultural Accidents

S. No.	*Activity*	*District*				*Total **
		Papum Pare	*Lower Subansiri*	*West Siang*	*East Siang*	
1.	Jungle clearing	5	4	6	8	23(26.47)
2.	Harvesting	5	1	4	1	11(12.94)
3.	Plouging	2	4	1	2	9(10.59)
4.	Going to field	2	2	1	2	7(8.24)
5.	Returning from field	2	1	1	2	6(7.06)
6.	Collecting firewood	0	2	1	2	5(5.69)
7.	Channel/bund preparation	2	1	1	1	5(5.69)
8.	Fencing	2	1	1	1	5(5.69)
9.	Weeding	2	1	0	2	5(5.69)
10.	Seed bed preparation	1	0	0	3	4(4.71)
11.	Searching animal	1	1	0	1	3(3.53)
12.	Hauling	1	0	0	0	1(1.18)
13.	Hit by animal	1	0	0	0	1(1.18)
14.	Puddling using power tiller	0	0	1	0	1(1.18)
	Total	26	18	17	24	85(100.00)

*Number in parenthesis indicates the percentage

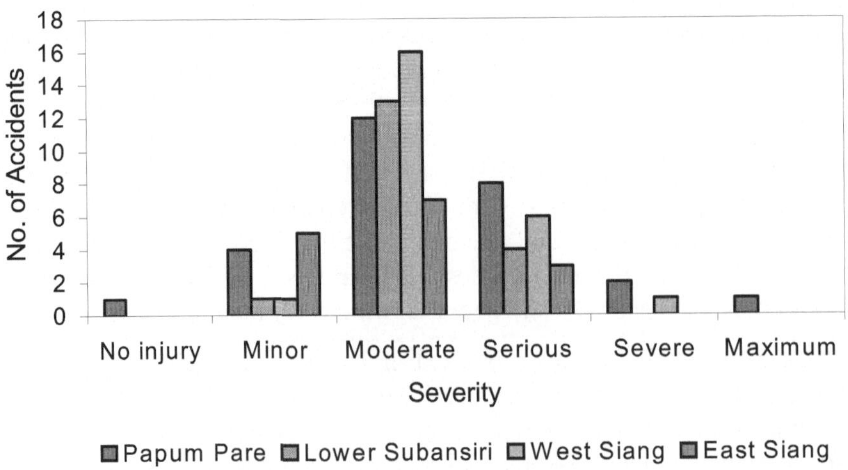

Fig. 2: Details of Severity of Agricultural Accidents

Table 4: Details of Type of Agricultural Accidents

Type of Accident	*Districts*				*Total*
	Papum Pare	*Lower Subansiri*	*West Siang*	*East Siang*	
Cut	13	8	11	7	39(45.88)
Snake/honey bee bite	4	6	7	4	21(24.71)
Falls	3	1	1	2	7(8.24)
Slip	2	1	3	0	6(7.06)
Hit	3	1	1	1	6(7.06)
Piercing of stone/throne	1	2	0	1	4(4.71)
Crushing	1	0	0	0	1(1.18)
Sucking	1	0	0	0	1(1.18)
Total	28	19	23	15	85(100.00)

*Number in parenthesis indicates the percentage.

Table 5: Details of Causes of Agricultural Accidents

Causes	*Districts*				*Total*
	Papum Pare	*Lower Subansiri*	*West Siang*	*East Siang*	
Ignorance	7	9	9	6	31(36.5)
Over speed Operation	10	7	9	5	31(36.5)
Lack of safety devices	3	0	1	1	5(5.9)
Sloppy surface	2	1	1	1	5(5.9)
Slippery ground	1	1	2	0	4(4.7)
Poor light	2	0	1	1	4(4.7)
Tiredness	2	0	0	0	2(2.4)
Imbalance of body	0	0	0	1	1(1.2)
Loose garments	1	0	0	0	1(1.2)
Overloading	1	0	0	0	1(1.2)
Total	29	18	23	15	85(100.00)

CONCLUSIONS

- The agricultural accident incident rate in Arunachal Pradesh was found to be 5.58/1000 workers/year and it was higher in districts, where jhum cultivation is extensively followed.
- About 72.7% of the total agricultural accidents occurred when the agricultural workers were using various hand tools.

- About 35.6% of implement related accidents occurred during the use of dao. Plough, sickle, spade and power tiller involved respectively 14.1%, 11.4%, 8.4% and 1.2% of the total accidents due to hand tools and implements.
- About 95.3% of the agricultural accidents were non-fatal and 4.71% of the agricultural accidents were fatal in nature. About 1.18% of the accidents were minor and did not result in any injury, 56.47% of the accidents were moderate, 24.71% were serious, 3.53% were severe and 1.18% was maximum type.
- Most of accidents resulted in cut (45.88%) followed by snake/honey bee/spider/insect bite (24.71%).
- The ignorance and over speed of operation (36.5%), lack of safety devices and sloppy surface (5.9%), slippery ground and poor light (5.9%), tiredness, imbalance of body and loose garments (1.2%) are the major causes of accidents.

REFERENCES

[1] Dewangan, K.N., Prasanna, Kumar, G.V. and Datta, R.K. (2004a). "Scope of Mechanisation in Arunachal Pradesh, India—Part–1: Traditioal Hand Tools of Shifting Cultivation." *Agricultural Engineering Today* 28(3–4), pp. 60–67.

[2] Dewangan, K.N., Prasanna, Kumar G.V. and Datta, R.K. (2004b). "Scope of Mechanisation in Arunachal Pradesh, India—Part–2: Present Status and Future Strategies." *Agricultural Engineering Today* 28(5–6), pp. 44–51.

[3] Gite, L.P. (2005). "Project Coordinator's Report." In: *Proceedings of the 3rd Workshop of All India Coordinated Research Project on Ergonomics and Safety in Agriculture,* CIAE, Bhopal, India, pp. 3–17.

[4] Mohan, D. and Patel, R. (1992). "Design of Safer Agricultural Equipment: Application of Ergonomics and Epidemiology." *Int. J. Indus. Ergon.* 10(4), pp. 301–309.

[5] Nag, P.K. and Nag, A., 2004. "Drudgery, Accidents and Injuries in Indian Agriculture." *Ind. Health* 42(4), 149–162.

[6] Tiwari, P.S., Gite, L.P., Dubey, A.K. and Kot, L.S. (2002). "Agricultural Accident in Central India: Nature, Magnitude and Economic Impact." *J. Agric. Safety and Health* 8(1), pp. 95–111.

Redesign of Pedal Operated Paddy Thresher

Nikhil Mantri and Gaur G. Ray

Industrial Design Centre, Indian Institute of Technology
Powai, Mumbai–400 076, India

ABSTRACT

Nowadays Pedal operated Paddy thresher is being extensively used by the farmers with small land holding. But when studied carefully it was found that the machine had number of design shortfalls like causing injury to the operator by rotating drum, uncomfortable for farmers with different body dimensions, dust and smaller straws which are inhaled by operator causing respiratory disorders, entanglement of body parts and clothes with moving objects, difficulty in transportation, no proper separation of grain and straws, etc. It was also found that this machine had number of good design features like portability, easy repairs at village level, possibility of operating by women, balanced gear ratios and use of easy rocker assembly.

Therefore, existing machine was redesigned to avoid injuries, making it comfortable for all type of farmers and enabling proper separation of grains and straws to reduce the effort made by people in overall process.

INTRODUCTION

Recently it was found that farmers with small land holding were moving away from cultivation of paddy crop, and choosing other alternative crops. Main reasons were increasing post harvesting, *i.e.*, threshing and winnowing losses (Figure 1), injuries to the body while operating machine, grain damage and heavy strain and fatigue during the operation. Proposed study was therefore undertaken to design a machine for farmers with small land holding which will help them to perform their post harvesting operation (mainly threshing and winnowing) easily.

Overall efforts on this project were to reduce total efforts in paddy post harvesting process, *i.e.*, threshing and winnowing, to produce fine and undamaged grain output.

(Ref b)

Fig. 1: Traditional Methods of Threshing and Winnowing

Reason for Selecting Redesign of Pedal Operated Paddy Thresher

1. Paddy crop is easy to thresh by beating but the losses are quiet high due to breakage of grains.
2. Pedal operated paddy threshers reduce drudgery.
3. This equipment is portable; it can be used in hilly areas and is easily operated by women.
4. Pedal threshing is a more strenuous activity than manual threshing by beating (ref a). Pedalling and holding of paddy plants at same time on the rolling drum result in high muscular strains. Ergonomic improvements in the pedal thresher may help to minimise postural strains.
5. Mohan and Patel (1992) recorded that the two machines caused 13% of total agricultural injuries, mainly (a) 11% by fodder cutting machines and (b) 2% by threshers.
6. Emphasis has been laid on farm mechanisation through provision of subsidy (25% of the cost of the equipment).

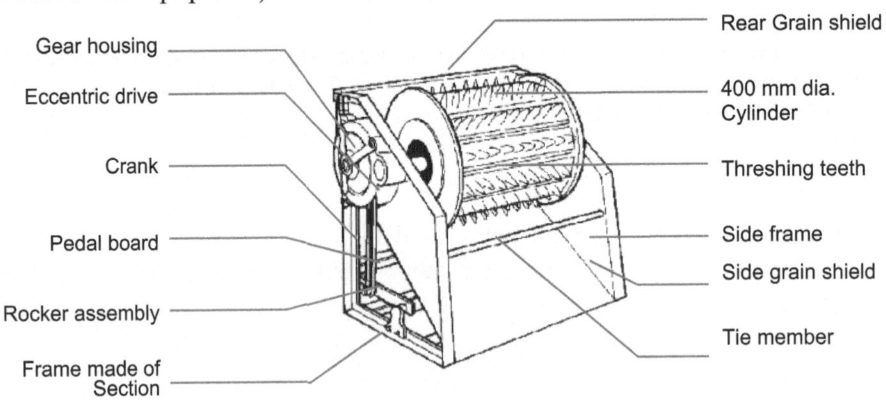

Fig. 2: Existing Pedal Operated Paddy Thresher

Method for Approaching the Problem

Paddy farming involves number of activities right from seedbed preparation to rice milling. All processes were deeply observed for a week and design input area was identified.

Entire paddy farming and processing can be divided into five sections:
1) Paddy farming, 2) Paddy harvesting, 3) Paddy Threshing, 4) Winnowing, 5) Rice milling

It was found that process of paddy threshing and winnowing had many problems like hand cutting, human fatigue, strain, injuries, hygiene and breathing problems. For this reason post harvesting process, *i.e.*, paddy threshing and winnowing was taken as design research area.

The study was conducted at Sunderwadi, Kolhapur district, Maharashtra-Karnataka border, South Maharashtra, India.

Detail of threshing operation was observed on three subjects of average age of 40 yrs (±5) with minimum experience in threshing machine operation of 5 years.

Design Areas for Existing Machine

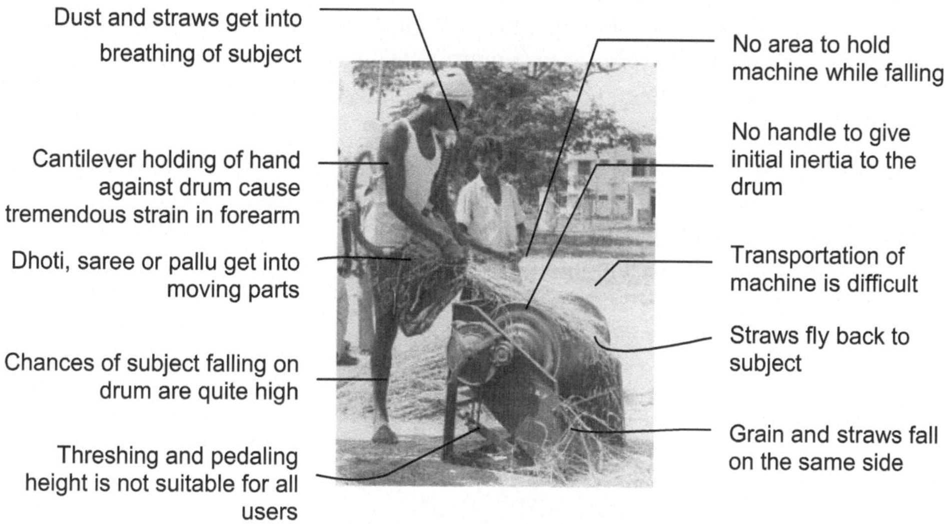

Fig. 3: User Interaction with Machine

Fig. 4: Machine is Placed on Windward Side for Straws to Fly Away

Fig. 5: Activity Analysis—First Person Moving Crop Bundles, Second Person Threshing and Third Person Separating Grain and Straws (Anon, 2007)

Design Requirements

1. *Safety:* Avoid injuries to the forearm caused during pushing the paddy bundle. Based on the information regarding factors associated with injuries, a safer design of thresher was prepared using anthropometric data of the Indian farmer (Chakrabarti).
2. *Ergonomics:* Drudgery and postural strain caused by pedalling and holding the crop bundle at same time should be reduced. Machine should be easily operated by users with all type of body dimensions (Figure 3).
3. *Hygiene:* Straws and dust particles those are blown in the air enter respiratory tract of farmer causing respiratory diseases (Figure 3). This is to be avoided.
4. *Feedback Mechanism:* Easy feedback and visibility of threshing operation should be possible.
5. *Portability and Transportability:* Machine should be easily transportable through farms and village roads.
6. *Stackable:* Since machine is used for only 2 months of the day, it should be portable for easy storage for next 10 months.
7. *Grain Separation:* Straws and grain should be properly separated in machine, which is not possible with the existing one.

Final Concept

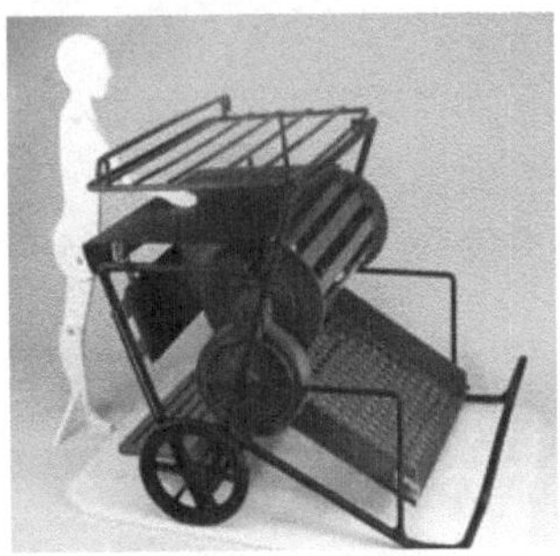

Fig. 6: The Machine

Injury Prevention

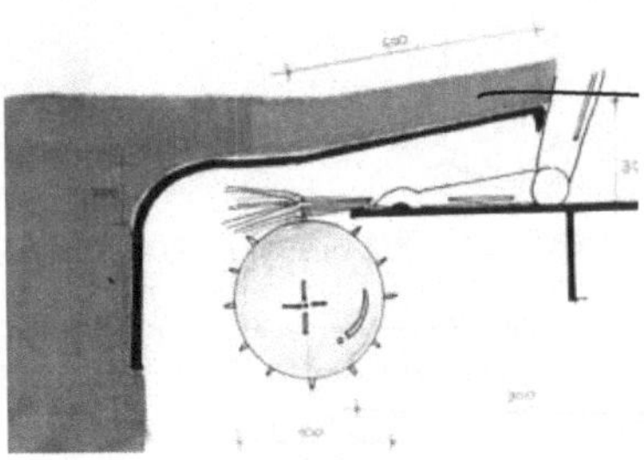

Hand is blocked at edge from reaching inside

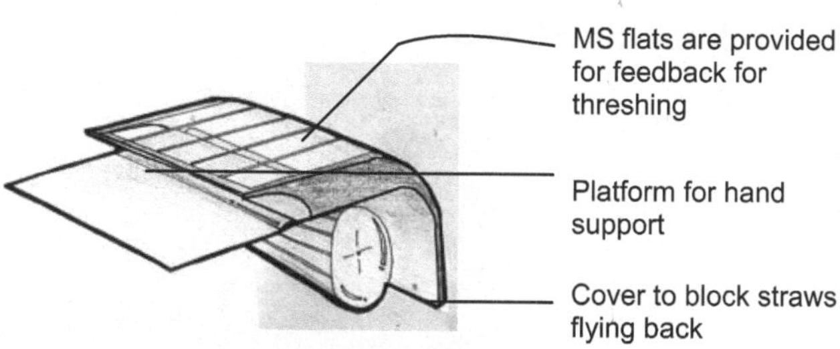

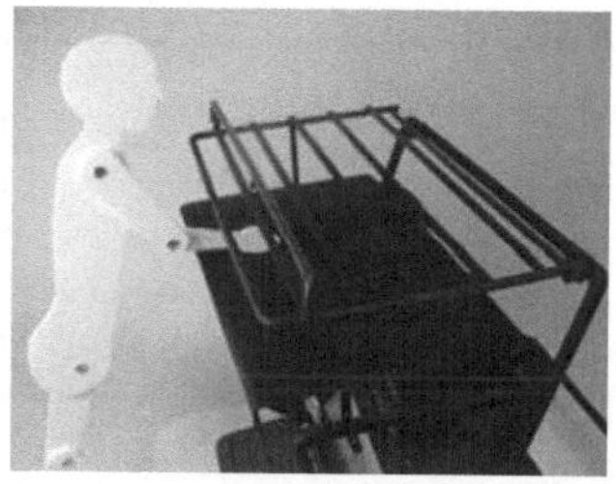

Hand contact with rotating drum is avoided in any condition. Platform cover is designed considering forearm length of 95th percentile of user. The pipes are provided as platform cover for the user to know if threshing is been done properly.

Hand Warning Signal

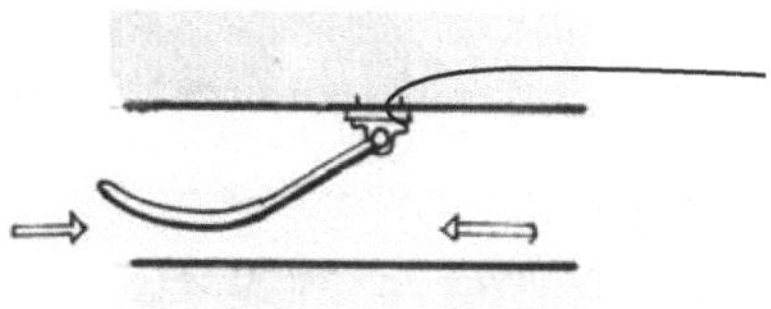

The unit gives proper warning for stopping hand reaching drum and hand can be easily removed after threshing.

Hand warning signal is provided to the stop hand at specific distance from the drum while feeding to avoid injuries.

Hand Brakes: To Stop Drum in Case of Accidents

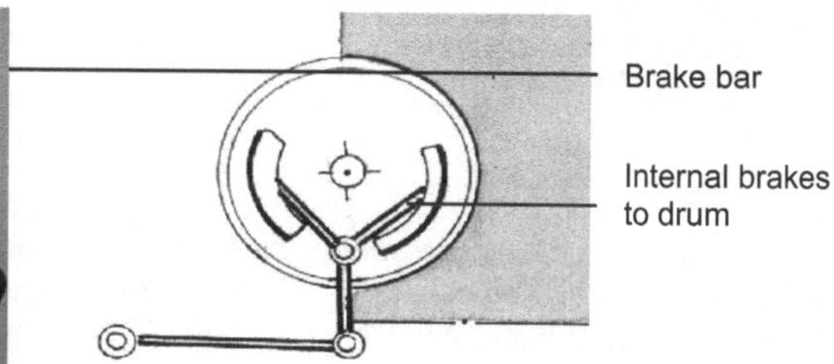

In case of accident drum should be stopped instantly. So hand brakes are provided on top which are easily accessible to worker.

Vibrating Sieve: For Grain Separation from Straws

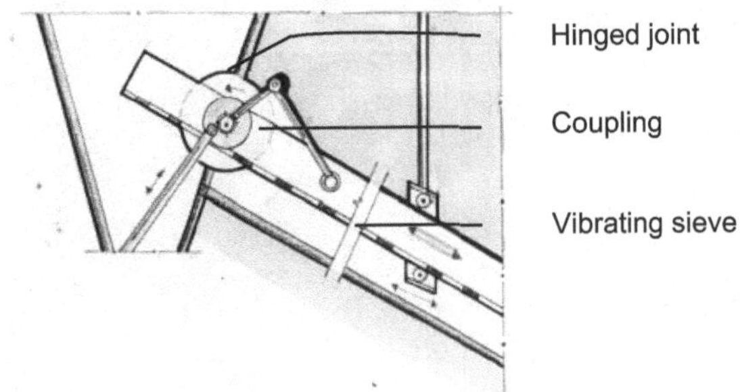

Grains fall down from sieve and straws are carried over it. Vibrating movement of sieve helps in overall process. This reduces work of one complete person of separating grain and straws after threshing is complete.

Ergonomic Considerations

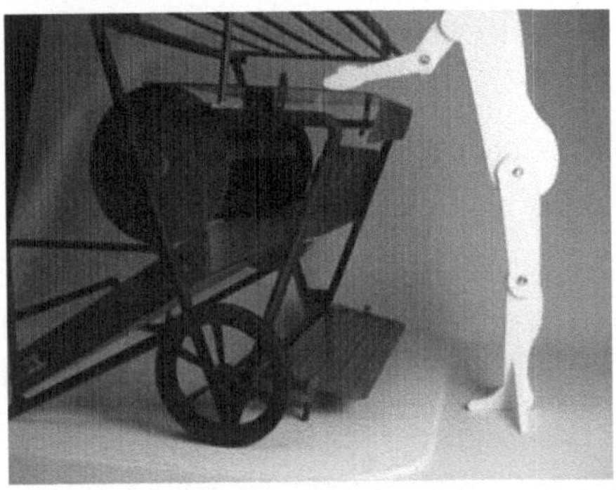

Pedaling height is at 120 mm from ground level for comfort of all size of users. Pedaling board is always horizontal with help of hinge at centre and connected to gear assembly through crank

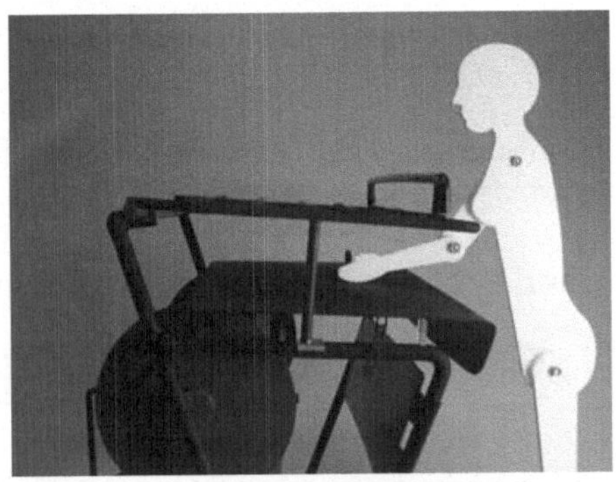

Threshing height (platform height) is designed for shortest *i.e.* 5th percentile of woman (elbow height-736 mm) and is adjustable for all further heights upto 95th percentile of man *i.e.* tallest man.(elbow height-939 mm)

Transportability

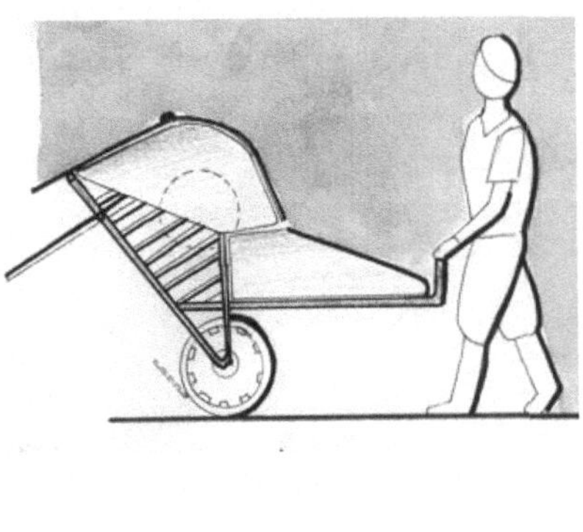

Machine can be lifted like wheel barrow to transport it from one farm to other. The whole weight of assembly comes on axis on which wheel is placed and not on the person. Mean height of user (1614 mm) is considered while designing the handle.

To carry the machine over bunds, Handles are provided, which enables complete machine to be lifted by two persons.

Cover Design

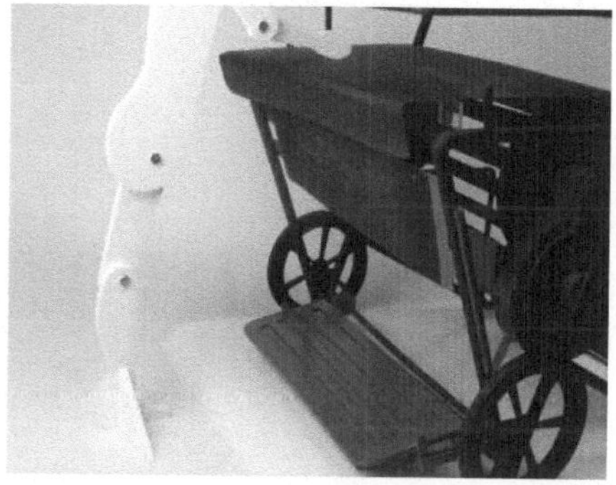

Cover made of FRP avoids straws getting into air. It also covers the moving parts to avoid accidents, which are caused by entanglement of body parts and clothes into moving parts of machine. Light weight of FRP also reduces weight of machine.

Rear grain shield is provided such that it doesn't obstruct pedalling movement but help in avoiding any material from flying back from drum.

REFERENCES

[1] Anon (2007). http://en.wikipedia.org/wiki/Threshing (Date: 20 October 2007).

[2] BIS (1982). "Indian Standard Specification for Pedal Operated Paddy Thresher (IS 3327–1982)." Bureau of Indian Standards, New Delhi.

[3] Chakrabarti, Debkumar. "Indian Anthropometric Data." NID Ahmedabad.

[4] Mohan, D. and Patel, R. (1992). "Design of Safer Agricultural Equipment: Application of Ergonomics and Epidemiology." *Int. J. Ind. Ergonomics*, 10, pp. 301–309.

SECTION–VI

Ergonomics in Industry

Ergonomic Interventions in Malaysian Small and Medium Industries

S. Darius Gnanaraj[1], Abdul Rahman Omar[2], Md Fuad Bahari[2] and Isa Halim[2]

[1]Directorate of Interdisciplinary Research, Karunya University
Coimbatore–641 114, India

[2]Faculty of Mechanical Engineering, Universiti Teknologi MARA
Shah Alam, Selangor D.E., Malaysia

E-mail: diridr@karunya.edu

ABSTRACT

Ergonomic interventions have been carried out in small and medium industries in Malaysia. Sheet metal stamping industries' workers are exposed to higher body vibrations as they stand and work throughout their shifts. Ergonomic interventions were made in two small industries involved with sheet metal stamping works. A Larson Davis HVM 100 (Human Vibration Meter) was used to measure the whole body vibration of an operator working on a 65-ton stamping machine. A 3-axis pad with built in accelerometer was placed under the feet of the worker and whole body vibration was measured based on ISO 2631-1. Thick rubber platforms interposed between the worker and the floor of the industry absorbed the vibration and removed the discomfort experienced by the worker. It was found that the magnitude of whole body acceleration reduced by about 78 per cent with the inception of the rubber anti-vibration pad under the worker's feet. Ergonomic interventions were made in another metal stamping company, which used a lot of manual material handling activities while producing their products. The production shop floor consisted of stamping machines having capacities ranging from 25 to 250 tonnes. Ergonomic redesign of workplaces with suitable lift tables and alternate sit/stand postures were found to remove the discomfort from workers to a greater extent.

INTRODUCTION

In stamping machines, sudden forces are applied on the material by the punch to make sheet metal parts having different geometry. The frequency of operation of stamping machines is about 2 Hz. If a person stays near a stamping machine in a small scale industry, he can experience heavy blows entering through his feet. Every time the punch exerts force on the material, a blow is felt in the feet of the worker. The frequency of this blow is also same as the frequency of operation of stamping machine, which is about 2 Hz. This paper examines the effect of such blows on the comfort level of the working environment and the amount of discomfort that can be removed by using an anti-vibration rubber pad through the reduction of the amount of whole body vibration experienced by stamping machine operators. This paper also reports the effect of ergonomic interventions made using an adjustable table, sit-stand stool and by raising the position of the receiving tray.

EXPERIMENTAL PROCEDURE

To measure the whole body vibration of a stamping machine operator, a 65-tonn stamping machine was chosen in a small and medium industry located in Shah Alam, Malaysia. A Larson Davis HVM 100 (Human Vibration Meter) was used to measure the whole body vibration of a stamping machine operator. A 3-axis pad with built-in accelerometer was placed under the feet of the worker and whole body vibration was measured based on ISO 2631-1. Table 1 shows the comfort/discomfort levels based on the acceleration values of whole body vibration values.

Table 1: Comfort/Discomfort Levels Based on the Acceleration Values of Whole Body Vibration Values (ISO 2631-1)

Acceleration Values in m/s^2	*Comfort/Discomfort Levels*
Less than 0.315	Comfortable
0.315 to 0.63	A little uncomfortable
0.5 to 1	Fairly uncomfortable
0.8 to 1.6	Uncomfortable
1.25 to 2.5	Very uncomfortable
Greater than 2	Extremely uncomfortable

RESULTS

The weighed r.m.s. acceleration values were obtained for all three axes by measuring the whole body vibration of an operator while he worked on a stamping machine. Experiments were conducted to collect the whole body vibration magnitudes while the operator worked under normal working conditions. A set of anti-vibration pads was introduced between the floor and the feet of the operator and the resulting whole body vibration magnitudes were collected by repeating the experiments. Figure 1 shows a worker on the stamping machine by standing over the anti-vibration rubber pads. Figure 2 shows the graphical view of the magnitude of the whole body vibration magnitudes before and after placing the anti-vibration pad. The average vibration magnitudes have been reduced from 0.767 m/s^2 to 0.170 m/s^2. When we refer to Table 1, we see that the worker was 'fairly uncomfortable' before placing the rubber anti-vibration pad and the workstation has become 'comfortable' after placing the rubber anti-vibration pad. The worker also agreed that he could feel that the anti-vibration rubber pad removes discomfort to some extent. It is known that the anti-vibration mount fixed between the machine and the floor removes part of the impact due to stamping operation. It can be seen from Figure 2 that the magnitude of whole body vibration is causing discomfort to the stamping machine operator when he does not use the rubber anti-vibration pad. We see that the magnitude of whole body acceleration reduces by about 78 per cent with the inception of the rubber anti-vibration pad under the worker's feet. We need to make this fact known to all stamping machine operators so that they would use rubber anti-vibration pads regularly and avoid problems due to occupational health hazards.

Fig. 1: An Operator Standing over the Anti-vibration Rubber Pads while Working on the Stamping Machine

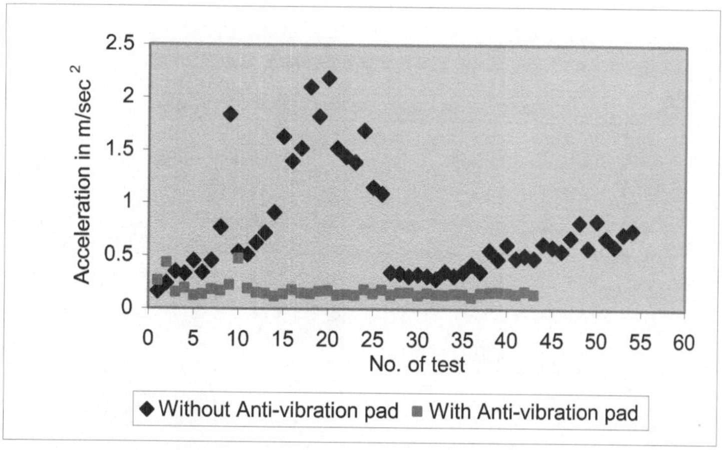

Fig. 2: Effect of Anti-vibration Pad on the Whole Body Vibration

Surface Electromyography (SEMG) and Oxygen Consumption Measurements

SEMG and Oxygen Consumption measurements were made on workers working in another Malaysian Small and Medium Industry (SMI). Since the workers of this company are few in number, only three male workers aged from 19 to 23 years old were selected. These workers were routinely working on stamping machines and were familiar with the task. The selected workers who have been chosen to participate in the measurements had excellent physical health, active and did not have any occupational disease. The workers were informed that the measurement involved two phases of assessments: (i) assessment at existing workstation (ii) assessment at improved workstation design. They were requested to participate in both assessments.

Instrument and Software for Surface Electromyography (SEMG)

The measurements of muscles activity were conducted based on real time monitoring basis. All electrodes were connected to data logger and the signals from data logger can be monitored from the screen of PC monitor using wireless networking. The aim of this system is to ensure that the assessment is under control throughout the whole measurement processes. MegaWin Software was used to store and interpret the data related to muscle activity.

Instrument and Software for Oxygen Consumption Measurement

During the assessment, the worker wore a chest-mounted light-weight oxygen analyser (Metamax Cortex, Germany) on his body. This portable device is an integrated telemetry system that measures minute ventilation and oxygen consumption. Figures 3a to 3d show one cycle of work performed by the worker before intervention.

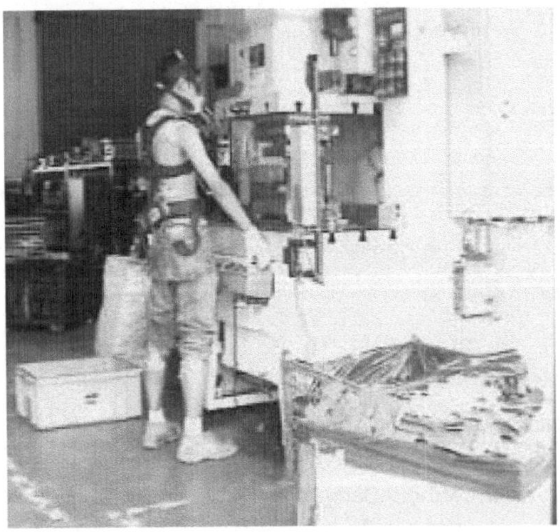

Fig. 3a: Worker Stands in Front of Stamp Machine

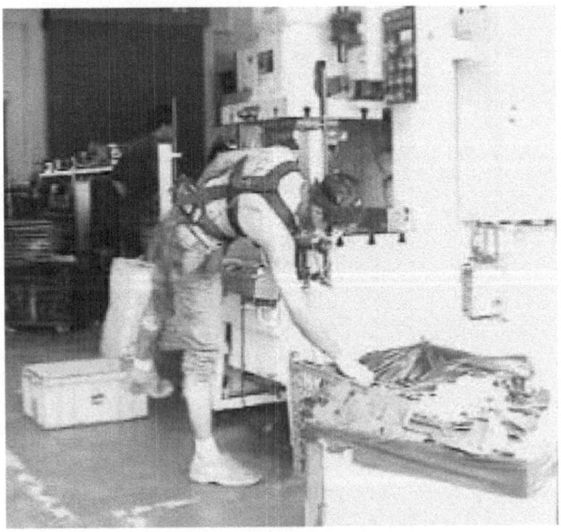

Fig. 3b: Worker Reaches the Material in Bending Posture

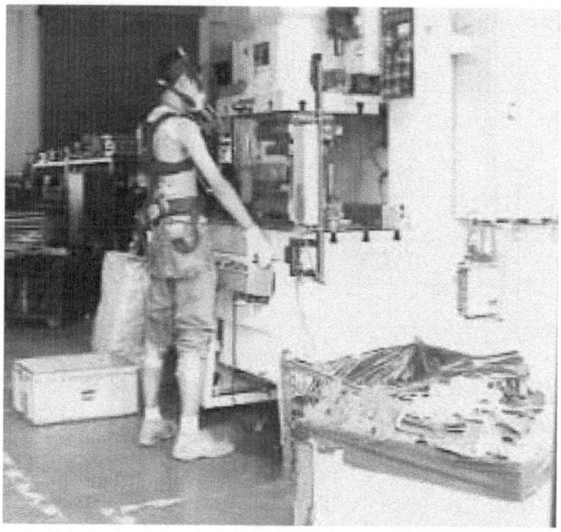

Fig. 3c: Worker Feeds the Material Into Machine Die

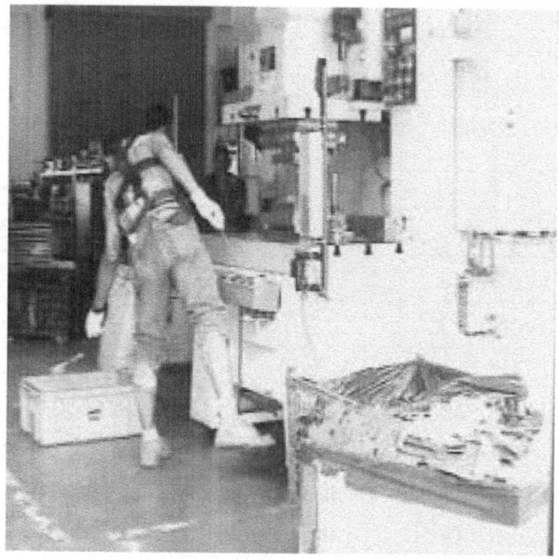

Fig. 3d: Worker Throws the Finished Part into Empty Bucket on His Left

Improved Workstation

Figure 4 shows the improved workstation provided with adjustable table, sit-stand stool and a raised container for placing the used metal plates. Table 2 presents the results of oxygen consumption before and after intervention. It can be observed that oxygen consumption and energy expenditure are reduced by 30 to 60% due to ergonomic intervention.

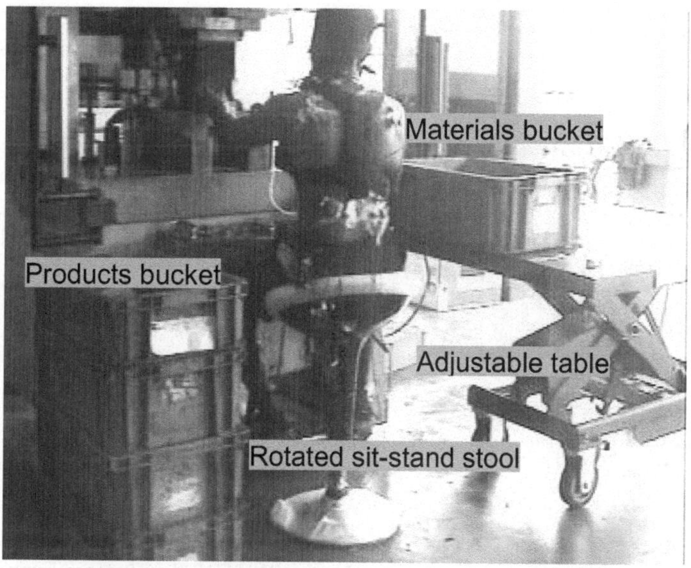

Fig. 4: Improved Workstation Design Provided with an Adjustable Table and a Rotated Sit-stand Stool

Table 2: Results of Oxygen Consumption Measurement

Worker	*Before Intervention*		*After Intervention*	
	Oxygen Consumption (l/min)	*Energy Expenditure (kJ/min)*	*Oxygen Consumption (l/min)*	*Energy Expenditure (kJ/min)*
Worker 1	0.735	15.582	0.291	6.169
Worker 2	0.465	9.858	0.336	7.123
Worker 3	0.295	6.254	0.131	2.777

Figure 5a shows the total load exerted on muscles present in different parts of the body of the third worker before the intervention. It is seen that the left neck muscle receives more load in this posture (12486 μVs out of 463076 μVs). Both legs are experiencing considerable amount of load. After the intervention, the legs receive negligible loading as seen in Figure 5b and the total value reduces to 415625 μVs from 463076 μVs; it amounts to a reduction of about 10.25%. Even though the left muscle appears to receive more loading, the value of loading has increased only by about 4.8%. But the right muscle receives a high load, more than double of the load it received before intervention. The heights of adjustable table, sit-stand tool and also the tray for placing the used metal plates are to be adjusted properly and some more study is required to reduce the loads experienced by the neck muscles.

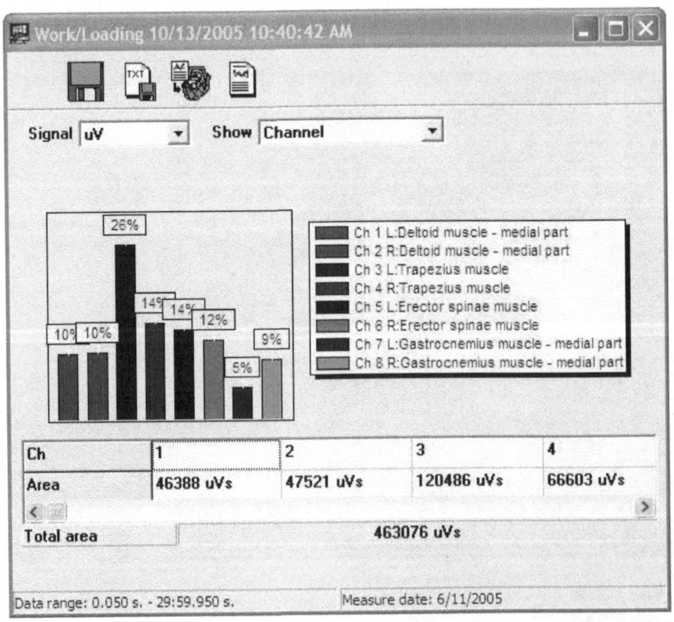

Fig. 5a: Total Load Exerted on Muscles for the Third Worker (before intervention)

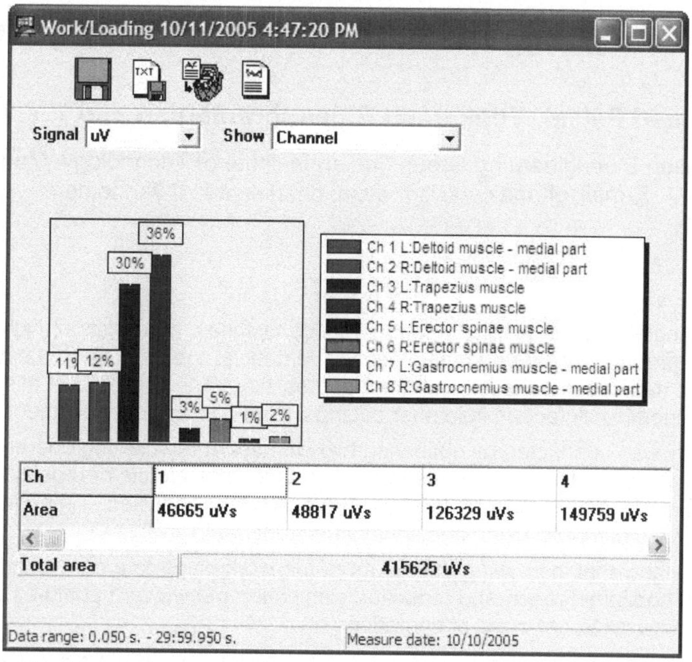

Fig. 5b: Total Load Exerted on Muscles for the Third Worker (after intervention)

CONCLUSIONS

Ergonomic interventions carried out in two Malaysian Small and Medium Industries (SMIs) show considerable decrease in the discomfort experienced by the workers operating stamping machines.

It is found that the magnitude of whole body acceleration reduces by about 78 per cent with the inception of the rubber anti-vibration pad under the worker's feet.

By providing an adjustable table, a sit stand stool and a raised tray for placing the used metal pieces, there is a decrease of 10.25% in the total load exerted by the worker working on a stamping machine.

When we make a standing worker to sit and work, the legs receive negligible loads but the neck muscles experience more loading. Some more refinements in the workstation design are required to make the worker more comfortable.

ACKNOWLEDGEMENT

The authors thank the Ministry of Science, Technology and Innovation (MOSTI), Government of Malaysia for funding this research project. The two Malaysian small and medium industries who granted permission for conducting these ergonomic studies are thanked. We thank one and all who contributed to this work in one way or other.

A Study of Ergonomics in the Leather Cutting Workplace

A. Mohammed Faisal, Venkatesh Balasubramanian and T.T. Narendran

Rehabilitation Bioengineering Group, Indian Institute of Technology, Madras, India
E-mail: chanakya@iitm.ac.in; chanakyaaa@yahoo.com

ABSTRACT

Leather cutting is a major activity in leather goods factories. It consists of highly repetitive operations, producing several small pieces of leather at the workplace. Leather cutting involves two main operations, *viz.*, clicking and splitting. The objective of this paper is to identify ergonomic risk factors in leather cutting operations and quantify them.

Motion Study was conducted by observing two workers at the clicking operation and one at the splitting operation for extended periods of time. The effects of repetition, height of work place, horizontal reach and awkward posture were analysed. Times taken for the operations were compiled using traditional time study.

The results show that increasing the height of the work place to a convenient level, and decrease in horizontal reach and reduction/elimination of awkward postures significantly reduce the time taken per cycle of operation.

Reduction of ergonomic risk factors and redesigning the workplace as a safe and comfortable one results in substantial saving of operational cycle times and thus contributes to increase in productivity.

INTRODUCTION

Manufacture and export of leather products is a major economic activity in India. Leather products are classified broadly as leather footwear and leather goods such as bags, handbags, gloves, wallets, brief cases, mobile covers and straps, files, belts and saddlery goods.

Leather industry is a sector which depends heavily on manual labour. Factors such as styling and customising the products and vagaries of the natural raw material make human intervention a necessity. In such a setting, it is evident that traditional industrial engineering concepts such as Time and Motion Study (Gilad and Elnekave, 2006) and Ergonomics (Dul and Neumann, 2005) will have tremendous scope for improving the workplace, improving productivity (Shikdar and Das, 1995) and cutting costs of production in order to retain the competitive edge of Indian leather industry in the global market.

Production of leather goods involves skilled and strenuous operations. They are broadly classified as cutting, assembly and finishing operations. Of these, leather cutting is a key activity that is required by all classes of leather products. It is a manual operation that needs high skill and is repetitive in nature. Leather cutting consists of two generic operations, *viz.*, clicking (cutting of leather pieces to required shape and size) and splitting (cutting of leather pieces to required thickness). Despite the labour intensive operations in the leather industry, the awareness of ergonomics in the sector is quite poor at present.

This paper reports a study that seeks to identify ergonomic risk factors in leather cutting operations. Recommendations for improvement, based on the analysis, are also provided.

METHODOLOGY

Three operators engaged in clicking and splitting were selected and recorded about their motions. Of these, one skilled and the other semi-skilled worker were involved in clicking while the third performed the splitting operation. A step-by-step recording of the clicking and splitting operations accompanied by a classification of the work elements constituting the operation was done. The duration of each work element was recorded using Time Study (Gilad and Elnekave, 2006). The work elements were also analysed for ergonomic risk factors (Resnick and Zanotti, 1997) in terms of height, reach and awkward postures involved in their execution (Table 1).

Table 1: Steps of Leather Cutting Operations

Operations	*Steps*
Clicking	Loading of Leather
	Picking up of die
	Set up of die with Leather
	Processing Pieces
	Removal of Leather Pieces
	Unloading of Leather Pieces
	Arrange the Leather Pieces
Splitting	Set the machine
	Picking up of Leather Pieces
	Processing Pieces
	Collecting of Leather Pieces
	Arrange the Leather Pieces

RESULTS AND DISCUSSION

The evaluation of ergonomic risk factor, *viz.*, height, horizontal reach and awkward posture and the time required to perform each step is shown in Table 2.

Each work element was observed in order to determine number of the repetitions of clicking and splitting operations. In clicking (Tables 3 and 4), we see that 71% non-productive elements are needed to perform 29% of productive elements. In a cycle of 5 minutes, value adding activities need only 87 seconds while non-value adding activities account for 213 seconds. The unnecessary repetition of non productive elements should be reduced or eliminated in order to improve the operation performance. From the ergonomic perspective, inconvenient work height, awkward posture and long reach are factors that affect the time taken to perform the non-productive element. For example, inconvenient level of work height and long horizontal reach for loading, unloading and arranging the leather pieces lead to increase in operation time.

Table 2: Steps of Leather Cutting Operations with Ergonomic Risk Factors and Time

Steps		*Average Time (s)*	*Ergonomic Risk Factor*		
			Height	*Reach*	*Awkward*
Clicking	Loading of Leather	8	•	•	
	Picking up of die	8			•
	Set up of die with Leather	1	•	•	
	Processing Pieces	2	•	•	
	Removal of Leather Pieces	1			•
	Unloading of Leather Pieces	2	•	•	
	Arrange the Leather Pieces	1	•	•	
Splitting	Set the machine	4	•	•	
	Picking up of Leather Pieces	4			•
	Processing Pieces	1	•	•	
	Collecting of Leather Pieces	4			•
	Arrange the Leather Pieces	2	•	•	

Table 3: Repetition and Time of Clicking Operation with Skilled Operator

Clicking Operation		*Labor (Skilled)*					
		Repetition			*Time*		
Step	*Average Time (s)*	*O*	*D*	⇨	*O*	*D*	⇨
Loading of Leather	8			3			24
Picking up of die	8			3			24
Set up of die with Leather	1		41			41	
Processing Pieces	2	41			82		
Removal of Leather Pieces	1		6			6	
Unloading of Leather Pieces	2			41			82
Arrange the Leather Pieces	1		5			5	
Total (5 minutes)		41	52	47	82	52	130
Percentage		29	37	34	29	37	34

264 seconds + 36 seconds for personal allowances = 300 seconds

In splitting operation (Table 5), there was less repetition of non productive elements. However, it has more of awkward postures in the pick and place of leather pieces resulting in an increase in the operation time. The overall split, in this case, is 151 seconds for productive elements and remaining 149 seconds for non-productive elements. The observed time was calculated based on the average taken over extended periods of production. The normal time (Barnes, 1980) of each operation was calculated from the observed time along with rating factor (based on the performance of worker) and standard time (Barnes, 1980) of the normal and allowance (2 to 5% of normal time for light work) as shown in Table 6.

Table 4: Repetition and Time of Clicking Operation with Semi Skilled Operator

Clicking Operation		*Labor (Semi-skilled)*					
		Repetition			*Time*		
Step	*Average Time (s)*	*O*	*D*	⇨	*O*	*D*	⇨
Loading of Leather	9			3			27
Picking up of die	9			3			27
Set up of die with Leather	2		33			66	
Processing Pieces	2	33			66		
Removal of Leather Pieces	1		10			10	
Unloading of Leather Pieces	2			33			66
Arrange the Leather Pieces	1		5			5	
Total (5 minutes)		41	52	47	66	81	120
Percentage		29	37	34	28	40	32

267 seconds + 33 seconds for personal allowances = 300 seconds

Table 5: Repetition and Time of Splitting Operation

Splitting Operation	*Average Time (s)*	*Repetition*			*Time*		
		O	*D*	⇨	*O*	*D*	⇨
Set the machine	4		2			8	
Picking up of Leather Pieces	4			12			48
Processing Pieces	1	151			151		
Collecting of Leather Pieces	4			13			52
Arrange the Leather Pieces	2		8			16	
Total (5 minutes)		151	10	25	151	24	100
Percentage		81	5	14	81	5	14

275 seconds + 25 seconds for personal allowances = 300 seconds

Table 6: Standard Time for Leather Cutting Operations

Operation	*Average Units Produced for 5 minutes*	*Observed Time (s/unit)*	*Rating Factor*	*Normal Time (s/unit)*	*Standard Time (s/unit)*
Clicking – Skilled	41	7.32	1.2	8.78	9.05
Clicking – Semiskilled	33	9.09	1.3	11.82	12.18
Splitting	151	1.98	1.3	2.58	2.65

Increasing the height of the work place to a convenient level, and reducing the horizontal reach and reduction or elimination of awkward postures can significantly reduce the time taken per cycle of operation.

RECOMMENDATIONS

The workplace should be redesigned such that height of workplace is approximately at elbow level of workers. Similarly, the distance moved by the arm for horizontal reach in clicking operation can be reduced by placing the leather and the dies closer to the operation. Removal of the leather piece during clicking operation involves awkward posture at present. This can be avoided by training the worker and by improving the design of the die. At present the unwanted leather pieces tend to stick with the die owing to improper clearance. At the splitting operation, the leather pieces intended for loading should be kept on the table and in front of the worker. At present, the worker has to turn back every time to pick up the leather pieces.

CONCLUSIONS

This study reports the analysis of a leather cutting workplace. Based on an evaluation of ergonomic risk factor and on a time and motion study of the workplace, we recommend redesign in order to achieve reduction in operational cycle times. The reduced operation time and ergonomic risk factors will lead to an increase in productivity.

REFERENCES

[1] Ashraf, A. Shikdar and Biman, Das, "A Field Study of Worker Productivity Improvements", *Applied Ergonomics,* Vol. 26, No. 1, pp. 21–27, 1995.

[2] Barnes, R.M. "Motion and Time Study: Design and Measurement of Work", 7th edn, John Wiley, 1980.

[3] Dul, Jan and Neumann, W. Patrick, "Ergonomics Contributions to Company Strategies" *10th International Conference on Human Aspects of Advanced Manufacturing: Agility and Hybrid automation*, USA, 2005.

[4] Gilad, I. and Elnekave, M. "Time-based Approach to Obtain Quantitative Measures for Ergonomics Hazard Analysis", *International Journal of Production Research*, Vol. 44, No. 23, pp. 5147–5168, 1 December 2006.

[5] Resnick, M.L. and Zanotti, A. "Using Ergonomics to Target Productivity Improvements", *Computers and Industrial Engineering,* Vol. 33, No. 1–2, pp. 185–188, 1997.

Computer Controlled System for Operating Machines in Modern Indian Ginneries

Jyoti M. Nath[1], S.K. Shukla[1], P.G. Patil[1] and K.M. Paralikar[2]

[1]Ginning Training Centre of CIRCOT, Nagpur–440 023, India
[2]CIRCOT, Matunga, Mumbai–400 019, India

ABSTRACT

Indian cotton is harvested in 4 to 5 pickings. Last two picks are high in trash content and require more than one pre-cleaner and lint-cleaner to bring its trash to acceptable level. As pre-cleaning and lint-cleaning requirements vary for last two picks, it is not desirable to use all available cleaners on first two picked cotton. Excessive cleaning could decrease bale weight, degrade fibre length and lower spinning quality of cotton. It is a very laborious job for altering sequence of cleaning machines for every fed cotton samples. Gin automation with proper management practices assists in decreasing level of contamination due to less human interventions, energy savings and minimisation of productivity loss. A programmed switching system for operating machines in ginnery through handheld keypad control or personal computer has been developed. It consists of infrared controlled system interfaced to computer. Through interactive computer programming, machine sequence can be altered as per requirement. Here, pre-cleaners and lint-cleaners are equipped with bypasses and used selectively according to cotton cleaning requirements. In addition, gin machines get started and stopped in definite sequence to avoid chockage of cotton as well as overloading of gin machines. Such automation saves human labour as there is no requirement of manual operations of the machines. Gin machines can be linked to central computer and all activity can be monitored and remotely controlled through a control room. Developed system assists in proper gin management with less labour involvement, energy conservation, minimisation of productivity loss and production of quality bales which in turn adds to monetary gain.

Keywords: Programme, Switching, Automation, Ginnery.

INTRODUCTION

International Textile Manufacturers Federation (ITMF) brands the Indian cotton under most contaminated cotton category followed by Pakistan, Turkey and Turkmenistan (ITMF, 2005). Gin processing is an important step in production of cotton lint with quality fibre attributes. Indian cotton is entirely hand picked unlike developed countries where cotton is mechanically picked. Hand picked cotton is considered as the best picked cotton which contain less than 2% trash content at the time of picking. Whereas machine picked cotton contains around 12 to 25% trash content depending on the type picking machines used. It has been found that processed cotton coming out of Indian ginneries contain around 4 to 6% trash content. Whereas in developed countries processed cotton is almost trash free mainly due to use of more than one pre-cleaning and post-cleaning machines. It has been observed that Indian cotton is harvested in 4 to 5 pickings whereas first two picked are relatively clean cotton's which require only one pre-cleaning and one post-cleaning machines for cleaning whereas last two to three picked cotton is more trashy, which requires more than one pre-cleaning and post-cleaning machines. Requirement of gin machinery varies as per the trash content present in raw cotton. As pre-cleaning and lint-cleaning requirements vary for different cotton samples, it is not desirable for ginnery to use all available cleaners on all cotton samples (Anthony & Wesley, 1982). Mangalardi (1972) reported that excessive cleaning could decrease bale weight, degrade fibre length and lower cotton spinning quality. Several types of cylinder pre-cleaners and post-cleaners are used to remove dust, sand, trash particle, immature locks, etc. from cotton (Shukla *et al.*, 2001). Pre-cleaners and lint cleaners should be equipped with bypasses and used selectively according to cotton cleaning requirements. It is also required that gin machines must get started and stopped in definite sequence to avoid chokage of cotton as well as overloading of machines in modernised ginning system. Energy expenditure is also one of the major processing heads that may affect the profitability of a ginnery. Anthony and Eckley (1994) reported that the cost of power consumption comprises about 15% for ginning cotton in modern gins of USA. Judicious use of gin machines with proper management practices assists gin entrepreneurs in energy savings and minimisation of productivity loss. For this, a programmed switching system to control gin operation through handheld keypad control/personal computer has been attempted.

MATERIAL AND METHODS

Machine sequence for processing cotton samples in a modernised ginnery is shown in Figure 1. Most of the modernised Indian ginneries have pneumatic conveying system for handling of raw cotton and lint. To avoid chockage and for smooth distribution of cotton through the entire system gin machines are equipped with PLC control system. Machine sequence for processing different picked cotton depends on the nature and percentage of trash content in the sample.

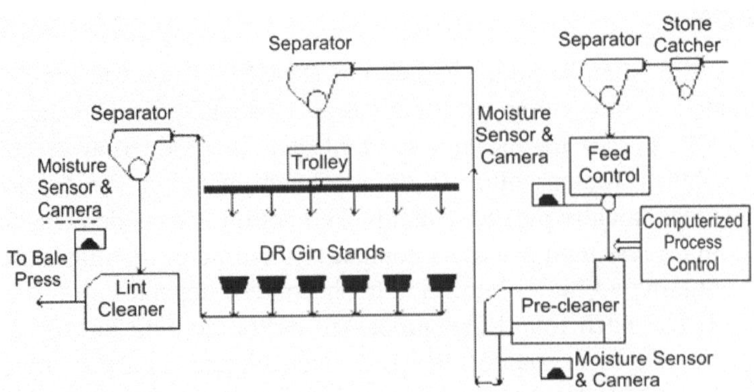

Fig. 1: Machine Sequence in Indian Modern Ginneries

For controlled switching of gin machines a programmed system to operate gin machines through handheld keypad controller/personal computer has been developed. Gin machines can be remotely operated and controlled with the help of infrared control. Infrared receivers are connected at each machine ends to control switching of their respective relays. Infrared transmitters of respective gin machines receive signal from the computer end. Accordingly user can control gin machine switching operation by giving commands through computer programming or by operating switches of a hand held keypad.

Block diagram of the infrared remote control transmitter and receiver sections are shown in Figure 2 and Figure 3. Transmitter section consists of a power supply, an oscillator, an output stage, whereas the receiver section comprises power supply, an infrared detector module, time delay circuit with noise filter, bistable flip flop, and an output section.

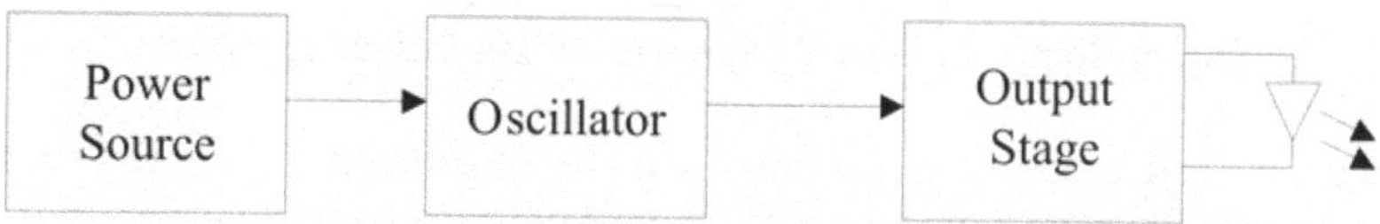

Fig. 2: Block Diagram of Infrared Remote Control Transmitter

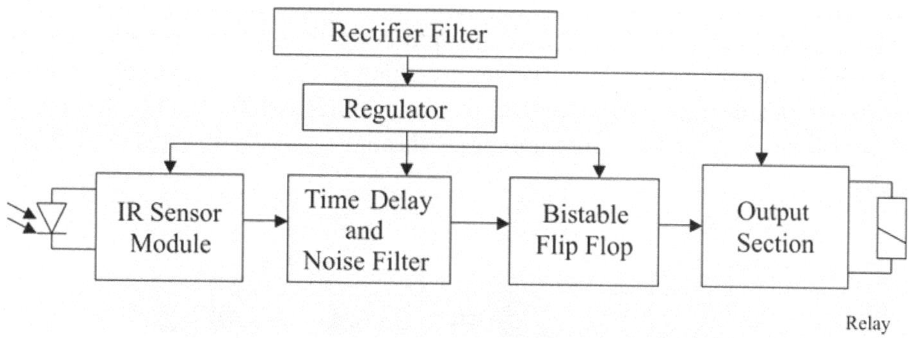

Fig. 3: Block Diagram of Infrared Remote Control Receiver

Remote control transmitter block with eight transmitters and remote control receiver block with eight receiver circuits were fabricated separately. Infrared receivers are activated after receiving signal from infrared LED of respective transmitters. Gin machines are directly connected through relays to infrared receiver modules such that electrical starters operate on/off depending on the state of relays connected to the receiver circuits. Thus the heavy-duty three phase motors of gins will run and operate the machines only if the receiver circuit relays activate the starters and make the electrical contacts (Figure 4).

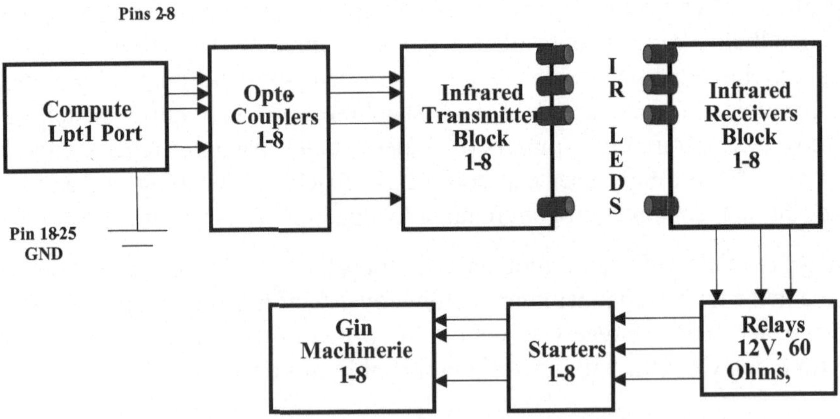

Fig. 4: Complete Block Diagram Gin Plant Switching System

The designed infrared transmitter circuitry can be interfaced to parallel printer port of personal computer (Figure 5). Through LPT1 computer port one can control up to eight machineries without much additional hardware. Pins 2–9 of LPT1 port are eight output terminals and they give 5V output. Thus, one can control switching of relays through computer programming. IC MCT2E is an opto-coupler used for isolation of external hardware from computer. Similarly, infrared technique assists in remotely controlling a number of equipment from a distance with the help of a handheld push button type on/off keypad controller or control room arrangement through centrally situated computer.

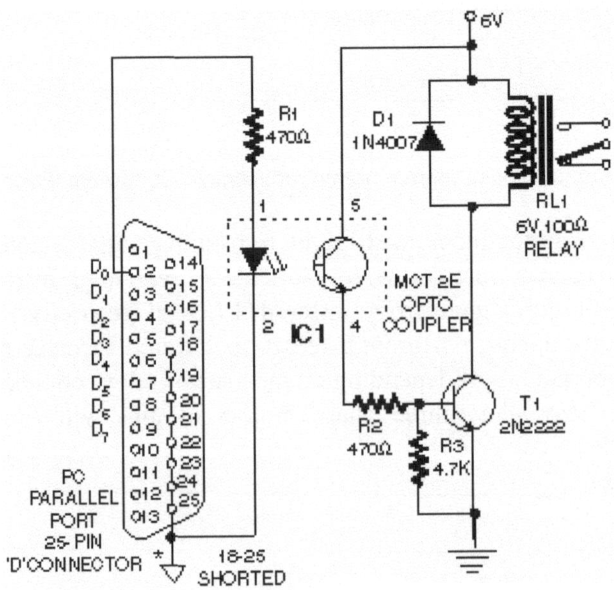

Fig. 5: Schematic Circuit Diagram of Interfacing Infrared Remote Control Transmitter to LPT1 Port of Computer

Each printer port consists of three port addresses: Data, status and control port. These addresses are in sequential order. That is, if the data port is at address 0x0378, the corresponding status port is at 0x0379 and the control port is at 0x037a. We need inpout32.dll to write codes for accessing computer port. Thus by user interactive computer programming (Figure 6) one can remotely control and operate gins in a particular sequence, time delays can be introduced, assigned machines can be switched off, all this can be done by mere click of buttons through computer keyboard. Ease of operation with additional cost benefits in terms of power saving is an added advantage of such automation.

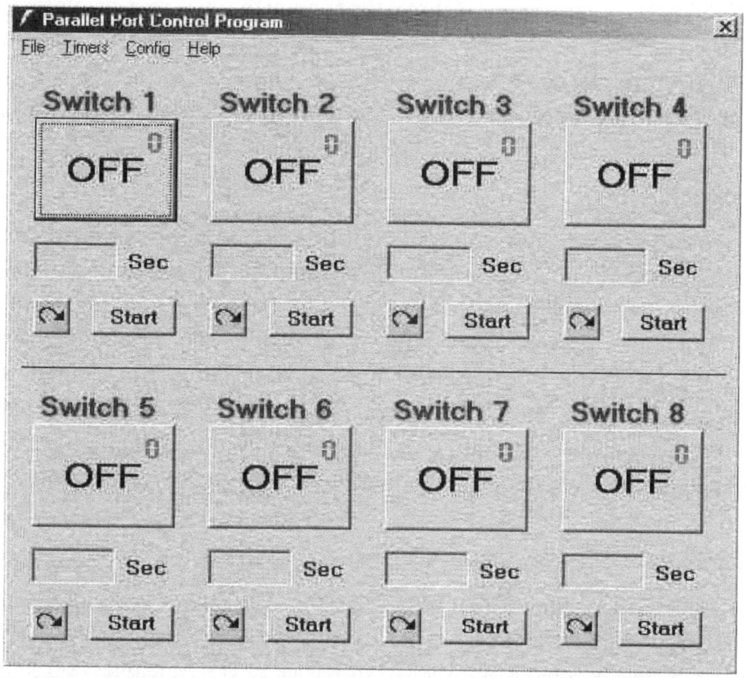

Fig. 6: Computer Program Developed to Control Gin Machineries

RESULTS AND DISCUSSION

Prototype of programmed switching system for controlling gin machines employing infrared transmitters/receiver modules has been developed. Designed system was tested, by operating double roller gin machines through developed remote control system. Gin machines were directly connected through relays to infrared receiver modules such that electrical starters operate on/off depending on state of relays connected to receiver circuits. Thus, heavy-duty three-phase motors of gins will run and operate machines only if relays connected to IR receivers activate respective starters and make electrical contact. Infrared transmitter circuitry can be interfaced to parallel port of personal computer as shown in Figure 5. It is possible to control eight machines through LPT1 computer port without much additional hardware. Pins 2–9 of LPT1 port are eight output terminals and they give 5V output.

A user interactive computer program was developed in visual basic programming language as shown in Figure 6. With this, ginners can remotely control and operate gin machines in particular sequence, time delays can be introduced and assigned machines can be switched off. Infrared technique assists in remotely controlling number of machines from a distance with the help of handheld push button type on/off keypad controller or control room arrangement through centrally situated computer. Transmitter block is powered from a 9V battery for a handheld keypad operation. The infrared remote control described here is free from ambient light interference and provides control range of about ten meters without the use of any focusing lens. Current consumption of IR transmitter is 162 mA. When operated with a 9-volt; 3.4 ampere-hour rating battery, keypad type remote controllers can give continuous operation of 30–40 hours.

CONCLUSION

A programmed switching system for controlled operation of gin machines was successfully developed. By user interactive computer programming, it is possible to remote control and operate gins in a particular sequence, time delays can be introduced and assigned machines can be bypassed or switched off. Designed infrared remote control is free from ambient light interference and provides control range of about 10 m without use of any focusing lens. Ease of operation with additional cost benefits in terms of power saving is an added advantage of such automation. Designed system will aid in proper gin management, gin energy conservation and productivity loss minimisation.

REFERENCES

[1] Anthony, W.S. and Eckley, R.C. (1994). "Energy Utilization and Conservation in Cotton Gins." Cotton Ginners Handbook, USDA Agricultural Handbook, 503, pp. 225–236.

[2] Anthony, W.S. and Wesley, R.A. (1982). "Influence of Gin Machinery on Monetary Returns to Cotton Producers." *Ginner's journal and year book,* 50 (1), pp. 7–18.

[3] ITMF (2005). Cotton Contamination Surveys 1999–2001–2003–2005. International Textile Manufacturers Federation, Am Schanzengraben 29, CH-8002 Zürich.

[4] Mangialardi, G.J. (1972). "Multiple Lint Cotton Cleaning; Its Effect on Bale Value, Fibre Quality, and Waste Composition." *Tech. Bull.* No. 1456, 69.

[5] Shukla, S.K., Manojkumar, T.S. and Murugesan, R. (2001). "Performance Evaluation of CIRCOT Inclined Type and Extractor Type Precleaners on Trashy Seed Cotton." *Souvenir: XXXVth ISAE held at OUAT,* Bhubaneswar during, Jan, pp. 22–24.

Ergonomic Evaluation in the Engine Calibration Section of an Engineering Industry

Rauf Iqbal[1] and P.C. Ghosh[2]

[1]National Institute of Industrial Engineering, Powai, Mumbai–400 087, India
[2]Central Labour Institute, Sion, Mumbai–400 022, India

ABSTRACT

The study was conducted in the calibration section of the shop floor of an engineering industry. The workstation for calibration is made for standing posture and one calibration operation takes 25 minutes (approx.) with moderate to high arm and wrist movement. The workers have complained high physiological fatigue in the thigh and calf muscles. Heart rate and blood lactate response of a total of 114 workers in the calibration section was recorded in three different shifts, namely morning shift (n = 43), afternoon shift (n = 45) and night shift (n = 26). Mean and SD of age, height and weight of the workers in the morning shift are 24.5 ± 3.19 years, 170.5 ± 5.11 cm and 60.4 ± 6.91 kg; afternoon shift are 23.6 ± 3.17 years, 168.3 ± 6.20 cm and 58.8 ± 6.67 kg; night shift are 23.6 ± 3.29 years, 170.9 ± 4.77 cm and 57.5 ± 5.61 kg, respectively.

Strength profile, *i.e.*, grip strength and back strength of the workers have also been recorded. Mean and SD of the hand grip strength of the workers of morning shift [37.6 ± 10.27 kg (R), 36.7 ± 10.59 kg (L)], afternoon shift [38.0 ± 13.49 kg (R), 39.2 ± 11.31 kg (L)] and night shift [38.6 ± 8.33 kg (R), 36.6 ± 10.37 kg (L)] are found to be similar and there is no significant difference among these strength levels. Mean and SD of the back strength of the workers of morning shift, afternoon shift and night shift are 77.0 ± 28.56 kgf, 78.2 ± 21.42 kgf and 76.0 ± 25.98 kgf respectively. Working heart rate of the workers of 3 different shifts are 102.0 ± 16.42 beats/min, 98.5 ± 11.92 beats/min and 104.6 ± 13.06 beats/min, respectively.

Blood lactic acid level during rest (pre work), work and recovery (15th minute) of the morning shift workers are 1.5 ± 0.27 mM/l, 3.9 ± 1.42 mM/l and 2.7 ± 1.27 mM/l, respectively. In case of afternoon shift workers the values are 1.8 ± 0.20 mM/l, 3.2 ± 0.39 mM/l and 2.1 ± 0.18 mM/l, respectively; whereas in case of night shift workers the values are 2.4 ± 1.13 mM/l, 4.3 ± 2.51 mM/l and 3.3 ± 2.03 mM/l, respectively.

Keywords: Calibration, Automobile, Hand Grip Strength, Back Strength, Lactic Acid and Ergonomic Intervention.

INTRODUCTION

Safety, health and productivity of any industry largely depend on critical analysis of human factors involved in Industrial operations. The study of human factors in industrial set up is termed as Human Factors Engineering or Ergonomics. The basic concept of Ergonomics stipulates that in any industrial set up problems should be identified, solved, and it should be followed-up periodically for the improvement of occupational health, safety and productivity in the industry (Baker, 1989). Many developed countries have practised the principles of ergonomics which resulted in higher productivity, lower accident rate and excellent health profile of the industrial workers. The practice of ergonomics depends on the knowledge of human biology of the worker in relation to the machine being used on shop floor. The compatibility among man and machine is the key factor to be considered on the shop floor for achieving proper man machine interaction based on the characteristics of both man and machine (David and Mustafa, 1985). One of the important criteria for ergonomics evaluation is total consideration of man, machine and his working environment both at Macro and Micro levels where exactly the operation is carried out. The ergonomics principles have long been adopted by the Industrial setup to ensure safety and health of the workers as well as to improve productivity. On the shop floor ergonomics deals with work load, work environment, energy cost of the work, health profile of the workers, etc., and evaluates the comfort and safety level of the workers with improved productivity.

Many industrial operations are being carried out in awkward postures, repetitive movements and repetitive bending that triggers various Musculo-skeletal disorders (Genaidy, 1995; Mallory and Bradford, 1989). Such unnatural work practice causes low back pain and neck pain etc among the workers. The ergonomics intervention addresses various human factors on the shop floor thereby helping in improvement of health status, lower accident rate and higher productivity (Marras and Schoenmarken, 1991). This is beneficial for both employee and the employer. The benefit of employee is safe, healthy and accident free working condition whereas the benefit of employer is the higher productivity, lower medical expenses, lower compensatory costs, etc. Quite a large number of industries now-a-days insist on Ergonomic Intervention Program for maintaining statutory safety standards and improving health and productivity at work (DG, FALSI, 1948).

In this context, a study was initiated and carried out in an engineering industry in India. The study was conducted in the calibration section on the shop floor. The calibration work is done in standing posture and the operators have complained of developing back pain because of prolonged standing posture. Therefore, the study was carried out to evaluate work station, working posture and work stress among the workers, so that suitable ergonomic intervention can be recommended.

Aims and Objectives

The Ergonomic study in the calibration section was initiated and conducted with the following aims and objectives:

- To study the Physiological cost of selected shop floor operations.
- To study the postural stress among the workers during the operation.
- To suggest some ergonomic intervention—if required, for making operations more user friendly.

The findings of the study will benefit the workers on the shop floor to maintain ergonomically recommended workstation arrangement and work practice; so that, proper posture during work can be maintained and work stress can be minimised. This will prevent any permanent injury/disability to musculo-skeletal system and maintain proper safety, health and productivity at work.

MATERIALS AND METHODS

Place of study and investigation subjects:

The study was conducted in the calibration section on the shop floor of an engineering industry.

The human volunteers covered in the study were the workers and supervisors working in the calibration section on the shop floor.

Methodology

Following techniques have been adopted for the study conducted on shop floor:

- *Observation:* The process of work and the workers working at the workstations were observed thoroughly to understand the problems faced by the workers during calibration work.
- *Questionnaire:* A questionnaire consisting of subjective feelings of the workers in relation to the workload, work station, work environment and occupational health hazard has been studied.
- *Work Station Analysis:* Details of the work station anthropometry like, working heights, reaches, width and depth of the work surface have been recorded.
- *Physiological Cost of Work:* Following physiological parameters were measured:
  - *Heart Rate:* Heart rate during rest (pre-work), work and recovery period was monitored telemetrically by using Polar Heart Rate Monitors (Polar HR Monitors, S 810i, Finland) to assess the cardio-vascular response at work.
  - *Blood Lactic Acid:* Blood lactic acid level during rest (pre work), working and recovery period was monitored by using a Portable Lactate Analyser (YSI 1500 Sport, USA) to evaluate the level of fatigue during and after work.
  - *Back Strength:* Back strength level of the workers was measured by using a Portable Back Strength Dynamometer (TKK 5402, Takei Scientific Instruments, Japan) to assess musculo-skeletal hazards, if any.
  - *Grip Strength:* Hand grip strength level of the workers was measured by using a Portable Grip Strength Dynamometer (Jamar, Sammons Preston Rolyan, USA) to assess musculo-skeletal hazards, if any.
- *Photography:* Photographs of the workstation as well as the working postures of the workers at the workstation were taken elaborately to study and analyse the postural stress.

RESULTS AND DISCUSSION

Physical profiles of the workers in the calibration section have been presented in Table 1. Workers are quite young in age. Mean of age of the workers in three different shifts varies from 23.6 y to 24.5 y. Mean of the height of the workers in the morning shift, afternoon shift and night shift are 170.5 cm, 168.3 cm and 170.9 cm respectively. Body weights of the workers in three shifts are 60.4 kg, 58.8 kg and 57.5 kg respectively (Table 1).

Table 1: Physical Profile of the Workers in the Calibration Section

Shift	*Age (y)* *Mean ± SD*	*Height (cm)* *Mean ± SD*	*Weight (kg)* *Mean ± SD*
Morning (n = 43)	24.5 ± 3.19	170.5 ± 5.11	60.4 ± 6.91
Afternoon (n = 45)	23.6 ± 3.17	168.3 ± 6.20	58.8 ± 6.67
Night (n = 26)	23.6 ± 3.29	170.9 ± 4.77	57.5 ± 5.67

The operation in the calibration section is being done in standing posture. Various components of the operations of calibration work are fixing of engine in the calibration station, monitoring of various parameters like, pressure, flow of oil, etc., which involves considerable degree of arm movement and bending, twisting of wrist. One calibration takes approximately 25 minutes and an individual worker performs 7–8 calibrations in a shift. Therefore, the work imposes static load on the back (standing posture) and stress on the arm and wrists due to high arm and wrist movement. Grip strength (Rt and Lt) and back strength of the workers in the calibration section were recorded. Table 2 shows the strength profile of the calibration workers. Mean of grip strength (both Rt and Lt) varies from 36.6 kg to 39.2 kg. Mean of back strength of the workers varies from 76.0 to 78.2 kgf. Grip strength and back strength levels in the present study are somewhat lower than the other industrial occupations like, VDT operators, industrial assembly workers and safety inspectors (Chandra *et al.*, 2007), in which ranges of grip strength Rt and Lt are 42.6–48.4 kg and 42.8–45.2 kg respectively. The range of back strength in the study of Chandra *et al.* (2007) is 108.6–130.6 kgf. This indicates that there is decrease of strength level in the calibration workers because of static standing posture. A significant decrease in hand grip strength was detected in metal industrial workers who used hand-held grinders, in a 4-year follow up study, indicating that prolonged occupational exposure should be considered as a risk for work-related MSDs (Mirbod *et al.*, 1999).

Job severity of the workers in the calibration section has been evaluated from prevailing work load. Workload on the shop floor was evaluated from physiological parameters, like heart rate and blood lactic acid level during work and recovery. Working heart rate in the night shift workers is higher (104.6 beats/min) than the morning shift (102.0 beats/min) and afternoon shift (98.5 beats/min) workers. 3rd min recovery is also poor in case of night shift workers (96.5 beats/min) in comparison to the morning shift (88.4 beats/min) and afternoon shift (84.8 beats/min) workers (Table 3).

Table 2: Strength Profile of the Workers in the Calibration Section

Shift	*Grip Strength (kg)*				*Back strength (kgf)*	
	Right		*Left*			
	Mean	*±SD*	*Mean*	*±SD*	*Mean*	*±SD*
Morning (n = 43)	37.6	10.27	36.7	10.59	77.0	28.56
Afternoon (n = 45)	38.0	13.49	39.2	11.31	78.2	21.42
Night (n = 26)	38.6	8.33	36.6	10.37	76.0	25.98

Table 3: Heart Rate Response of the Workers in the Calibration Section

Shift	*HR Rest (pre work) (beats/min)*		*HR Work (working) (beats/min)*		*3rd min HR Recovery (beats/min)*	
	Mean	*±SD*	*Mean*	*±SD*	*Mean*	*±SD*
Morning (n = 43)	81.4	5.19	102.0	16.42	88.4	6.91
Afternoon (n = 45)	79.6	7.17	98.5	11.92	84.8	6.67
Night (n = 26)	82.7	7.29	104.6	4.77	96.5	5.67

Working blood lactic acid level also in the night shift workers is higher (4.3 mM/l) than the morning shift (3.9 mM/l) and afternoon shift (3.2 mM/l) workers. 15th minure recovery lactate is also comparatively poor in case of night shift workers (3.3 mM/l) than the morning shift (2.7 mM/l) and afternoon shift (2.1 mM/l) workers (Table 4).

Table 4: Blood Lactic Acid Responses of the Workers in the Calibration Section

Shift	*Blood Lactate at Rest (pre work) (mM/l) Mean ± SD*	*Blood Lactate at work (working) (mM/l) Mean ± SD*	*15th Minute Recovery Blood Lactate (mM/l) Mean ± SD*
Morning (n = 43)	1.5 ± 0.27	3.9 ± 1.42	2.7 ± 1.27
Afternoon (n = 45)	1.8 ± 0.20	3.2 ± 0.39	2.1 ± 0.18
Night (n = 26)	2.4 ± 1.13	4.3 ± 2.51	3.3 ± 2.03

It is a well known fact that those who stand for a long time experience discomfort due to static effort and subsequent muscle fatigue (Grandjean, 1985; Jorgensen *et al.*, 1993 and Miedema *et al.*, 1997). Prolonged standing has also been associated with discomfort in the lower back (Jorgensen *et al.*, 1993; Ryan, 1980). The findings from the physiological parameters in the present study show that workload on the workers is light to moderate in nature but the recovery is comparatively poor. This is because of persistent stress among the workers due to working in standing posture. This leads to loss of strength level of the workers, particularly back strength (Hill, 1989).

RECOMMENDATIONS

Prolonged standing posture has developed persistent strain on the back and leg muscles as evident from the reduced back strength of the workers. This has been reflected as back pain and leg strain as reported by the workers. Since, the calibration work can not be done in sitting posture, a waist support working stool has been recommended for the workers, so that during calibration work in standing posture, the hip and waist can be rested as and when required.

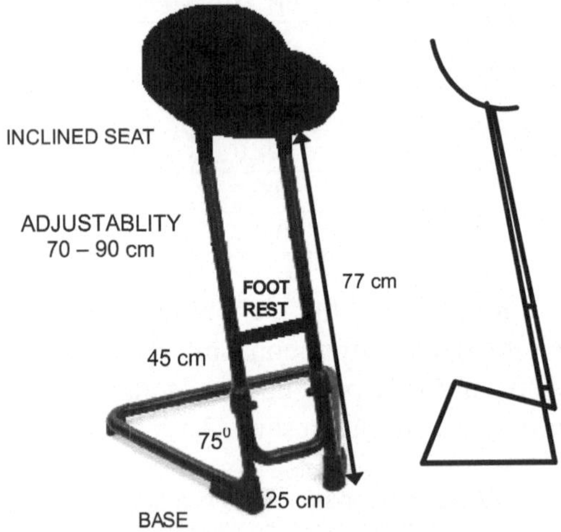

Fig. 1: Waist Support for the Workers in the Calibration Section

Recommended design of the Waist Support stool is as under:

(a) *Fixed Height Stool:* Stool would be made of steel pipe (hollow). Height of the stool would be 77 cm (Chakrabarti 1997) in case of fixed height stool. Triangular broad base (45 cm each) with backward tilt (750 with the base) for stability, foot rest for resting of one foot alternately as and when required.

(b) *Height Adjustable Stool:* In case of height adjustable stool, the range of adjustment of height would be 77–90 cm. Rest of the features are like fixed height stool.

CONCLUSION

The concept of total safety, health and productivity, is directly related to various factors like physical, physiological, environmental and postural, etc. prevalent in the work place. Thus, the ergonomic evaluation in industrial situation is needed for making the workers healthy and productive. Good ergonomic design reduces tension and stress at workplace. It creates an atmosphere of concern between management and staff. The effects on absenteeism, job turnover, productivity, and quality of work will show positive results.

The application of ergonomics is a complex process that requires careful management. To ensure the standards and guidelines drawn from human factors and anthropometric data are safe and effective for individuals, they must be validated through progressive demonstrations and case studies in real life learning environments.

REFERENCES

[1] Baker, W.E. (1989). "Human Factors, Ergonomics, and Usability: Principles and Practice." In E.T. Klemmer (Ed.), *Ergonomics harness the power of human factors in your business* (pp. 79–196). Norwood, Jersey: ABLEX Publishing.

[2] Chakrabarti, Debkumar (1997). "Indian Anthropometric Dimensions for Ergonomic Design Practice." National Institute of Design, p. 161.

[3] Chandra, A.M., Ghosh, S., Iqbal, R. and Sadhu, N. (2007). "A Comparative Assessment of the Impact of Different Occupations on Workers' Static Musculoskeletal Fitness." *International Journal of Occupational Safety and Ergonomics (JOSE).* 13(3), pp. 271–178.

[4] David, C.A. and Mustafa, B.P. (1985). "Industrial Ergonomics. A Practitioner Guide." Industrial Engineering and Manufacturing Press. Institute of Industrial Engineers. Norcross. G.A. 30092.

[5] Directorate General Factory Advice Services and Labour Institute, Mumbai 1948. Model Rules under the Factories Act.

[6] Genaidy, A., Barkawi, H. and Christensen, D. (1995). Ranking of Non-neutral Postures Around the Joints of the Upper Extremity and the Spine." *Ergonomics,* 38, pp. 1851–1858.

[7] Grandjean, E. (1985). "Fitting the Task to the Man: An Ergonomic Approach." 3rd ed. London, UK: Taylor and Francis.

[8] Hill, C. (1989). "Getting to Grips with Repetitive Strain Injuries." Personal Management (UK), pp. 32–35.

[9] Jorgensen, K., Hansen, L., Lundager, K. and Winkel, J. (1993). "Low Back Muscle Reaction to Constrained Standing in Relation to Shock Absorbing Properties of Floor and Shoes." In: Nielson, R., Jorgensen, K., editors. *Advances in industrial ergonomics and safety.* V London, UK, Taylor and Francis, pp. 279–283.

[10] Mallory, M. and Bradford, H. (1989). "An Invisible Workplace Hazard Gets Harder to Ignore." Business Week, Industrial Technology Edition, pp. 92–93.

[11] Marras, W. and Schoenmarken, R. (1991). "Improving Posture While Keyboarding will Reduce the Incidence of CTS." *Paper presented at International Ergonomics Association, Paris, France.*

[12] Miedema, M., Douwes, M. and Dul, J. (1997). "Recommended Maximum Holding Times for Prevention of Discomfort of Static Standing Postures." *Int J Ind Ergon,* 19, pp. 9–18.

[13] Mirbod, S.M., Akbar-Khanzadeh, F., Onozuka, M., Jamali, M., Watanabe, K., Inaba, R. *et al.* (1999). "A Four-year Follow up Study on Subjective Symptoms and Functional Capacities in Workers using Hand-Held Grinders." *Ind. Health,* 37(4), pp. 415–25.

[14] Tyan, A.G. (1989). "The Prevalence of Musculoskeletal Symptoms in Supermarket Workers." Ergonomics, 32(4), pp. 359–71.

Development of a Bench Marking Tool—Safety Management System in Industry

Amitabha De, T. Mogare, J. Saharia and N.S. Naik

National Institute of Industrial Engineering, Powai, Mumbai–400 087, India
E-mail: amitabhade@nitie.edu

ABSTRACT

In India, Bureau of Indian Standards published, "Code of Practice on Occupational Safety and Health Audit" to carry out audit of safety systems as a part of statutory requirements. This audit system provides overall scenario of safety, health and environment of an Industry. However, industry would like to have a quantitative SWOT analysis tool which enables to compare with the best available practices. The process is known as benchmarking. The present paper aims at providing a methodology for developing such a tool. A survey of questionnaire for seeking intention to take part in the study was prepared and sent to 220 industries, 20 responses were received. Based on the above, a detailed questionnaire mentioning corporate information, plant safety profile, plant safety history, safety policy and safety information, safety accountability and responsibility, training programme, action plan and performance appraisal was prepared and sent to the industries for validation. Simultaneously, a team comprising experts having prior knowledge of safety and scaling methodology, experts with no knowledge of safety but having knowledge of scaling methodology and experts having neither the knowledge of safety nor the knowledge of scaling methodology were formed. This team developed relative importance of each HSE management criterion mentioned above using Analytical Hierarchy Process (AHP). Responses received were validated. Questionnaire was modified and again sent to safety professionals and the industries. The scaling method developed had provided the overall HSE performance with 51–60% as 1 star, 61–70% as 2 star, 71–80% as 3 star, 81–90% as 4 star and greater than 90% as 5 star. It also provides information about overall and individual criterion best safety management practices.

Keywords: Occupational Safety and Health Audit, SWOT, Bench Marking.

INTRODUCTION

Professor Robert M Solow, Professor of Economics, MIT while delivering his address on the occasion of receiving his noble prize in economics in 1987, had said, "Bulk of economic growth is the result of technological progress". It is not only in economic growth wherein technology has left its imprint but also in every sphere of human civilisation. However, technology is a double-edged sword; if it not handled properly, it can cause more damage and harm than the benefits accrued from it. A study by United Nations (UN) has estimated that occupational health cases worldwide were around six millions (ILO, 1998). Since Second World War, industries have become increasingly multi-national. Communications and transportations have advanced rapidly. World organisation has been set up to monitor this trade; different cross-country trade groups have been formed. However, issues like health and safety in such organisations are addressed indirectly (ILO, 1998). There are trade blocks who feel' "Life must be preserved for labour and labour must be harmless to life". European Union in 1974 formed the advisory committee in safety, hygiene and health protection at work, as a single European Act in 1987 and CEFIC guidelines on transfer of technology in 1991. This suggests that a beginning has been made to monitor and compare safety and health issues as a common base.

To stay in business and to remain globally competitive, it is necessary to have a proactive approach and to follow the best practices in industry. The process is called bench-marking. Safety benchmarking or safety audit is not a new activity in various industrially developed countries and is being practised for more than three decades. Some of the well known audit systems are Pallenberg Audit System, Responsible Care, Process Safety Management, International Safety Star Rating System (ISRS) and Occupational Health and Safety Assessment Series (OHSAS) 18000.

In India, Bureau of Indian Standards in 1988 published, "Code of practice on Occupational Safety and Health Audit"—Indian Standard IS 14489. The Directorate General Factory Advise Service and Labour Institutes (DGFASLI), National Safety Council and various mutual aid organisations over the years have been rewarding various industries for their safety and health performance.

However, the above organisations, while carrying out an excellent job in promoting safety, health and environment in Indian industries, in most cases have focused on the overall performance and the statistics. An industry would like to have a quantitative SWOT analysis of its safety, health and environmental performances in every sphere of its activities and practices and would like to compare and achieve the best available. Such methodology suitable to Indian industries is not available. Hence, the study titled, "Development of a benchmarking tool-safety management system in Industry" is proposed. The present study is designed for large scale industries and focuses on organisational and operational safety.

METHODOLOGY

A survey questionnaire for obtaining the intention to take part in the study had been prepared and sent to 220 industries. 20 organisations had responded positively. Based on the above response, a detailed questionnaire comprising corporate information, plant safety profile, plant safety history, safety policy and safety organisation, safety responsibility and accountability, safety training programme, safety action plan, safety performance appraisal, and safety reports was prepared and sent to the above industries for their feedback. Simultaneously, based on the replies of the initial survey, an expert group was formed for assessing the relative importance of the above factors and their contribution on the overall, safety performance. Analytical Hierarchical Process (AHP) (Satty, 1980) was used for the above purpose. Even though various other benchmarking methods like paired comparison, Delphi technique are available, AHP was used as it provided relative importance of each safety performance measures and the reliability of the above measures. Based on the above, a five star rating system was arrived at, with ≤ 50% has no score, 51–60% one star, 61–70% two star, 71–80% three star, 81–90% four star, and > 90% five star. The responses received from the above companies along with their documents were then evaluated against this scale.

RESULTS AND DISCUSSION

Results are presented in Tables 1 to 8.

Table 1: Weightage of the Factors Contributing to Overall Safety Management System

Factors	*% Weightage of Each Factor*
A. Factory/plant safety profile	4.5
B. Plant safety history	19.9
C. Safety policy and safety organisation	11.0
D. Safety responsibility and accountability	17.1
E. Safety training programme	10.0
F. Safety action plan	12.2
G. Plant safety performance appraisal	12.3

Table 2: Hierarchy of Factory/Plant Safety Profile

Factors	*% Weightage of each factor*
A. Capital outlay/assets	5
B. Total no. of employees	7
C. ISO certification	13
D. Strength of safety department	14
E. Capital investment in safety department	24
F. Safer technology process change	37

Table 3: Weightage of the Factors Contributing to Plant Safety History

Factors	*% Weightage of Each Factor*
A. Accident information over the past five years	31
B. Amount of compensation paid during the last 5 years	10
C. Environmental consequences during the last 5 years	20
D. Accidents in other plants/locations which have affected the working of your plant during the last 5 years	05
E. Any recognised performance awards	14
F. International safety rating	20

Table 4: Weightage of the Factors Contributing to Safety Policy and Safety Organisation

Factors	*% Weightage of Each Factor*
A. Safety policy, policy document with organisational set up	27
B. Details of organisational set up for safety	19
C. Safety department acts as a...	13
D. Composition and activities of the safety committee in the last financial year	29
E. Does safety policy cover contract labourers	12

Table 5: Weightage of the Factors Contributing to Plant Safety History Safety Responsibility and Accountability

Factors	*% Weightage of Each Factor*
A. Job descriptions are available for...	10
B. Safety responsibility	29
C. Is specific accountability fixed for safety	45
D. Is there any safety-linked incentive/promotion scheme	16

Table 6: Weightage of the Factors Contributing to Plant Safety History Safety Training Programme

Factors	*% Weightage of Each Factor*
A. Training for all employees other than safety personnel	13
B. Training for safety employees	33
C. Types of safety training	31
D. Safety training programme rating	23

Table 7: Weightage of the Factors Contributing to Plant Safety History Safety Action Plan

Factors	*% Weightage of Each Factor*
A. Does a safety plan exist?	17
B. Does the organisation have an emergency plan	17
C. Has job safety analysis been carried out?	34
D. Are Standard Operating Procedures (SOP) specified	24
E. Safety meetings	08

Table 8: Weightage of the Factors Contributing to Plant Safety History Plant Safety Performance Appraisal

Factors	*% Weightage of Each Factor*
A. Safety inspection	12
B. Safety measures	23
C. Safety programme appraisal	42
D. Safety tools used in appraisal	23

The result of the hierarchy of various safety functions is presented in Table 1. The reliability of the hierarchy was found to be 0.016 against the accepted value of 0.1. This showed that the experts were consistent in their judgment. The plant safety performance appraisal was found to be the top most priority area with a contribution of 25.3%, followed by plant safety history, accountability and responsibility, safety action plan, safety policy, safety organisation and safety training programme. Even though the Factories Act 1948 have emphasised the size of a factory (in terms of labour force), the present study showed that factory-plant safety profile is of least importance.

While analysing the details of the above factors, it was observed that in appraisal of safety programme; safety audit, analysis of existing safety related data, hazard analysis and safety measures were found to be more important than that of safety inspection activity. This suggests a need for shift in the approach in safety appraisal programme. Similarly, in responsibility and accountability, fixation of accountability had 45% relative importance while safety linked incentive scheme, promotion scheme and job description together contribute half of the fixation of accountability. In plant safety history, as expected, accident records were found to be a more important factor.

An overall result of the evaluation based on the above scaling technique had shown that only two companies had secured three stars, rest of the four companies had less than three star grading. Results were compared with areas safety reward systems as awarded to these companies and values were not found to be different.

The areas of concern even for the best practices were found to be safety action plan, plant safety history, safety policy and safety organisation and quality of the training programme.

CONCLUSION

The study has showed that the methodology developed may be used for benchmarking safety management practices.

REFERENCES

[1] International Labour Organisation, Encyclopedia of Occupational Health and Safety, 4th Edition, 1998 (ILO/98/39).

[2] Responsible care: http://www.responsiblecare.org/page.asp?p=6341&l=1.

Evaluation of Heat Stress and Ventilation Level in a Chemical Factory in West Bengal, India

Durgesh Chakrabarty[1], Rauf Iqbal[2] and P.C. Ghosh[3]

[1]Tata Chemicals Ltd, Haldia, West Bengal, India

[2]National Institute of Industrial Engineering, Powai, Mumbai–400 087, India

[3]Central Labour Institute, Sion, Mumbai–400 022, India

ABSTRACT

The study was conducted in a chemical factory in West Bengal. Environmental conditions like, Dry Bulb Temperature (DBT), Wet Bulb Temperature (WBT), Globe Temperature (GT), Relative Humidity (RH), Air Velocity (AV), etc., on the shop floor as well as inside the office areas were monitored using a Portable Heat Stress Monitor. Heat stress indices like Effective Temperature (ET), Corrected Effective Temperature (CET) and Wet Bulb Glob Temperature (WBGT) were calculated and compared with the norms. Physiological responses at work like, working heart rate, 3rd min recovery heart rate and working oral temperature of 15 workers working at old ware house and new ware house were monitored. Mean and SD of the age, height and weight of the workers are 47.9 ± 5.18 y, 162.3 ± 5.09 cm and 64.3 ± 6.27 kg, respectively. Their work experience is 12.4 ± 6.97 years. Mean and SD of the working heart rate, 3rd minute recovery heart rate and working oral temperature are 97.2 ± 8.27 beats/min, 91.9 ± 8.93 beats/min and 36.8 ± 0.36°C respectively.

Mean and SD of the RH and air velocity in the Administrative block and OHS & HR section were 57.2 ± 12.06%, 49.7 ± 54.65 cm/s and 63.6 ± 10.74%, 51.0 ± 52.72 cm/s, respectively. These areas in the office are temperature controlled and the heat stress indices have been found well within the normal range. It was found that in all the important locations various heat stress indices were well within the normal range. Only in two locations namely Old warehouse and New warehouse, the environmental stress needed proper attention. Mean and SD of the RH, AV, CET and WBGT inside the Old warehouse are 55.2 ± 1.96%, 22.9 ± 4.65 cm/s, 30.5 ± 0.82°C and 31.6 ± 2.14°C respectively. Whereas in case of New warehouse the values are 50.1 ± 2.77%, 22.5 ± 4.53 cm/s, 31.0 ± 0.72°C and 31.8 ± 0.82°C, respectively. Based on the findings of the study various ergonomic interventions have been provided.

Keywords: Heat Stress Monitor, DBT, WBT, GT, RH, AV, ET, CET and WBGT.

INTRODUCTION

The combination of environmental as well as process heat in industry sometimes makes the working condition stressful to the worker. One of the most prominent factors for developing human fatigue is the level of heat in tropical countries like India. There are various heat stress disorders faced by the workers in industries (Liethead *et al.*, 1958). Frequently appeared problems are heat cramp, dehydration and heat exhaustion, etc. However, the most fatal is the "Heat Stroke" (Lind, 1960; Leithead *et al.*, 1961). The heat stress and the concomitant heat disorders are one of the major problems during summer (Leithead and Wilkinson, 1958) when ambient temperature goes in excess of 40°C. It is one of the most vibrant problems the industry is facing even after the last amendment of Factory Rules (1987). Nowaday majorities of the management are showing positive attitude towards the improvement of physical working environment realising its scientific background and physiological merits for higher productivity. More and more industries are coming forward for evaluation and control of adverse working condition.

Human heat balance system is governed by the climatic factors as well as non-climatic factors (Clifford *et al.*, 1959). The climatic factors are air temperature, air velocity, humidity and radiant heat in and around the workplace. The non-climatic factors are metabolic heat, evaporated heat loss, the conducted heat loss or gain, etc. The consequence of adverse working conditions will affect all the non-climatic factors, if those are not taken care of. The heat disorders are one of the major causes of low industrial productivity. Frequently faced problems are heat cramp, dehydration and heat exhaustion, etc., which are directly associated with poor fitness and low productivity. Some of the workers visited health centre were found of having the problem of dehydration and cramp.

Thus, a need was felt to evaluate the level of Heat Stress Indices and ventilation inside the plant (shop floor and office area) of a chemical industry in West Bengal, India, so that suitable measures can be initiated to reduce heat stress, if found necessary. The study was planned with the following aims and objectives:

- To monitor the selected heat stress indices at various locations on shop floor as well as in office areas.
- To monitor the level of air velocity present in the working environment.
- To monitor the physiological parameters of the workers during work.
- To provide appropriate remedial measures for reducing prevailing heat stress, if any.

MATERIALS AND METHODS

The environmental condition on the shop floor as well as in side of the offices has been monitored in detail. The working environment was evaluated using a Portable Heat Stress Monitor (Questemp°36 NMH), made by QUEST Technologies, USA.

Locations of the Study

Office areas:
(i) Canteen, (ii) Admin block, (iii) OSH & HR, and (iv) Store & Electrical Department.

Shop floor:
(i) Old ware house, (ii) New ware house, (iii) DAP 1 & 2 field, (iv) STPP Bagging, (v) Soda Ash godown, and (vi) Sulphonation plant.

Following parameters have been measured by the Portable Heat Stress Monitor:

- Dry & wet bulb temperature
- Globe temperature as a measure of radiant heat
- Humidity
- Air velocity.

All the physical measurements have been converted into various physiological indices of heat stress such as Effective Temperature (ET), Corrected Effective Temperature (CET), and Wet Bulb Globe Temperature (WBGT) index in degree centigrade (°C) from the standard nomograms (Hatch, 1963). The outside physical measurements also had been carried out in similar manner and used for comparison of radiant heat present on the shop floor.

Physiological Parameters

Working heart rate was monitored by Polar HR Monitor (S 610i) and working oral temperature was monitored by clinical thermometer. Energy cost of work was calculated indirectly from the heart rate responses (David and Mustafa, 1985; McAddle *et al.*, 1996).

RESULTS AND DISCUSSION

Evaluation of Environmental Heat Load

Environmental Heat Load has been calculated in terms of Relative Humidity (RH), Effective Temperature (ET) [Yaglou, 1927; Smith, 1952], Corrected Effective Temperature (CET) and Wet Bulb and Globe Temperature (WBGT) of specific locations both inside and outside so that a comparison can be made. ET, CET and WBGT are the indicators of the heat stress level. Effective Temperature (ET) is a sensory scale of warmth. The ET is equivalent to the temperature corresponding to temperature of still air saturated with water vapour which is supposed to give the same sensation of warmth as that is experienced in the environment in question. Different environmental conditions in different places are converted into ET to make proper comparison. In situations where there is radiant heat source, radiant heat is incorporated with the ET and the product is Corrected Effective Temperature (CET) [Bedford, 1946]. WBGT is another heat stress index which corresponds to radiant heat, ambient air temperature and humidity (Yaglou and Minard, 1957). ET, CET and WBGT are compared with the permissible level and corrective measures are taken if required.

Existing level of heat stress indices on all the locations (shop floor and office areas) have been evaluated and compared with the norms. Physical and physiological profiles of the selected workers on the shop floor have been recorded and presented in Table 1 and Table 2. Mean of age, height and weight of the workers taken into consideration was 47.9 y, 162.3 cm and 64.3 kg respectively. Average work experience of the workers was 12.4 y.

Table 1: Physical Profiles of the Workers Covered in the Study (n = 15)

Variables	*Age (y)*	*Height (cm)*	*Weight (kg)*	*Work Exp. (y)*
Mean	47.9	162.3	64.3	12.4
±SD	5.18	5.09	6.27	6.97
Min	42	153	55	2
Max	54	168	75	25

Table 2 shows that average working pulse is 97.2 beats/min with a corresponding energy cost of 13.5 kJ/min (David & Mustafa, 1985; McArdle *et al.*, 1996). Mean of 3rd min recovery pulse is 91.9 beats/min. This indicates that recovery of the workers is not up to the mark. Mean value of working oral temperature is 98.3°F with a range of 97.0°F to 99.0°F (Table 2). Maximum oral temp is 99°F (Table 2) which indicates that heat is absorbed by the body.

Table 2: Physiological Cost of the Work done on Shop Floor (n = 15)

Variables	*Working Pulse (beats/min)*	*Oxygen Uptake (l/min)*	*Energy Cost of Work (kJ/min)*	*Recovery Pulse (beats/min)*	*Oral Temperature (°F)*
Mean	97.2	0.6	13.5	91.9	98.3
±SD	8.27	0.17	3.48	8.93	0.66
Min	80	0.3	6.3	76	97.0
Max	108	0.9	18.1	100.2	99.0

Office Areas

Various environmental parameters and heat stress indices in the office areas have been presented in the Tables 3–6. It is found from Table 3 that CET in the canteen area is slightly higher than upper permissible limit (29–29.5 °C) as recommended by the Chief Advisor Factories (1957). The range of difference of globe temperature and dry bulb temperature in the canteen is 0.3°C (32.4–32.1) to 1.4°C (36.8–35.4). This indicates that the radiant heat load is preset in the canteen though it is not significantly higher to augment fatigue among the workers.

The areas like Admn block, OSH & HR and Store & Electrical departments are temperature controlled and various heat parameters and indices are found within the recommended range.

Table 3: Heat Stress and Ventilation at Canteen Area

Variables	*DBT (°C)*	*WBT (°C)*	*RH (%)*	*GT (°C)*	*AV (cm/ s)*	*ET (°C)*	*CET (°C)*	*WBGT (°C)*
Mean	34.1	29.1	65.5	34.7	49.7	29.5	29.9	29.3
±SD	1.44	1.04	4.51	1.80	29.39	1.25	1.22	4.13
Min	32.1	27.7	59	32.4	19.9	27.8	28.2	23.1
Max	35.4	30.2	69	36.8	79.6	30.6	30.8	31.6

Table 4: Heat Stress and Ventilation at Admn Block

Variables	*DBT (°C)*	*WBT (°C)*	*RH (%)*	*GT (°C)*	*AV (cm/s)*	*ET (°C)*	*CET (°C)*	*WBGT (°C)*
Mean	26.7	21.4	57.2	27.7	49.7	22.4	22.8	23.3
±SD	3.31	3.70	12.05	2.94	54.65	3.09	2.97	3.49
Min	21.6	16.6	44	23.9	9.69	17.5	17.4	18.8
Max	33.1	27.9	86	33.1	159.6	27.8	27.8	29.4

Table 5: Heat Stress and Ventilation at OHS & HR

Variables	*DBT (°C)*	*WBT (°C)*	*RH (%)*	*GT (°C)*	*AV (cm/s)*	*ET (°C)*	*CET (°C)*	*WBGT (°C)*
Mean	29.3	24.1	63.6	29.8	51.0	24.9	24.9	25.8
±SD	1.96	2.95	10.74	2.08	52.72	2.90	3.25	2.65
Min	27	20.9	50	27.4	6.6	22.5	21.7	23.2
Max	31.5	27.5	76	32.1	129.5	28.1	28.6	28.9

Table 6: Heat Stress and Ventilation at Store and Electrical Deptt.

Variables	*DBT (°C)*	*WBT (°C)*	*RH (%)*	*GT (°C)*	*AV (cm/s)*	*ET (°C)*	*CET (°C)*	*WBGT (Indoor) (°C)*
Mean	30.7	24.6	59.0	31.1	29.6	26.0	26.3	26.5
±SD	4.97	4.82	4.90	4.80	0.00	4.68	4.55	4.81
Min	25.2	18.2	53	25.6	29.6	20.3	20.6	20.4
Max	36.6	29.3	65	36.5	29.6	30.6	30.6	31.8

Shop Floor

The areas covered in the study on the shop floor are Old warehouse, New warehouse, DAP 1 & 2 Field area, STPP bagging, Soda ash godown and Sulphonation plant. The results have been shown in the Tables 7–12.

Tables 7–8 show that CET and WBGT in the Old warehouse and New warehouse are higher than the upper permissible limit (29 to 29.5°C) as recommended by the Chief

Advisor Factories (1957). Difference of mean values of GT and DBT in case of Old warehouse is 1.5°C and in case of new ware house is 1.8°C. This indicates the presence of radiant heat load in the warehouses. Air speed is also significantly low in both the warehouses.

Table 7: Heat Stress and Ventilation at Old Warehouse

Variables	*DBT (°C)*	*WBT (°C)*	*RH (%)*	*GT (°C)*	*AV (cm/s)*	*ET (°C)*	*CET (°C)*	*WBGT (Indoor) (°C)*	*WBGT (Outdoor) (°C)*
Mean	35.7	28.3	55.2	37.2	22.9	30.0	30.5	31.6	30.9
±SD	1.27	0.56	1.96	1.35	4.65	0.63	0.82	2.14	0.74
Min	32.3	27.5	52	35.6	19.9	28.6	28.6	29.9	29.8
Max	37.6	29.5	59	39.5	29.6	30.8	31.9	38.2	32.3

Table 8: Heat Stress and Ventilation at New Ware House

Variables	*DBT (°C)*	*WBT (°C)*	*RH (%)*	*GT (°C)*	*AV (cm/s)*	*ET (°C)*	*CET (°C)*	*WBGT (Indoor) (°C)*	*WBGT (Outdoor) (°C)*
Mean	36.8	28.9	50.1	38.6	22.5	30.6	31.0	31.8	31.2
±SD	0.97	0.56	2.77	1.44	4.53	0.72	0.72	0.82	0.76
Min	35	28.1	44	36.1	19.9	29.2	30.0	30.3	30
Max	38.1	29.7	55	40.1	29.6	31.7	31.9	32.8	32.1

Tables 9–12 shows the heat parameters and ventilation level in the DAP 1 & 2 field area, STPP bagging, Soda ash godown and Sulphonation plant. Though the mean values of CET and WBGT are within the recommended range or marginally high but the maximum values of CET and WBGT in all the locations are beyond the recommended range. However, this will not effectively cause fatigue to the workers since the level of exposure is only for a short period of time during the shift and it can be controlled by applying various ergonomic techniques.

Table 9: Heat Stress and Ventilation at DAP 1 & 2 Field Area

Variables	*DBT (°C)*	*WBT (°C)*	*RH (%)*	*GT (°C)*	*AV (cm/s)*	*ET (°C)*	*CET (°C)*	*WBGT (Indoor) (°C)*	*WBGT (Outdoor) (°C)*
Mean	35.0	28.3	64.1	35.2	73.3	29.1	29.1	30.3	30.1
±SD	1.03	0.66	6.49	0.92	56.74	1.17	1.23	0.72	0.74
Min	34.1	27.2	59	34.1	9.7	26.7	26.7	29.3	29.3
Max	36.5	29.1	82	36.7	209.6	31.1	31.1	31.4	31.3

Table 10: Heat Stress and Ventilation at STPP Bagging

Variables	*DBT (°C)*	*WBT (°C)*	*RH (%)*	*GT (°C)*	*AV (cm/s)*	*ET (°C)*	*CET (°C)*	*WBGT (Indoor) (°C)*	*WBGT (Outdoor) (°C)*
Mean	33.3	27.8	63.3	33.7	52.1	28.4	28.5	29.5	29.5
±SD	2.06	1.42	3.77	2.31	22.14	1.23	1.25	1.59	1.58
Min	31.6	26.4	59	31.7	29.6	27.2	27.5	28.2	28.1
Max	35.9	29.1	68	36.8	79.6	29.7	30.0	31.4	31.3

Table 11: Heat Stress and Ventilation at Soda Ash Godown

Variables	*DBT (°C)*	*WBT (°C)*	*RH (%)*	*GT (°C)*	*AV (cm/s)*	*ET (°C)*	*CET (°C)*	*WBGT (Indoor) (°C)*	*WBGT (Outdoor) (°C)*
Mean	34.6	28.1	58.3	35.6	43.2	29.4	29.8	30.3	30.2
±SD	2.69	1.18	7.09	3.33	25.16	1.37	1.53	1.82	1.76
Min	31.5	26.7	52	31.8	19.9	27.8	28.1	28.2	28.2
Max	36.2	28.8	66	37.8	69.9	30.3	30.8	31.4	31.3

Table 12: Heat Stress and Ventilation at Sulphonation Plant

Variables	*DBT (°C)*	*WBT (°C)*	*RH (%)*	*GT (°C)*	*AV (cm/s)*	*ET (°C)*	*CET (°C)*	*WBGT (Indoor) (°C)*	*WBGT (Outdoor) (°C)*
Mean	32.5	28.4	62.4	34.9	63.8	27.7	29.3	30.4	30.4
±SD	5.50	0.73	4.39	2.23	67.92	3.54	1.83	1.05	1.02
Min	22.9	27.6	55	32.1	9.7	22.2	27.2	29.3	29.2
Max	36.7	29.1	66	37.9	179.5	31.4	31.1	31.7	31.6

RECOMMENDATIONS

Based on the findings of the study, it is evident that level of heat stress and ventilation is well within the normal range in all the office areas (except canteen) and most of the shop floor locations. Following recommendations have been provided to improve the working conditions and making workers more productive, and to ensure health, efficiency and safety at work:

1. There is not any exhaust system in the new warehouse. This is the main cause of heat stress and discomfort among the workers. Wall exhausts and overhead air extractor in the ceiling has been recommended, so that the air circulation can be improved and level of heat would be lowered.
2. Most of the workers in the new and old warehouses have reported extreme weakness and body ache. Their work is mainly loading and handling of heavy loads. It was observed that they used to sweat profusely during work. This caused loss of

electrolytes (Sodium, Potassium and Calcium ion) from the body. Frequent intake of water will not replace the electrolytes into the body. Electrolyte replacement is needed. Therefore,

(a) Workers in the new and old warehouses have been recommended to provide supplement ORS or any other electrolyte solution, so that electrolyte loss can be compensated.

(b) Green vegetable salad with working meal has been recommended since green vegetable contains Sodium, Potassium, Calcium, etc.

3. Water cooler has been recommended in specific areas in the plant, so that cold drinking ware becomes easy access to all the workers in the plant.
4. Source of heat in the canteen for contract workers are oven and asbestos top. It was found that there is only one exhaust present in the kitchen. False ceiling has been recommended, so that radiant heat can be reduced. Additional exhaust has been recommended in the kitchen.
5. Periodical educative programme has been recommended for the workers to make them aware of the adverse effect of hot environment and its impact on the body. Workers should be educated through display of posters in and around the workplace and also through demonstrations, so that they can learn to protect themselves from the adverse effect of heat, since maintenance of safety is the combined effort of the management as well as the workers.

CONCLUSION

The concept of total safety, health and productivity in industrial situation is directly related to the environmental conditions like heat, humidity and ventilation, etc. Thus, monitoring and controlling of environmental condition is essential for a safe and healthy work place and making the workers productive.

REFERENCES

[1] Bedford (1946). "Environmental Warmth and its Measurement." Med. Res. Counc. War Memo No. 17, HMSO, London.

[2] Chief Advisor of Factories, Govt. of India (1957). "Thermal Stress in Textile Industry." Report no 17, Ministry of Labour and Employment, Govt of India, New Delhi.

[3] Clifford, J., Kerslake, McK, D. and Waddell, J.L. (1959). "The Effect of Wind Speed on Maximum Evaporative Capacity in Man." J. Physiol. (Lond.), 147, 253.

[4] David, C.A. and Mustafa, B.P. (1985). "Industrial Ergonomics. A Practitioner Guide." Industrial Engineering and Manufacturing Press. Institute of Industrial Engineers. Norcross. G.A. 30092.

[5] Hatch, T.F. (1963). "Assessment of Heat Stress." In Hardy, J.D. [ed.], Temperature: Its Measurement and Control in Science and Industry. Vol. 3, Pt, Reinhold Publishing Corporation, New York, p. 307.

[6] Leithead, C.S., Gunn, E.R. and Chung, C. (1961). "Muscle Cramps in Cane Cutters in British Guiana." (Mimeograph). Liverpool School of Tropical Medicine.

[7] Lind, A.R. (1960). "The Effect of Heat on Industrial Worker." Ann.occ.Hyg., 2, 190.

[8] Leithead, C.S., Leithead, L.A. and Lee, F.D. (1958). "Salt Deficiency Heat Exhaustion." Ann. Trop. Med. Parasit., 52, 456.

[9] Leithead, C.S. and Wilkinson, W.M. (1958). "Tetany in Heat Stroke." Trans. Roy. Soc. Trop. Med. Hyg., 52, 19.

[10] McArdle, W.D., Katch, F.I. and Katch, V.L. (1996). "Exercise Physiology: Energy, Nutrition and Human Performance." Williams & Wilkins. 4th ed, USA.

[11] Yaglou, C.P. and Minard, D. (1957). "Control of Heat Casualties at Military Training Centers." A.M.A., Arch. Ind. Hlth. 16: 302.

Human Computer Interactions and their Influence on High Performance Manufacturing Systems—A Simulation Study

T. Aravind Krishna, Venkatesh Balasubramanian and T.T. Narendran

Rehabilitation Bioengineering Group, Indian Institute of Technology, Madras, India
E-mail: chanakya@iitm.ac.in; chanakyaaa@yahoo.com

ABSTRACT

High Performance Manufacturing Systems (HPMS) are systems in which factors such as products and their variants, demand, economical conditions and available technologies, undergo frequent changes over time. Considering the ways in which HPMS affect jobs, the perceptual, judgmental and cognitive abilities of humans are likely to be more important than their manipulative skills. This is a major reason for the perception that humans will retain central roles in such manufacturing systems. Hence there is a need for overcoming the inherent physical limitations of the humans and enhance the capabilities in performing a task of interest.

Traditional ergonomic methods and human system interfaces are no longer adequate. This study examines how reliance on human-computer interactions influences performance and how sharing of information related to interaction mediates this influence in the context of an HPMS.

Simulations were conducted based on Decision Field Theory. The simulation results show that reliance on human computer interactions and sharing of interaction related information across the system create synergy between human capabilities and the manufacturing system capabilities. This synergy can be used effectively to enhance system performance under dynamic and uncertain conditions. Sharing both performance and reliance information enhances cooperation between the human operators and the system in an additive manner.

Keywords: High Performance Manufacturing Systems, Human Computer Interaction (HCI), Simulation, Decision Field Theory.

INTRODUCTION

The relentless pressure on modern organisations to stay lean and trim has induced them to keep production, increase production efficiency, respond quickly to customer demands and meet dead lines for delivery.

In this scenario, managements need real time estimation of the current state, rapid decision making, and quick execution of decisions (Kaplan and Norton, 1992; Kumar *et al.*, 2005). The consequent need is good coordination of different management activities and information support for these activities since these plays an important role in effective decision making. Forza and Salvador (2001) underscore the need for information flow to achieve efficient production. They require the operators to possess knowledge of their domains and become up-to-date data on current production.

In the environment of working with online information this paper analyses the effect of Human Computer Interaction (HCI) on decision making in dynamic and uncertain conditions. We used simulations based on decision field theory to analyze the impact of HCI on decision making.

Human Computer Interaction

Human Computer Interaction (HCI) is the study of how people design, implement, and use interactive computer systems and how computers affect individuals, organisations, and society. Human Computer Interfaces are special devices in man machine (human-computer) systems. They serve as a bridge between man and machine (between human and computer) and by their basic function, they allow information exchange between human and machine participants in performing specific operations (Forza and Salvador, 2001).

When investigating human-computer interfaces, with respect to machines and humans as transmitters, receivers and processors of information in a man-machine system, the machine (the computer) is designed to perform well-defined operations, and interaction is served by proper hardware components (keyboard, mouse, visual display unit, etc.), and by specific software components, databases, knowledge-bases, etc.

The human reacts non-deterministically, based on properties such as: physiological attributes (eyes, ears, fingers, etc.), intellectual characteristics (capacity, recognition. learning. decision, etc.), knowledge basis (knowing the environment, the system himself/herself, etc.), and psychological states (concentration, vigilance, fatigue. patience, etc.) (Bannon, 1991; Bannon and Bodker, 1991). Interfaces should allow adequate interaction between the different participants of the man-machine system. Through the interaction processes they should: facilitate satisfactory monitoring of machines by humans, support human intervention in machine operations, aid human decision-making by providing information on system state diagnosis and intervention possibilities. The aim is to establish quick, error-free or error tolerating operation of the entire system, and hence guarantee efficient and reliable system performance.

The Role of the Human Operator in Manufacturing Systems

The role of a human operator in a manufacturing system can be defined by identifying the operator's responsibilities and the services required to fulfill those responsibilities (Figure 1). As an example, the shop floor operator's role may include the activities of controlling and monitoring the machines, taking real-time scheduling decisions, handling materials, etc. These responsibilities are fulfilled by simultaneous performance of several activities which include collecting and processing information, reasoning, executing commands, etc.

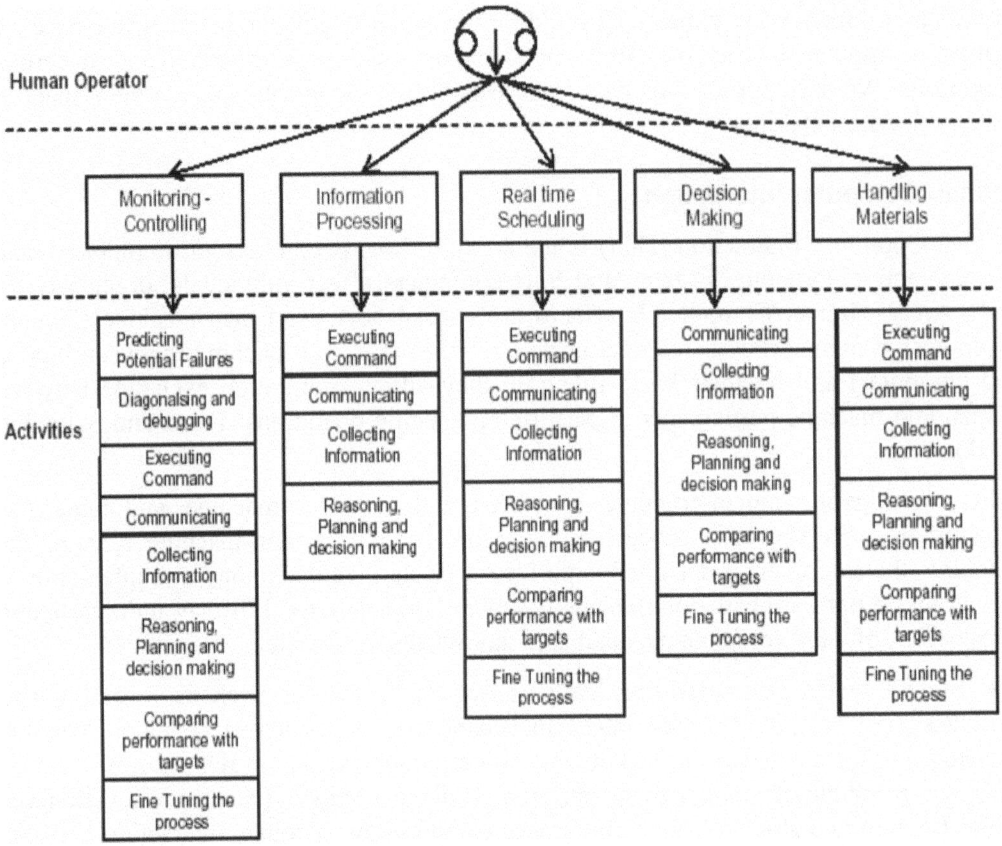

Fig. 1: The Role of the Human Operator in Manufacturing Systems

Automated Shop Floor Control System

An automated shop floor control system places a higher demand on the human operator in terms of the volume of data and work involved in automated control systems. Under such circumstances the operator has to monitor, detect and resolve conflicts to ensure the continued operation of the shop floor leading to increased cognitive fatigue on the operator.

Implications

The above two models of operator (under a conventional and automated shop floor control system) explain the nature of the environment and activities which the shop floor operators are performing. They explain the cognitive fatigue involved in the human operator while making decisions in maintaining the shop floor system under dynamic situations of uncertainties and disturbances. This cognitive fatigue exists in addition to the physical fatigue which normally exists and is quite well explained by number of research works in occupational ergonomics pertaining to physical problems in working environments such as manual material handling, postures to be maintained during performance of physical work in various environments, etc.

Problem Definition

In this paper we examine the scheduling decisions in real time environments which involve uncertainty in demand, frequent changes in product requirements, machine availability, etc. We examine the effect of HCI on decision making pertaining to scheduling jobs in dynamic and uncertain conditions.

Nature of Real Time Scheduling Environments

A lot of research has been carried out in job scheduling and majority of the research work concentrate on mathematical scheduling techniques and algorithms. They make simplified assumptions to remove complexity during computations. Further here shops are perceived as deterministic and manufacturing processes, and are assumed to be stable with predictable and reliable machines, setup times and processing times, etc. This makes the problem less relevant to actual practice. However in the real world, working environments are quite unpredictable. Unplanned events occurring frequently during the scheduling period are common. Operators have to schedule in circumstances where situations are stochastic with unpredictable changes in machines (availability, break downs, etc.), material availability (jigs and fixtures, tools, raw material, etc.), job attributes (priorities, arrival rates, processing rates, etc.) are quite common. Schedulers have to make timely and quick alterations in plans which may be in the following forms:

- Changing the batch size.
- Assigning priorities to jobs.
- Interrupting operations to run urgent jobs.
- Splitting operations between machines and overlapping operations to speedup work.
- Using machines in non standard ways to increase the short term capacity.
- Renegotiating due dates with customers to spread work load.

Uncertainty in Scheduling

Dealing with uncertainty is a key challenge in scheduling. New jobs arrive, before previously scheduled jobs in the system have been processed. The new arrivals may make the existing plan suboptimal. The state of the shop may restrict choices available for amending the schedule. It may be impractical to alter the priority of some jobs in a queue. Often those jobs for which processing is imminent have already been planned for specific resources and materials. Reversing these plans may be difficult. Also frequent changes to the processing order of jobs that are currently on the shop floor may cause chaos and confusion. To limit these effects a scheduler may decide to restrict changes.

Uncertainty extends beyond making changes in resource allocations *viz.* arrival times, rework, scrap rates and machine breakdowns. Several existing constraints such as transportation, raw material, type of work, work force absenteeism, labour rates, etc., need to be considered while altering schedules. Further the constraints and hence the decision taken depends on several factors such as the situation and time pressure. A good schedule generated at this hour may be considered to be a bad schedule if generated a few hours later.

Implications

The above nature of the scheduling requirements demands that the real time schedulers need to have deep knowledge of the working environment. This tacit knowledge needs to be harnessed in the areas of availability of multiple resources and routings, operation precedence, job priorities, random failure of machines, availability of material, changes in production goals, etc. While it is possible to frame rules for many of these factors, frequent changes in products, methods and resource inhibit from spending time and money on creating rule based systems comprehensive enough to make timely decisions.

An alternative approach is for humans and computers to make interactively decisions. Nakamura and Salvendy have proposed and shown that human and computer intelligence can combine to solve the scheduling problem (Paul and Vincent, 1996). The human handles the unstructured part of the scheduling problem, while the computer handles the structured parts of the problem. As an illustrative example one could consider that computer applies procedures for job scheduling using a standard set of heuristics such as FCFS, SPT, etc., while the user exercises the local knowledge in the form of quantitative information (such as order size, due dates) and subjective information (such as customers, operators, management priorities). This enables the system to function effectively as the humans can cope with information that is incomplete, ambiguous, biased, outdated and erroneous.

Interactive Scheduling

Sheridan offers ten variants in sharing responsibility between humans and computers. At one extreme humans act as the principal controller taking advice from the computer while the opposite extreme has computer as the principal controller with humans performing corrections and adjustments (Sheridan and Verplank, 1978). Between these extremes we have the hybrid systems where both humans and computer put forward solutions. Computers carry out data processing displaying results graphically. Humans evaluate the proposed solutions and modify them to suit the situations. Here humans take the primary role in achieving total system performance objectives.

Simulation and Discussion

As the problem definition states, we examine the effect of Human Computer Interactions (HCI) on decision making (scheduling) in dynamic and uncertain conditions. The act of scheduling under uncertainty is studied as an activity of decision making in an uncertain environment. Here task of scheduling is modelled as a dynamic decision task where the choice and decisions of future depends upon the choice and decisions made in earlier stages.

Decision Field Theory

Decision Field Theory (DFT) is a dynamic cognitive approach to human decision making (Busemeyer and Diederich, 1993). It is based on psychological rather than economic principles. The decision processes proposed by DFT have been applied successfully across a broad range of cognitive tasks including sensory detection, perceptual discrimination, memory recognition, etc. DFT proposes the usage of data in the form of a matrix with the options available for selection in the rows and the states with the values representing all the possible evaluations of each option on each attribute under each state of nature in the column.

In this simulation experiment a model based on DFT was created with two options considered for the scheduling activity, (a) with HCI and (b) without HCI. Two states considered were (a) situation under time pressure and (b) situation with no time pressure.

Two scenarios were created; situations where the operator wants to (i) maximize machine efficiency and (ii) give priority for job completion.

Simulated Environment

A "simulated environment" was created in our lab where computers replicated machines in the shop floor. Ten subjects representing operators performing scheduling were directed to schedule jobs to computers. Situations of uncertainty and dynamic job arrivals were created by randomly introducing urgent jobs and altering priority of jobs. The performance of the scheduling activity with HCI was simulated by displaying complete information of machine and job status both in numbers and graphical form (Gantt chart) infront of the operators, while the scheduling activity without HCI was replicated by appraising the operators every now and then about the status of the machines and jobs. The shop floor environment was replicated by defining certain machines as identical machines capable of performing similar jobs; randomly creating situations of pre-emptions, machine breakdowns, etc.

The performance of the system in terms of machine efficiency and meeting due dates (each with complete information and with briefing) was recorded for each operator pertaining to each option and under each state. In addition the time taken by the operator for decision making was also recorded. The range of variations of each parameter was recorded.

Simulation Using Pro-model

ProModel software was used to simulate the scheduling environment. The decision making time of the operator under each scenario was randomly chosen within the range noted down in the simulated experiment. The simulation was run for 720 hours (1 month) for each operator. The performance of the shop in terms of machine efficiency and number of late jobs per day and the performance of the operator in terms of average time taken for each decision making were measured under each scenario.

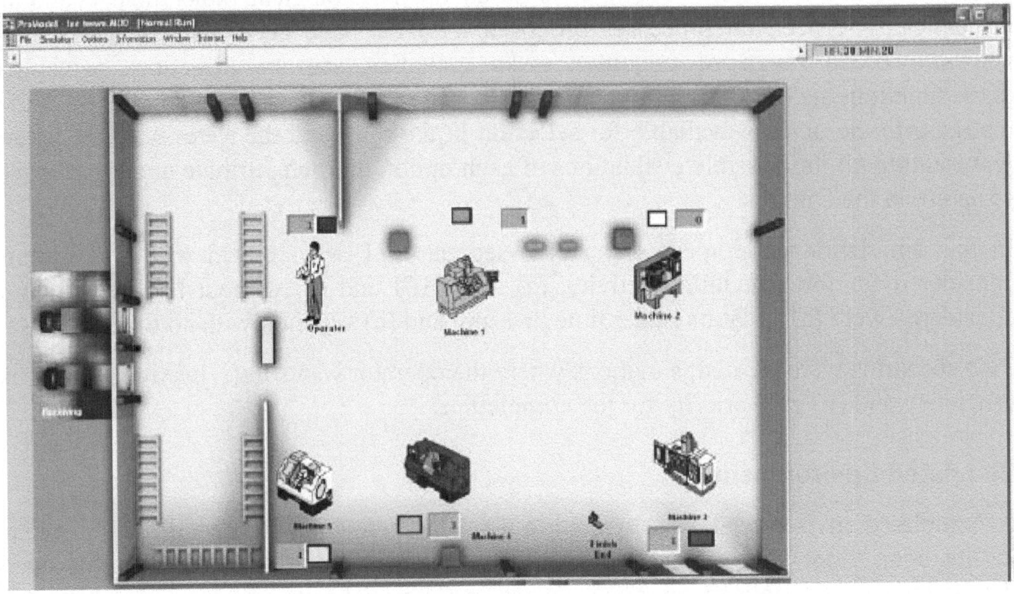

Fig. 2: Snapshot of Simulation done using ProModel

RESULTS AND IMPLICATIONS

Statistical analysis was performed using one-way ANOVA on decision making time for each condition maintaining machine efficiency criterion and ensuring job completion criterion under time pressure and no time pressure circumstances. These results showed that HCI interventions were significant ($p < 0.05$) under all circumstances. The results indicate that human computer interactions improve the performance of the operator. It supports the belief that human processing speed is higher with complete information than with piecemeal information. It also supports the view that the difference between decision making under time pressure and without time pressure are significant. The outcome will be more pronounced with higher number of jobs and machines. This indicates that human computer interaction aids in human decision making by providing clear and relevant information. This will be particularly effective under time pressure, since it improves operator's awareness of the situation and reduces cognitive fatigue.

CONCLUSIONS

The study examines how reliance on human-computer interactions influences performance and how availability of information mediates this influence in the context of an HPMS. The simulation results show that reliance on human computer interactions and sharing of related information across the system create a synergy between human capabilities and the manufacturing system capabilities by reducing the work pressure on the operator. The operator can be more effective and efficient in decision making thus improving the performance of the entire system. This partially supports Bannon's recommendation (Bannon, 1991; Bannon and Bodker, 1991) that it is more important to design systems that support workers in their activities than construct fully automated systems.

ACKNOWLEDGEMENT

We would like to express our sincere thanks to all RBG members and our friends who participated in our study.

REFERENCES

[1] Bannon, L.J. (1991). "From Human Factors to Human Actors: The Role of Psychology and Human-Computer Interaction Studies in System Design." In J. Greenbaum & M. Kyng (Eds.), *Design at Work: Cooperative Design of Computer Systems* (pp. 25–44). Hillsdale, NJ: Lawrence Erlbaum.

[2] Bannon, L.J. and Bodker, S. (1991). "Beyond the Interface: Encountering Artifacts in Use." In J.M. Carroll (Eds.), *Designing Interaction: Psychology at the Human-Computer Interface* (pp. 227–252). Cambridge, UK: Cambridge University Press.

[3] Forza, C. and Salvador, F. (2001). "Information flows for high-performance manufacturing", *International Journal of Production Economics,* 70, 21–36.

[4] Jerome, R. Busemeyer and Adele, Diederich (1993). "Survey of decision field theory", *Psychological Review,*Vol. 100. No. 3.432–459.

[5] Kaplan, R.S., Norton, D.P. (1992). "The Balanced Scorecard—Measures that drive performance", *Harvard Business Review*, January–February, pp. 71–79.

[6] Kumar, K.D., Karunamoorthy, L., Roth, H. and Mirnalinee, T.T. (2005). "Computers in manufacturing: towards successful implementation of integrated automation system", *Technovation,* 25, pp. 477–48.

[7] Paul, P.M. Stoop and Vincent, C.S. Wiers (1996). "The complexity of scheduling in practice", *International Journal of Operations & Production Management,* Vol. 16, Number 10 1996, pp. 37–53.

[8] Sheridan, T.B. and Verplank, W.L. (1978). "Human and Computer Control of Undersea Teleoperators," MIT Man-Machine Systems Laboratory, Cambridge, MA, Tech. Rep.

Ergonomic Issues Related to Computer-based Devices Used in Intensive Care Unit

Ganesh Bhutkar[1], N.R. Rajhans[1], Dinesh Katre[2] and M.L. Dhore[1]

[1] VIT, Pune

[2] C-DAC, Pune

ABSTRACT

This paper is evolved during the research work on usability engineering issues with computer-based products used in the Intensive Care Units (ICUs). The main objective of the research work was to study some selective computer-based products used in ICUs and identify the problems with designs of these products.

The research work involves major knowledge domains like Human Computer Interface (HCI), Ergonomics, Medical Science, Electronics and Computer Science. Medical usability is an interesting field of research. Internationally, organisations such as Usability Professional's Association (UPA), Human Factors and Ergonomic Society (HFES), US Department of Health and Human Services and few American and European Universities are working in this field. But as hospitals and ICUs are going for more automation with sophisticated equipment, machinery and software such as Intelligent Systems or Decision Support Systems (DSS), Central Monitoring Station, Electrocardiogram (ECG) Analysers, etc., medical usability is becoming more relevant. The automated ICU systems can result in the wrong medication being dispensed to patients. Therefore, medical usability, especially in ICU, is often a matter of life or death.

Whenever user resorts to sticky notes or other paper-based workarounds, it is almost certain that a failed user interface has occurred. If the interface fails to provide adequate feedback, users might not even realize that they have committed an error. The software person, equipment manufacturer and the user generally work independently. This creates a big gap between these three agencies. To some extent, working in unison or standardisation of equipment displays will help in achieving the target.

Keywords: Ergonomics, Intensive Care Unit (ICU), Usability Issues, Ventilator Systems and Multi-Parameter Systems.

INTRODUCTION

Ergonomics is a discipline that seeks to improve human performance in use of devices by means of hardware, software or architectural design that is compatible with the abilities of the users. Although many manufacturers of medical devices now integrate ergonomics or usability design into their devices, there exists a wider need for improvement in medical devices, especially those that are used in the Intensive Care Units (ICUs). Device-induced errors in the ICUs can lead to patient injuries or deaths. The behaviour of physicians and their assistants is directly influenced by operating and physical characteristics of the devices. User interfaces that are misleading or illogical can induce errors even when operated by skilled users like ICU physicians in emergencies (Nielson, 2005). Therefore, medical devices can be used safely and effectively only if the interaction among the different elements like operating environment, user capabilities and stress levels are addressed in the design.

Our research cuts across multiple disciplines like Medical Science, Human Computer Interface (HCI), Ergonomics, Electronics and Computer Science. Medical usability is all the more relevant, as hospitals are going for more automation with sophisticated equipment, machinery and computer software such as Intelligent Systems (Melles *et al.*, 2002), Decision Support Systems (DSS) (Thursky and Mahemoff, 2007), Ventilator Systems, Multi-Parameter Systems, etc. ICU system interfaces must focus on efficiency, ease of use, accuracy and goal fulfilment. Otherwise, it can result in wrong medication being dispensed to patients. It is a matter of life or death.

Medical Devices in ICU

The study consisted of following computer-based devices in ICU:

Multi-Parameter Monitoring Systems

This device is shown in Figure 1A. It is used for monitoring of invasive or non-invasive pressures, ECG, EEG, arrhythmia analysis, arterial oxygen saturation, etc.

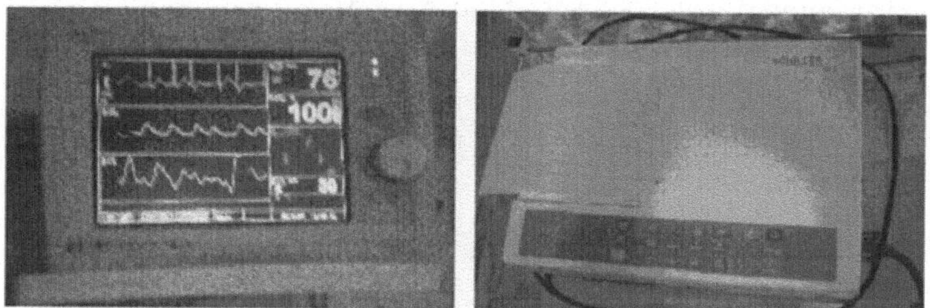

(A) (B)

Fig. 1: Medical Devices

ECG Analyzer

This device is shown in Figure 1B. It helps in monitoring and analyzing the ECG tracing to provide and store full diagnostic reports periodically.

Ventilator Systems (Thursky and Mahemoff, 2007): This device is shown in Figure 2. It gives respiratory support to critically ill patients.

These are most critical, commonly required computer-based devices in ICU, which provide extensive information; so their selection becomes a necessity. The research was conducted in few ICUs in and around the city of Pune to study and identify major ergonomic issues with computer-based products used in ICU.

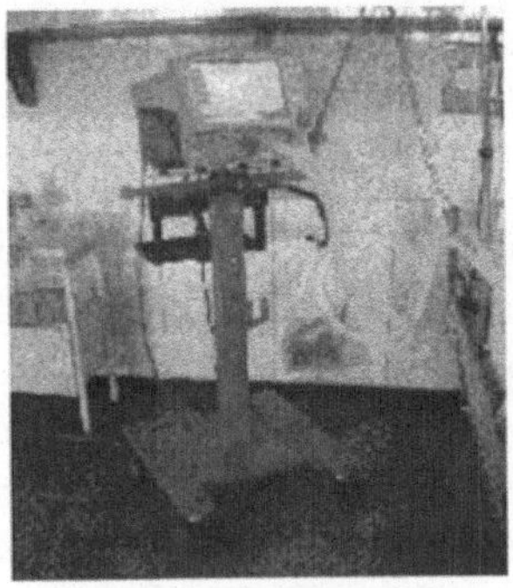

Fig. 2: Ventilator

Types of Users

Different types of users are considered in this research.

Primary Users: ICU physicians, specialised physicians, resident doctors, sisters and patients.

Secondary Users: Intern doctors, ward boys and biomedical technicians.

Tertiary Users: Administrators, relatives of patients, visitors.

We have tried sampling of some subset of these users.

METHODOLOGY

Following two research methods were used.

Questionnaire

The questionnaire had two sections. First section was about personal data like name, age, sex, height, left or right handed dexterity, education, organisation, products used, frequency of usage, time spent, failure rate, etc. Second section asked about user response on different ergonomic issues (Sawyer, 1996) related to selected computer-based devices used in ICU.

Fieldwork, Interviews and Observations

It included frequent visits to hospitals, rigorous interviews with ICU staff and observations while using the medical devices.

Ergonomic Issues

Some ergonomic issues (Sawyer, 1996) that were observed during this research with computer-based devices in ICU are elaborated below.

Controls

Some users complain about improper arrangement and spacing of controls and switches. They felt that there was enough scope for improvement. As per their experience, frequently used switches don't last long.

Right-handed Design: We observed that the positioning and layout of control panels, especially with ventilator systems, were designed only for right-handed users. As shown in Figure 3A, the right-handed design neglects the convenience of left-handed users by making it difficult for them to use.

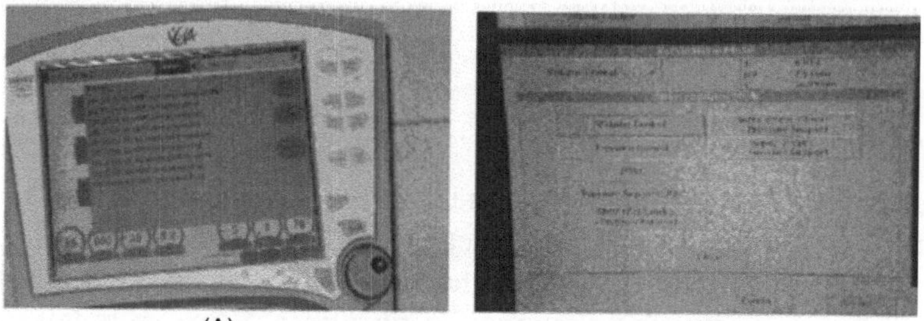

(A) (B)

Fig. 3: Controls and Functionality

Functionality

Most users felt that there were extensive functionalities provided with both ventilator and multi-parameter monitoring systems. One screen showing extensive functionalities with ventilator system can be seen in Figure 3B. Many of these functionalities were underutilised and some of them at times even cause cognitive load to the user.

Readings

Some users felt that multi-parameter monitoring systems had poor readability. Due to reflected light, the numerical readings were not visible in ventilator systems and monitors as observed in Figure 4A. Technicians felt that aging factor of both systems and the users can complicate the readability and contrast issue with monitors of these systems.

Training

All users felt that there was need for training for some users working in ICU. Almost all users learn by experience and not by training. Due to lack of training, many functions remain hidden from the users. Many physicians felt that nurses needed training as they lack reasonable understanding of graphical interfaces, even in the form of reports. For example, ECG reports generated by ECG analysers or multi-parameter monitoring systems (as shown in Figure 4B) should be understood even by intern doctors and nurses at times like emergencies. Physicians also felt that they were extremely busy and had lesser or no time for training.

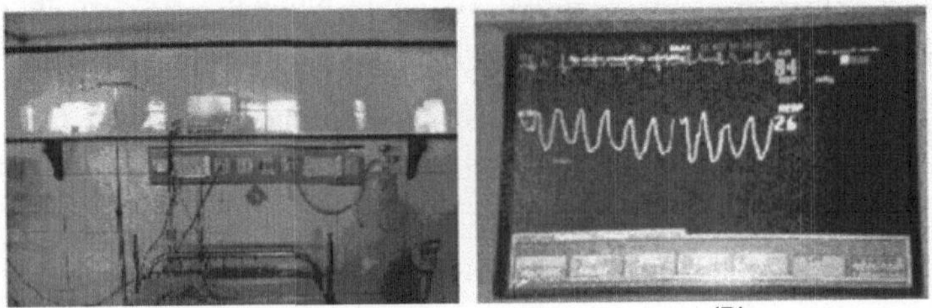

(A) (B)

Fig. 4: Ambient Light and Graphical Interface

Manuals

User manuals are very important element in use of medical devices. Users have observed that manuals were difficult to locate when needed or manuals were not accessible because they were in custody of some authority. Others felt that it was not frequently required as they were happy using routine functionalities. Some users felt that it should be available on CDs or as an online help. Textual complexity of user manuals (Vela Ventilator Systems, 2004) frustrates users. Many users complained that these manuals were available in many foreign languages like English, French or Dutch, but not in Indian languages like Hindi, Marathi or Tamil. Especially, the secondary users as mentioned in point 2, felt frustrated with foreign languages.

Interface Labels and Symbols

A few users felt that interface labels, symbols and abbreviations were cultural specific. They need to be localised. Some users felt that there was a need for some standardisation of symbols and abbreviations. As medical devices in ICU remain operational for about five to ten years, the labels and symbols disappear if hard solvent is used for cleaning the devices.

Data Entry

The devices like multi-parameter monitoring systems and ECG analysers have an interface to enter patient data. But most users found the entry steps to be cumbersome and time-consuming. So they rarely use data entry facility.

Self Diagnosis

Users felt that medical devices like ventilator systems should be more intelligent to support an error recovery and self-diagnosis. Medical devices should support use of defaults or templates to guide secondary users to handle some specific situations. The alarms informing different conditions should be differentiated.

Portability

Most of the users felt that ventilator systems need to be more compact with a robust design to support mobility, portability and storage. Especially, dismantling of ventilator system stand was time-consuming.

Accessories

Usually the accessories were complex. Many vendors have monopolistic approach about accessories and their consistent availability can be doubtful. The probes are delicate and they need to be sturdier. Figure 5 shows outlet tube, which is an important accessory of ventilator systems. The tubes can obstruct the view of monitors.

RESULTS

In this research, about ten users from five different hospitals gave feedback. They are classified as ICU physicians, specialized physicians, resident doctors, intern doctors, nurses and biomedical technicians.

Following graph shows the common trends in the user responses. It is shown in the form of commonality of opinion on user concerns specific to the issues discussed so far. The higher the height of bar, greater is the concern.

In this survey, users were also asked about their satisfaction level related with above-mentioned medical devices. Their response is recorded as a User Satisfaction Index (USI). USI is a response value (out of 10) provided by the user. The USI graph is shown in Graph 2. Here, red line conveys average value of USI, which is about 7.15.

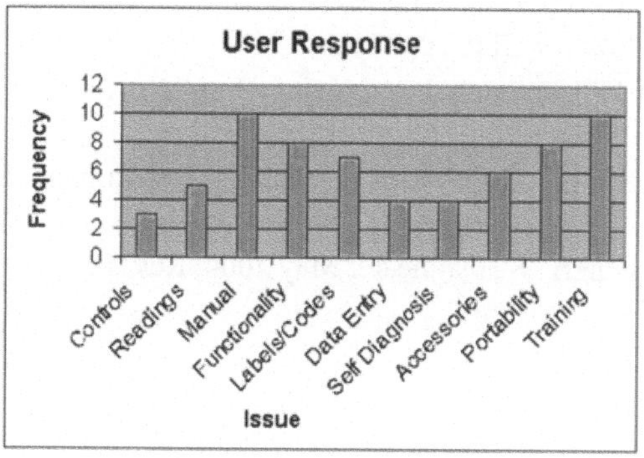

Graph 1: User Response

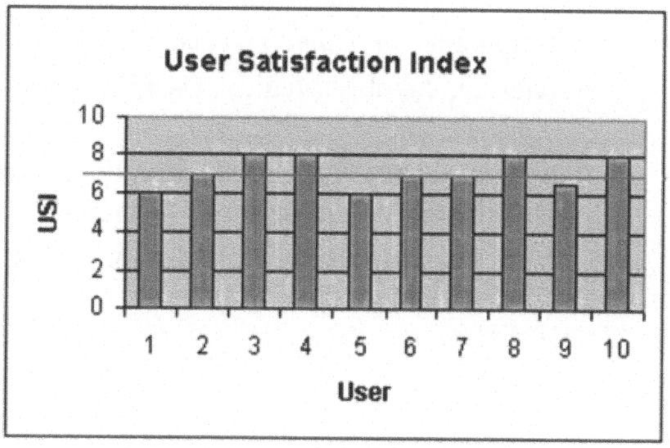

Graph 2: User Satisfaction Index

CONCLUSION

The study and observations discussed so far evidently show major scope for enhancing the medical devices, particularly the ventilator systems in ICU, from ergonomic and usability perspective. Considering the life criticality and the impact of design on ICU patients and doctors, the proposed enhancements must be accorded highest priority.

FUTURE WORK

The problems highlighted in this paper should be taken as challenges for further research. Our explorations henceforth will focus on identifying the usability guidelines and solutions to specific ergonomic problems. We also would like to explore the possibility of evolving universal guidelines for enhancing the usability of ICUs globally.

ACKNOWLEDGEMENT

Authors would like to express their gratitude to physicians and staff at Bharati Hospital, Sahyadri Hospital, Siddhi Hospital at Pune, Giriraj Hospital at Baramati and Jogalekar Hospital at Shirwal for their involvement and support in the research work for development of this paper. We would specially like to thank Dr. Shahaji Deshmukh, Director, Bharati Hospital, Pune for his continuous encouragement during this research work. We also thank Dr. Mrs. Gayatri Godbole, Dr. Rajiv Doshi, Dr. Ramesh Bhoite and Dr. Harshwardhan Anturkar for their kind support in this work.

REFERENCES

[1] Karin, A. Thursky and Michael, Mahemoff. "User-centered Design Techniques for a Computerized Support Antibiotic Decision System in an Intensive Care Unit", *International Journal of Medical Informatics* 2007, 76, pp. 760–768.

[2] Melles, M., Freudenthal, A. and Bouwman, C.A.H.M. "Introducing Intelligent Systems into the Intensive Care Unit: A Human-centred Approach", *Proceedings of the 21st European Conference on Human Decision Making and Control,* 2002, pp. 110–117.

[3] Nielson, Jakob. "Medical Usability: How to Kill Patients through Bad Design", Alertbox, April 11, 2005.

[4] Sawyer, Dick. "An Introduction to Human Factors in Medical Devices", US Department of Health and Human Services, Dec 1996.

[5] Vela Ventilator Systems, Operator's Manual, VIASYS Healthcare, May 2004, Rev. E.

Demographic Characteristic of Upper Limb Discomfort in Computer Users: A Cross Sectional Study

Shweta Bodkhe, Kavitha Raja and John Solomon

Department of Physiotherapy MCOAHS, Manipal University, Manipal, India

ABSTRACT

A questionnaire was prepared consisting of 16 items that included: (1) anthropometric characteristics, (2) job description, (3) hours of computer use, (4) frequency and duration of computer use, and (5) characteristics of upper limb discomfort, if any.

Three hundred eighty four computer users of Udupi city participated in the study, of them 40.63% were positive for upper limb discomfort. Out of total subjects 181 subjects were female in which 44.75% were positive for discomfort and 203 subjects were male in which 36.95% were positive for discomfort. Further correlations with demographics and the work related factors were correlated to the discomfort and discussed in the paper.

Keywords: Upper Limb Discomfort, Computer Use, Demographic Characteristics.

INTRODUCTION

The upper extremity is a complex machine, intricate in its construction and function. The hand is an organ of grasp as well as of fine movements. It is an organ of sensation, fine discrimination, and exquisite dexterity. The large portion of brain controls the hand is an evidence of the importance of this organ. The primary role of the entire upper limb shoulder, arm, and forearm is to place the hand in its position of function. The daily activities involving repetitive movement may cause micro trauma to tendons controlling the upper limb and may leave muscles either in shortened or stretched positions which make the underlying soft tissues nerves vessels prone to compression leading to discomfort.

Upper extremity discomforts include a variety of musculoskeletal and nerve disorders that can be related to overuse of the upper limb. There is a tremendous increase in the number of reported cases of upper extremity discomfort in the last two decades (Werner *et al.*, 2005). It is now well accepted that repetitive work with the hand can lead to these complaints. The use of high forces and the need for fixed postures compound the effect on the clerical workers. The less well defined shoulder /neck disorder is seen in workers who undertake light repetitive work in fixed postures, such as computer operators. There has been an exponential increase in the number of "light" jobs and in the percentage of workers under taking these jobs. The causes of occupational upper extremity discomfort arising from these light white-collar jobs are important and particularly for office/ clerical worker performing keyboard tasks (Anon, 2005).

The high prevalence of computer work in almost every industry has given rise to many Musculoskeletal Disorders (MSDs) of the neck and shoulders. These disorders known as repetitive strain injuries or cumulative trauma disorders generally occur in the upper limbs, low back and neck regions (Lacerda *et al.*, 2005; Erus and May, 1999).

The resulting health consequences and the loss of productivity stemming from the modern office have encouraged some countries to adopt a reactive approach rather than a pro-active approach towards these problems. In India the individuals working, as clerical staff do not find ergonomically designed workstations. This worsens the scenario. Prevalence studies from India of discomfort due to use of VDT in clerical staff is difficult to find, but data collected from epidemiological studies provides sufficient evidence to find a relationship in other countries. However, data pertaining to India is crucial prior to embarking on any meaningful intervention strategy.

Need for the Study

Upper extremity discomfort is the first indicator of repetitive strain injuries of upper extremity. Upper extremity discomfort may be used, as a predictor of upper extremity disorder. To our knowledge there are no reliable estimates of incidence of upper extremity discomfort among clerical staff in Indian population, and this study will help to identify the population at risk.

Objective

To identify the demographic characteristics in computer users with upper limb discomfort

METHODOLOGY

A questionnaire to assess the upper limb discomfort was prepared. Principles were followed and questionnaire was designed so as to make it simple and easy to understand and address work related musculoskeletal symptoms.

Content validity of the questionnaire was ascertained by five experts in the field. Modifications suggested were incorporated which included adding questions specific to type of symptoms and simplifying the questions.

The completed questionnaire was checked for understandability on five subjects who were part of our target population and suggested changes were incorporated and the final questionnaire was prepared, which consisted of two domains (A) generic which had nine items in it that asked for personal details (1) anthropometric characteristics (2) job description, (3), Hours of computer use, (4) frequency and duration of computer use (5) health status (B) Specific domain asked about characteristics of upper limb discomfort if any, (1) type, (2) area of discomfort which was tabulated in a manner so that each type of discomfort could be rated for the specific site, using rating on VAS scale. Body chart was incorporated in the questionnaire to ascertain the area of discomfort and different color/shading could be used to demarcate the area.

All the establishments using computer in Udupi city were identified, and permission was sought to survey their employees using the formal letter/personal meeting notifying the intentions, questionnaire, presentation and data collection details.

Thirty-two institutions out of 42 gave permission for conducting survey. Employees in these institutions who fitted the inclusion criteria of computer usage for 4 h/day or 20 h /week were included in the survey. Subjects with pre-existing history of neurological or musculoskeletal complaints were excluded. For surveying first a formal introduction and type of data collection and informed consent was taken from the subjects. Then an introduction to computer work related musculoskeletal disorder was given and were told about how the symptoms may look like.

RESULTS

Data analysis of the questionnaire that was completed (384) showed that out of the total population 156 subjects (40.625%) were affected or suffered upper limb discomfort. The gender wise prevalence was 44.75% of women and 36.95% of men. Age range of the male and female subjects is tabulated in Table 1 which shows that 40.81% of male affected in age range 20 to 30 years are more than others, while 37.83% of male in age range 30 to 40 years, 34.92% of male in age range >40 to 50 years and 31.81% in age range >50 to 60 years were affected.

Table 1: Age and Gender Distribution of Affected Individuals

Age Range	*Total*	*Affected Male*	*Affected Males (%)*	*Age Range*	*Total Female*	*Female*	*Affected Females (%)*
20–30	49	20	40.82	20–30	76	53	69.74
>30–40	37	14	37.84	>30–40	51	24	47.06
>40–50	63	22	34.92	>40–50	43	15	34.88
>50–60	44	14	31.82	>50–60	15	10	66.67

Table 1 also shows that for female subjects the frequency of discomfort observed (69.75%) is more in age range 20 to 30 years while percentage of discomfort reported for age range >30 to 40, >40 to 50, >50 to 60 were 47.05%, 34.88%, 66.67% respectively.

The frequency of area for discomfort shows that upper back is the most common area for discomfort. Next to it is the shoulder, hand, finger, wrist, elbow, arm, and forearm in descending order (Figure 1).

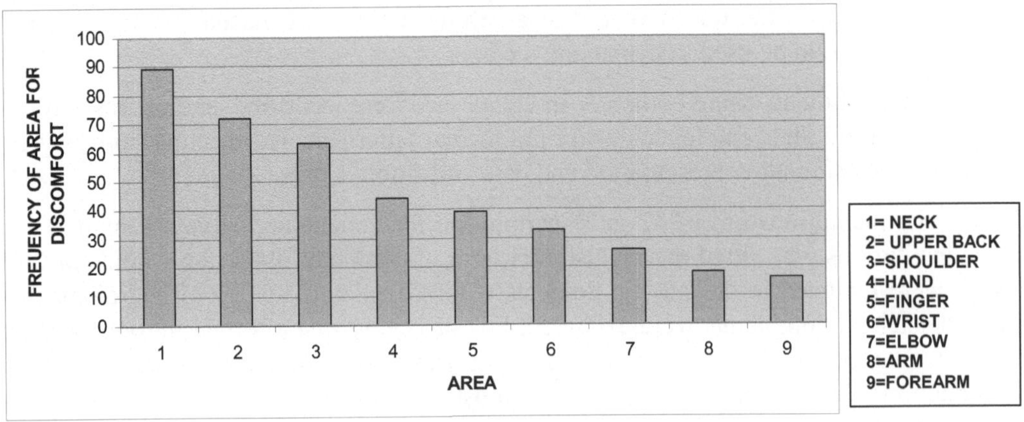

Fig. 1: Area Wise Distribution of Discomfort

DISCUSSION

The issues of health perception and reporting behaviour arises with the use of symptoms questionnaires. Perception and reporting of pain are often dependent on health belief systems, gender, age, culture and the respondent's level of trust about what will be done with the information provided. Symptoms questionnaire are not intended to be clinically diagnostic but instead allow standardisation of information for surveillance data collection and help to identify potential problems in the work environment.

We observed insignificant difference between men and women when overall population was considered. When the population between 20–30 years was considered, females were more affected than males (69.75% women *vs.* 40.82% men). Though most of the studies shows wrist as the commonly reported area for discomfort, we did not find a significant number of subjects reporting wrist discomfort. The most common areas reported for discomfort were neck, upper back and shoulder in descending order. This may be related to inappropriate furniture height, long working hours and inappropriate concern to person's height.

Here it is important to specify that none of the establishments paid much attention towards ergonomics, but they were using off the shelf furniture. In certain establishments they had made minor considerations to ergonomic workstations but these were rarely used by subjects because of lack of awareness.

Although results of this study are inconclusive due to geographic restriction to one area, it however indicates the need for preventive strategies and awareness programmes to forestall the occurrence of WMSD's which can reach epidemic proportions.

In future we need a larger scale prevalence study and to see if there is any correlation between working hours, working condition and discomfort so we need to do a longitudinal study from the time a person enters the work force and note the time when he develops the discomfort.

REFERENCES

[1] Anon (2003). "Cumulative Trauma Disordes in Office Workers." *Public Employees Occupational Safty And health Program VDT Guidelines,* N.J. department of Health and Senior Services.

[2] Erus, H. and May, S. (1999). "Employment Relations and Union Services Health and Safety-repetitive Strain Injuries." The charted society of physiotherapy 14 Bed ford row London WCIR 4 ed.

[3] Lacerda, E.M., Nacul L.C., Augusto, L.G.S., Olinto, M.T.A., Rocha, D.C. and Wanderley, D.C. (2005). "Prevalence and Association of Symptoms of Upper Extremities, Repetitive Strain Injuries (RSI) and RSI-like Condition." *A cross sectional study of bank workers in Northeast Brazil BMC public Health.* 5, p. 107

[4] Werner, Robert A., Franzablau, A., Gell, N., Ulin, S.S. and Armstrong, T.J. (2005). "Predictors of Upper Extremity Discomfort", *A Longitudinal Study of Industrial and Clerical Workers Journal of Occupational rehabilitation,* 15(1).

SECTION–VII

Occupational Health and Safety

Role of Anthropometry in Ascertaining Occupational Health Problems in VDT Operations in a Chemical Industry

Prabir K. Sarkar[1] and Amalendu Samanta[2]

[1]Factory Medical Officer, MCC PTA India Corp. Pvt. Ltd.
Vill & P.O. Bhuniaraichak Purba, Midnapore, West Bengal–721 635, India
[2]Asst. Research Officer, All India Institute of Hygiene and Public Health
110, C.R. Avenue Kolkata–700 090, India
E-mail: IDA1160@cc.m-kagaku.co.jp; [2]amalendusamanta06@gmail.com

ABSTRACT

Video display terminals have become an integral part of regular plant operations. The present study was initiated among the video display terminal operators in a chemical manufacturing industry to correlate the various anthropometric parameters with stress symptoms pertaining to the musculoskeletal and visual system. The occupational stress was evaluated through the administration of a questionnaire and its comparison with the measured anthropometric parameters from the working posture of the operator. The following six component angles namely eye-monitor top, neck-monitor, neck-twist, chair-backrest, trunk-thigh and monitor-tilt were selected for correlating the subjective musculoskeletal and visual stress. It was observed that out of the 168 operators, 94 complained of musculoskeletal and 91 complained of visual stress respectively. It was seen that there was a substantial deviation from the reference component angles ranging from 33%–79% with eye-monitor top angle surpassing all other five parameters. It was observed that 54.6% operators had deviation in terms of eye-monitor top, neck-monitor, neck-twist and monitor-tilt angles contributing for 53.8% visual stress. An estimated 45% of operators suffered from musculoskeletal stress out of the 48% whose workstation parameters namely eye monitor top, neck monitor, neck twist, trunk-thigh and chair backrest angles were not within the reference limit for an ideal computer workstation. The study has established a definite relationship among the occupational stress symptoms and the various ergonomic factors involved in an ideal video display terminal.

Keywords: Video Display Terminal.

INTRODUCTION

Computers have become an essential element in every type of office environment. Now many office users worry about ergonomic problems, including visual fatigue, poor sitting posture, repetitive motion injuries and exposure to radiation. Physical factors and their influence on health among computer or VDT operators have been studied since the 1970s, and a range of workstation and physical environment factors have been linked with musculoskeletal problems in VDT work (Cakir *et al.*, 1978). Considering the sedentary lifestyle compiled with prolonged time spent on VDT, there is the increased incidence of musculoskeletal and visual stress symptoms in the workforce. The continuous nature of computer work is responsible for a steep rise in cases of back pain and other morbid conditions affecting the general lifestyle of the worker. The current study was conducted amongst the various categories of employees in a chemical industry who were involved in VDT operations for a significant period of their working time. The study was performed to assess the existing office-seating scenario and its correlation with the associated health risks. The thorough assessment of the workplace aided by anthropometric measurement enabled the actual evaluation as to the deviations from the prescribed standards and their correlation with stress symptoms. The involvement of the total population and their average physical characteristics made this an ideal platform for assessment of occupational stress in context of Indian industrial scenario.

METHODS

The present study was carried out with an objective to assess the occupational health problems related to computer work using anthropometry as a tool to relate the deviations from the prescribed standards of ergonomic seating and workstation design. In consideration of the meagre number of subjects, 95% sampling was done for data collection. The study was performed for a period of two weeks in November 2005. The study subjects were administered a pre-tested, pre-designed semi-structured questionnaire covering the details like age, experience with computer work, duration of daily computer exposure and musculoskeletal and visual stress symptoms experienced by the operator. Considering the nature of job, two categories of employees namely control panel and video display terminal operators were selected for the study. A total of 109 video display and 59 control panel operators were included in the study. The working environment of each operator was assessed with the help of portable anthropometric set (GPM, Swiss made). The parameters included certain component distance and angles alongside equipment measurement. In the current study six component angles, *viz.* eye-monitor top, neck-monitor, trunk-thigh-chair backrest and neck twist along with monitor tilt have been considered for their correlation with musculoskeletal and visual stress symptoms. The basic physical characteristics of the subjects in relation to Body Mass Index (BMI) are comparable to the other working populations in eastern India. Each subject was investigated during one random habitual working shift while he was regularly engaged in work. The study was conducted in two phases with a brief understanding of anthropometry and the measurements alongside their involvement in the design process in the first phase followed by anthropometric measurements in the subsequent phase.

Participants (n) = 168			
Parameters	*Mean*	*SD*	*Range*
Age (years)	32	5.1	24–56
Height (cm)	169.1	6.5	147–186
Weight (kg)	70.6	10	41–99
BMI (kg/m^2)	24.6	2.8	17.97–32.7

RESULTS AND DISCUSSION

The anthropometric assessment of the workstation considered the following six component angles and the maximum deviation from prescribed reference (Table 1) was found for the eye monitor top angle which was in accordance with prevailing eye stress among the study population.

Table 1: Component Angles and Observed Deviations

Component Angles	*Prescribed Reference*	*Deviation from Reference*
Eye-monitor Top	900 (50 on either side)	133 (79.2%)
Neck-monitor	Between 850 and 950	82 (48.8%)
Trunk-thigh	Between 900 and 1000	68 (40.5%)
Chair-back rest	Between 1000 and 1050	55 (32.7%)
Neck-twist	00 (50 on either side)	60 (35.7%)
Monitor tilt	Between 1000 and 1050	86 (51.2%)

The average working hours on the computer at a stretch was found to be varying from less than an hour to a maximum of 8 h (Table 2). It was observed that the maximum number of subjects was on the computer for 1–3 h at a stretch. The maximum prevalence of musculoskeletal and visual stress was found to be 64% as observed with the individuals on VDT for 4–6 h at a stretch. It is worth pointing out that as the average daily working hours increase, the percentage of operators with complaints increase at a greater rate than for the years of computer use (Hochanadel, 1995).

Table 2: Relation of Working Time on Computer with Stress

Working at a Stretch on VDT	*No. of Employees*	*Musculoskeletal Stress*	*Visual Stress*
Less than an hour	16	10 (62.5%)	8 (50%)
1–3 hours	88	56 (63.6%)	46 (52.2%)
4–6 hours	25	16 (64%)	16 (64%)

The eye stress resulted from the deviation in the eye monitor top level, which was found to be below the reference for 79% of the subjects (Table 3). This low positioning of the display unit led to a deviation from the neck monitor angle for 82 operators. It appears that the optimal location of the monitor for the visual system requires a more downward gaze angle whereas the optimal location for the muscles of the head and neck requires a gaze angle closer to horizontal (Burgess-Limerick *et al.*, 2000). There was almost a 50% incidence for the visual stress for the individuals out of reference for the eye-monitor top and neck monitor angles. It was also observed that 60 operators had their monitor positioned away from their straight line of vision resulting in an unwarranted neck twist while working on the VDT. The problem of eye stress was compounded with the fact that there was the presence of substantial amount of glare and reflection on the VDT screen as because of wrong workstation orientation in reference to the overhead lighting. It was found that 51% of the study population had their monitors tilted at an angle, which deviated from the reference recommendation compounding to the problem of visual stress.

Table 3: Relation of Component Angles with Visual Stress

Component Angles	*No. of Subjects Out of Reference*	*Visual Stress*
Eye-monitor Top	133 (79.1%)	64 (48.1%)
Neck-monitor	82 (52.4%)	44 (53.6%)
Neck-twist	60 (40.5%)	36 (60.1%)
Monitor Tilt	86 (51.2%)	46 (53.5%)

A low monitor top also resulted in a bent posture for the neck resulting in musculoskeletal stress. The prevalence of neck pain was found to be 41% cumulatively for the three angles namely eye-monitor top, neck-monitor and neck-twist with 55.7% deviation from the reference (Table 4). Several authors have discovered physical troubles among VDT operators, which fit the description of cervicobrachial disorders (Laubli *et al.*, 1986). It was observed that 51% of the subjects complained of back pain as because of deviations from the reference backrest and trunk thigh angles (Table 5). In moving from standing to a seating posture (trunk to thigh angle of 900), the pelvis rotates about 600 and the remaining 300 is taken up by the reversal of the lumber spine. A rearwards inclined backrest removes the pressure on the inter-vertebral discs (Grandjean *et al.*, 1983). Mandal (1981) has stated that forward sloping seat pan would allow the pelvis to rock forward thereby restoring the lumber lordosis, maintaining the preferred trunk thigh angle. Since the chairs assessed in the current study didn't have a sloping seat pan, the trunk thigh angle was beyond reference for 68 subjects resulting in the discomfort to the back. ANSI - HFES 100/1988 specifies that backrest angles should have an inclination up to 150 behind the vertical. Larger inclination upto 300 behind vertical can be beneficial by reducing disc pressure and muscle activity (Andersson *et al.*, 1979).

Table 4: Relation of Component Angles with Neck Pain

Component Angles	*No. of Subjects Out of Reference*	*Neck Pain*
Eye-monitor Top	133 (79.1%)	50 (37.6%)
Neck-monitor Angle	82 (52.4%)	37 (45.5%)
Neck-twist	60 (40.5%)	24 (40.2%)

Table 5: Relation of Component Angles with Back Pain

Component Angles	*No. of Subjects Out of Reference*	*Back Pain*
Trunk-thigh angle	68 (40.5%)	31 (45.6%)
Chair-back-rest angle	55 (32.7%)	31 (56.4%)

CONCLUSION

There are substantial deviations from the reference VDT guidelines for the operators as observed from the anthropometric measurement of the workstation. A definite relationship with the prevailing musculoskeletal and visual stress symptoms was also evident in relation to the relevant component angles. Considering the fact that the subjects are in their early thirties, the prevalence of the stress symptoms are not as rampant. It gives a definite indication that such problems will be more common if proactive measures are not initiated immediately.

REFERENCES

[1] Andersson, G.B.J., Murphy, R.W., Ortengren, R. and Nachemson, A.L. (1979). "The influence of Back Rest Inclination and Lumber Support on Lumber Lordosis." *Spine*, 4, pp. 52–58.

[2] Burgess-Limerick, R., Mon-Williams, M. and Coppard, V.L. (2000). "Visual Display Height." *Human Factors*, 42(1), 140.

[3] Cakir, A., Hart, D.J. and Stewart, D.F.M. (1978). "Research into the Effects of Video Display Workplaces on the Physical and Psychological Functions of Persons." Federal Ministry for Work and Social Order, Bonn, West Germany.

[4] Grandjean, E. (1984). "Postural Problems at Office Machine Work Stations, in E. Grandjean (ED) *Ergonomics and Health in Modern Offices*", pp. 445–455.

[5] Hochanadel, C.D. (1995). "Computer Workstation Adjustment: A Novel Process and Large Sample Study." *Applied Ergonomics*. 26(5), pp. 315–326.

[6] Laubli, T., Senn, E., Fasser, W., Mion, H., Carlo, T. and Zeihr, H. (1986)."Klinische Befunde und Subjektive Klagen uber Beschwerden im Bewegungsapparat", Sozial—und Praventivmedizin, 31(2).

[7] Mandal, A.C. (1981). "The Seated Man (homo sedans)." The seated work position. Theory and Practice. Applied Ergonomics, 12.1, pp. 19–26.

Prevalence of Musculoskeletal Disorders among Young Students of Architecture and Interior Designing Courses

Manjit Kaur Chauhan

Department of Family Resource Management, SVT College of Home Science (Autonomous)
SNDT Women's University, Juhu, Santacruz (West), Mumbai–400 049, India
E-mail: manjit_k66@yahoo.co.in

ABSTRACT

Musculoskeletal pain is a major worldwide occupational health problem, with the most common pain sites being the neck/shoulder and low back. Objective was to determine the effect of work posture on musculoskeletal problems in the college students engaged in Architecture and Interior Designing courses.

Random sample of 60 students of Architecture and Interior Designing courses, with mean age 20 years, doing this course for more than 1½–2 years were investigated. The students adapted sitting, alternate sitting standing and forward bending postures/ positions while working on drawing boards. The standardized Nordic Questionnaire and Ovako Working Posture Analysis System (OWAS) were used to identify and evaluate harmful working postures. The prevalence of pain and discomfort within the past 12 months was very high. Musculoskeletal symptoms were localised primarily in the lower back, neck and shoulder. The respondents consistently identified factors such as forward bending, frequent twisting to grab work tools, adopting one posture for long duration and stools for sitting with no back support as the leading cause of the pain or discomfort. These results were supported by the posture analysis results, which also recommended that the corrective measures are required to improve the working postures. Other problem such as long working hours in the same position with inadequate breaks needs to be addressed. Modifications in workplace design will ensure that design concepts would achieve the ergonomics objective of reducing health problems and improving performance.

Keywords: Workplace Design, Students, Musculoskeletal Disorder, Inadequate Breaks.

INTRODUCTION

Musculoskeletal pain is a major worldwide occupational health problem, with the most common pain sites being the neck/shoulder and low back. Work related Musculoskeletal Disorders (WMDs) have become a major problem in many industrialised countries (Hagberg *et al.*, 1995). These disorders have caused a considerable human suffering and reduced the working capacity. Harkness *et al.* (2005) compared the prevalence of low back, shoulder, and widespread pain in two surveys conducted 40 years apart in England. For all three symptoms examined the prevalence increased from 2–4 fold between the two surveys. Szeto (2003) in study on health problems among young (12–13 years) population who use computers reported that eye, neck and shoulder discomforts related to computer use ranged from 94% for eyes, 86% for the neck and 71% for the shoulders.

Today there are various colleges, high schools, institutes, polytechnics, offering vocational courses like fashion designing, textile designing, interior designing and architecture, early childhood care education, mass communication, advertising, photography, etc. This vocational course puts lots of demand on time, energy and different postures that are adopted at their work set-ups. One of the courses where the work posture and health of the students are affected is interior designing and architecture. More and more number of students is enrolling for this course. These institutes are designed differently than the other traditional and conventional colleges. All interior designing and architecture institutes have well designed studios where the drawing work is done by the students. These students spend a large amount of time in these studios. The studio consists of drawing tables, drawing boards and stool/chair for sitting and drafting. Each and every institute varies in designs of the furniture's; there is no standard or guidelines. As per the knowledge, appearance, availability, budget, etc., the furniture is procured. No ergonomic consideration is taken into account for making purchases. Also there are very few studies on designing furniture for school students and studies related to mismatch between students and furniture designed. As the body dimensions of the students vary, there is a mismatch between the furniture and students due to which student's complaint about the musculoskeletal pain in different body parts (Diep, 2003). Szeto (2003) reported that both in school and at home, the young adolescents work with computers in very confined spaces using furniture that are not designed for the job. He added that in many schools, non-adjustable furniture are used which make it even more difficult to match the students to the computer workstation, thus, leading to poor postures and negative physical health effects which may have long term implications on their musculoskeletal growth and health in their adult lives. Programs and interventions countering the precursors of the MSDs may benefit future generations of interior designers and architects. This paper reports on the prevalence of musculoskeletal disorders and other health problems among young college girl students enrolled for interior designing and architecture courses in two environments—home and college. This may provide support to the notion that furniture design and the type of posture adopted while working on drawing boards is leading to MSDs among the younger population.

Objective

The main objective of the study is to determine the effect of work posture on musculoskeletal problems in college students engaged in interior designing and architecture courses.

METHODS

A simple random sampling method was used to select the subjects for the present study. The study was conducted in three colleges in Mumbai city. Only those students who were engaged in the course for more than 1½–2 years participated in the study. The study was limited to only female students as two colleges chosen were only for women. 20 subjects from each college were selected randomly. A total of 60 female subjects participated in the study.

Instrument and Administration

The standardied Nordic Questionnaire was used to study the musculoskeletal pain in different body parts. Keeping this in mind, the human figure/body marked with nine anatomical regions was used. The questionnaire also contained questions concerning student's physical characteristics and number of hours spent on working in classroom and working at home. Before administration of questionnaire, participants were instructed on the contents and importance of the study as well as the voluntary nature of the survey. If they wish to withdraw at that time they could, but no student declined participation.

Ovako Working Posture Analysis System (OWAS) is entire body assessment tool. It is a practical method for identifying and evaluating poor working postures. It determines the load caused by the most common postures of the back, arms and legs in different jobs. It is useful for directing improvements in the working methods.

- Back posture is identified by a number indicating whether the back is straight, bent or bent and twisted.
- Arm posture is identified by a number indicating whether one or both arms are at or below shoulder level.
- Leg posture is identified by a number indicating that the worker is sitting, standing on one or two legs, squatting on one or two bent legs, kneeling or walking.
- The load or use of force is characterized by the mass involved.

Observations were made randomly for few seconds at the time when students were performing the drawing activity in the classroom. The posture adopted at home was not studied, only questions were asked related to working posture at home. The posture was observed and studied using Sony cybershot camera. This results in a summary description of the work.

Postures are classified based on a risk assessment of musculoskeletal disorders and physical load (Karhu *et al.*, 1977).

- Each classified posture is recognised by a 4-digit code that indicates the back, arm and leg posture and the load or use of force.
- A fifth digit is used to indicate the work phase or task that the worker is working on.
- Back posture and movement score determined.

Postures are classified into four action categories indicating the urgency and priority for corrective measures required. The classes of action are determined based on postural data. The class of action determined will indicate whether an intervention is required or not.

RESULTS AND DISCUSSION

Table 1 shows the characteristics of the survey population. A total of three surveys in three different colleges, with 20 female students from each college were studied. The mean age (years), weight (kg) and height (cm) were recorded. The overall mean for age, weight and height was 21 (range: 18–26), 52.57 (range: 30–82) and 158.4 (142–170), respectively. Large variation was observed for weight and height in the students as they belonged to different ethnic groups.

Table 1: Characteristics of the Population Studied (n = 60)

Colleges	*Age (years)*		*Weight (kg)*		*Height (cm)*	
	Mean (Range)	*S.D.*	*Mean (Range)*	*S.D.*	*Mean (Range)*	*S.D.*
LBH (n = 20)	21.5 (18–26)	2.617	49.2 (38–71)	8.304	157.9 (152–167)	3.953
SVT (n = 20)	20.3 (19–23)	0.842	52.65 (39–74)	8.156	159.3 (154–170)	3.887
PVP (n = 20)	21.1 (18–25)	2.188	55.85 (30–82)	13.657	157.5 (142–167)	6.355
Total	20.97 (18–26)	2.089	52.57 (30–82)	10.710	158.4 (142–170)	4.882

Table 2 presents the average time spent by the students working on drawing board in classroom and at home. Mean time spent (in hours) in classroom and at home in LBH, SVT and PVP colleges is 3.35, 4.70 and 4.95; and 3.65, 5.05 and 5.60 respectively. On an average total time spent is quite high among PVP college students *i.e.*, 10.5 hours per day. These working hours include 1 long break (30 minutes) and 2–3 short breaks (5–10 minutes).

Table 2: Amount of Time Spent While Working on Drawing Boards in Classroom as Well as at Home

Name of the College	*Time Spent Per Day (hours)*		*Total Hours Spent Per Day (Range)*
	In College	*At Home*	
	Mean ± S.D. (Range)	*Mean ± S.D. (Range)*	
LBH	3.35 ± 1.014 (2–5)	3.65 ± 1.558 (2–8)	7.00 ± 1.975 (4–12)
SVT	4.70 ± 1.417 (3–8)	5.05 ± 1.600 (3–8)	9.75 ± 2.447 (7–16)
PVP	4.95 ± 1.802 (3–8)	5.60 ± 1.624 (3–8)	10.5 ± 2.766 (7–15)

The various sitting and standing postures adopted by the students working in the classroom on drawing boards are summarised in Table 3 and Figure 1. The scores for the assessment of postural stress were calculated. The scores are obtained according to OWAS worksheet. It could be seen from the table that most of the college students adopt back bent forward, with both arms below shoulder level in sitting and standing postures. On an average 50% of students stand with weight on one leg whereas 25% of LBH college students stand with both legs straight and 10% PVP students back is twisted/bent sideways/forward. In sitting, on an average 25% students back is bent forward whereas 25% of PVP students possess back twisted/bent sideways/forward position and 10% adopt posture with back twisted/bent sideways. Based on the assessment worksheet of OWAS, score 1 need no corrective measure, but score 2 needs action to be taken in near future. As seen from the table PVP has four times scored 2, which means some measures needs to be taken. This could be associated with the workplace (chairs used) and furniture design (height of the table, height of the chair with backrest), which needs that type of posture to be adopted. Therefore, careful consideration needs to be given to furniture design also.

Posture 's adopted

SVT College—Standing-back bend forward/ twisted with one leg straight

PVP College—Standing bending forward with legs straight

LBH College—Standing bending forward with one leg straight

LBH College—Sitting-back bend forward

SVT College—Sitting-back bend forward/twisted

PVP College—Sitting-back bend forward

Fig. 1: Sitting and Standing Postures Adopted in Classroom

Table 3: Description of Activity and Postural Analysis

S. No.	*Posture*	*Description of Posture*	*LBH*	*SVT*	*PVP*	*Action Category* *
	Standing					
1.		Back bent forward, Both arms below shoulder level, standing with weight on one leg, load <10 kg.	55%	50%	40%	2
2.		Back bent forward, Both arms below shoulder level, standing with both legs straight, load <10 kg.	25%	–	–	2
3.		Back twisted/bent sideways, Both arms below shoulder level, standing with weight on one leg, load <10 kg	5%	–	–	1
4.		Back twisted/bent sideways/forward, Both arms below shoulder level, standing with weight on one leg, load <10 kg.	–	–	10%	2
	Sitting					
5.		Back bent forward, Both arms below shoulder level, sitting, load <10 kg.	15%	50%	15%	2
6.		Back twisted/bent sideways/forward, Both arms below shoulder level, sitting, load <10 kg.	–	–	25%	2
7.		Back twisted/bent sideways, Both arms below shoulder level, sitting, load <10 kg.	–	–	10%	1

*are the scores for the actions to be taken for any harmful posture adopted;
Score 1—no corrective measures are needed
Score 2—corrective measure needed in near future.

Interviewing each student about musculoskeletal pains revealed that the percentage of students having MSDs were almost close to each other in all three colleges studied. Table 4 lists the various incidences of health problems among Interior Designing and Architecture students which are based on the Nordic questionnaire having human figure marked with back, arms and legs related problems. It could be seen from the table, these students suffer from low back (58%), neck pain (48% and shoulder pain (40%) respectively. As observed in Table 3, these students adopt varied sitting and standing postures for long hours because of which they develop these problems.

Table 4: Common Health Related Problems among Students

Regions	*LBH (n = 20)*	*SVT (n = 20)*	*PVP (n = 20)*	*Total (n = 60)*
Neck	18 (90%)	13 (65%)	17 (85%)	48 (80%)
Shoulders	14 (70%)	11 (55%)	15 (75%)	40 (66%)
Elbow	5 (25%)	–	1 (5%)	6 (10%)
Wrist/hands	5 (25%)	7 (35%)	5 (25%)	17 (28%)
Upper Back	12 (60%)	9 (45%)	11 (55%)	32 (53%)
Lower Back	20 (100%)	19 (95%)	19 (95%)	58 (96%)
Hips/thighs/ buttocks	1 (5%)	6 (30%)	1 (5%)	8 (13%)
Knees	8 (40%)	4 (20%)	3 (15%)	15 (25%)
Ankles/feet	4 (20%)	2 (10%)	2 (10%)	8(13%)

Note: Figure in parenthesis indicate percentages.

The prevalence of pain, discomfort and injury within past 12 months and past seven days from the time of interview is given in Table 5. The results revealed that 92% students complaint about low back pain, 73% complaint of neck pain, 63% and 62% complaint of upper back and shoulder pain respectively for past 12 months. Similar results were observed for last 7 days also. Here, 80% complained of low back pain, 68% neck pain, 60% and 55% shoulder and upper back problems respectively. The other problem which was observed was that of whether these MSD problems affect the normal working life of the students, at home/outside. The results showed that low and upper back pain and shoulder pain do affect the other activities performed by the students. This again could be due to long working hours in varied postures, using improper/unergonomically designed furniture.

Alongwith MSDs problems the students also complained about other problems they face, such as headache, eye irritation and feeling of fatigue. It is seen from Table 6 that 25% of students frequently felt fatigued whereas 20% felt fatigued sometimes. Also incidents of headache were 20% frequently and 25% sometimes. The students also complained about eye irritation like watering of eyes, pain in the eye and swelling of eye. 13% students frequently had eye problem whereas 17% felt discomfort sometimes.

Table 5: Prevalence of Musculoskeletal Disorders during Past One Year and Past One Week (n = 60)

Period >	*Last 12 months*				*Last 7 days*				*Other activities affected for last one year*			
Regions of Pain	*LBH (n = 20)*	*SVT (n = 20)*	*PVP (n = 20)*	*Total (n = 60)*	*LBH (n = 20)*	*SVT (n = 20)*	*PVP (n = 20)*	*Total (n = 60)*	*LBH (n = 20)*	*SVT (n = 20)*	*PVP (n = 20)*	*Total (n = 60)*
Neck	15 (75%)	15 (75%)	14 (70%)	44 (73%)	13 (65%)	9 (45%)	19 (95%)	41 (68%)	6 (30%)	7 (35%)	4 (20%)	17 (28%)
Shoulders	12 (60%)	14 (70%)	11 (55%)	37 (62%)	13 (65%)	8 (40%)	15 (75%)	36 (60%)	9 (45%)	7 (35%)	5 (25%)	21 (35%)
Elbow	4 (20%)	4 (20%)	1 (5%)	9 (15%)	4 (20%)	2 (10%)	1 (5%)	7 (12%)	3 (15%)	3 (15%)	–	6 (10%)
Wrist/hands	10 (50%)	13 (65%)	8 (40%)	31 (52%)	10 (50%)	6 (30%)	8 (40%)	24 (40%)	8 (40%)	4 (20%)	2 (10%)	14 (23%)
Upper Back	12 (65%)	13 (65%)	13 (65%)	38 (63%)	13 (65%)	11 (55%)	9 (45%)	33 (55%)	13 (65%)	11 (55%)	4 (20%)	28 (47%)
Lower Back	19 (95%)	19 (95%)	17 (85%)	55 (92%)	17 (85%)	13 (65%)	18 (90%)	48 (80%)	15 (75%)	14 (70%)	13 (65%)	42 (70%)
Hips/thighs/ buttocks	6 (30%)	9 (45%)	5 (25%)	20 (33%)	4 (20%)	5 (25%)	5 (25%)	14 (23%)	5 (25%)	4 (20%)	–	9 (15%)
Knees	9 (45%)	6 (30%)	4 (20%)	19 (32%)	7 (35%)	–	5 (25%)	12 (20%)	4 (20%)	1 (5%)	2 (10%)	7 (12%)
Ankles/feet	7 (35%)	5 (25%)	4 (20%)	16 (27%)	9 (45%)	3 (15%)	5 (25%)	17 (28%)	4 (20%)	2 (10%)	1 (5%)	7 (12%)

Note: Data is based on multiple responses.

Table 6: Other Health Related Problems among Students (n = 60)

Symptoms	*Frequently*	*Sometimes*	*Rarely*	*Never*
Fatigue	25%	20%	13%	2%
Headache	20%	25%	8%	6%
Eye problems	13%	17%	18%	11%

CONCLUSION

The traditional OWAS method used in this study showed that the students used some harmful postures while working on drawing boards placed on table and needed certain posture correction. This may lead to large number of complaints of work related musculoskeletal disorders. Drawing on drawing boards in the classroom required frequent forward bending and twisting.

ACKNOWLEDGEMENT

I am grateful to Dr. S. Chande, Principal, SVT college of Home Science, Ar. P. Bhatt, Co-ordinator and Head of Interior Designing Diploma in SVT College of Home Science and LBH College of Interior Designing and Architecture and Mrs. Manjiri Bhalerao,

Head-Department of Family Resource Management. My thanks are due to Ms. S. Pugalia for assisting in data collection and analysis. I am indebted to all the respondents for their valuable co-operation in the study.

REFERENCES

[1] Diep, N.B. (2003). "Evaluation of Fitness between School Furniture and Children Body Size in Two Primary Schools in Haiphong, Vietnam." Lulea University of Technology, Masters Thesis.

[2] Hagberg, M., Silverstein, B.A. and Wells, R.V. (1995). "Work Related Musculoskeletal Disorders." London: Taylor and Francis.

[3] Harkness, *et al.* (2005). "Is Musculoskeletal Pain More Common Now Than 40 Years Ago? Two Population—Based Cross-sectional Studies." *Rheumatology* (Oxford), 44, pp. 890–5.

[4] Karhu, O., Kansi, P. and Kuorinka, I. (1977). "Correcting Working Postures in Industry: A Practical Method for Analysis." *Applied Ergonomics*, 8(4), pp. 199–201.

[5] Szeto Grace, P.Y. (2003). "Potential Health Problems Faced by an Asian Youth Population with Increasing Trends for Computer Use."

Evaluation of Worker Exposure to Noise and Dust Pollution in Cotton Ginneries

V. G. Arude[1], S.K. Shukla[1], T.S. Manojkumar[1] and K.M. Paralikar[2]

[1]Ginning Training Centre, CIRCOT, Nagpur–440 023, India

[2]TOTD, CIRCOT, Mumbai–400 019, India

ABSTRACT

Airborne dust, noise, accidents and fire are potential sources of health hazards for gin workers. A study was conducted to evaluate dust and noise levels of different machinery and for different workstations in ginneries and to suggest appropriate measures. Noise levels of double roller gin, pre-cleaner, lint cleaner and baling presses and at different workstations in ginneries were measured with a decibel meter. The dust levels in mg/m^3 of PM 2.5 µm and PM 10 µm were measured at different workstations in different types of ginneries with the use of Haz Dust Monitor (EPAM 5000). The Time Weighted Average (TWA) for the particular particle size at each workstation was found out.

The noise levels for double roller gin, pre-cleaner, lint, cleaner and baling presses were found in the range of 79.3 to 97.7 dB(A). Gin house was found to be the noisiest workstation with an average noise level of 96.0 dB(A). The noise level for the workstations such as seed cotton suction fan, pre-cleaner, gin house, central platform, seed blower and press operator room were observed above the permissible exposure limit of 85 dB(A). In conventional and automatic ginneries TWA of PM 2.5 was found up to 2.30 and 0.20 mg/m^3 respectively and TWA of PM 10 was found up to 3.75 and 0.58 mg/m^3 respectively. Thoracic and respirable dust levels were observed above the permissible limits in all workstations. Dust levels were found above the permissible limits in all workstations. Workers in automatic ginneries were found exposed to lower levels of dust compared to conventional ginneries.

Exposure of workers to higher levels noise and dust adversely affect overall efficiency and safety. Appropriate technical and personal measures needs to be employed to limit dust and noise pollution to an acceptable level.

Keywords: Noise, Dust, Pollution, Worker, Ginnery.

INTRODUCTION

India has over 4000 ginneries which are dispersed all over the nine cotton-growing states. Out of these about 1000 ginneries are being modernised under Technology Mission on Cotton (TMC) Mini Mission IV (MM-IV). Ginning is the first important mechanical processing operation that cotton undergoes on its way from field to the textile mills. In these ginneries thousands of workers toil under intolerable conditions. Airborne dust, noise, accidents and fire are potential sources of health hazards for gin workers.

Cotton dust is the dust present in the air during the handling or processing of cotton. This dust contains a mixture of substances including ground up plant matter, short fibre, bacteria, fungi, soil, pesticides, non-cotton matter, and other contaminants. The production of dust in ginning operations is influenced by the dust generation potential of the cotton being processed (Anderson, 1979). Dust generated in ginneries originates from two major sources: cotton—machine interaction in which various machinery parts come in direct contact with the cotton as in ginning, pre-cleaning and post cleaning, as well as in cotton—human interaction in which the workers handle cotton, such as during seed cotton heaping, feeding cotton to the gins, and handling of lint and cotton seed. Besides these primary factors, generation of dust also occurs within factories as a result of strong winds blowing over the cotton heaps and in gin premises. (Eugene 1973).

Dust in cotton is a nuisance as well as a potential respiratory hazard. Gin operators and workers are exposed to respirable dust and very often get fever with unique symptoms. This fever syndrome is popularly known as Byssinosis (Antony, 1974). The inhaled dust causes cough, cold, breathing difficulties and may even lead to lung disorders if exposed for long term.

Environmental Protection Agency (EPA) has identified two sizes of Particulate Matter (PM) with which health risk is associated. These two types of PM are called as PM 10 and PM 2.5. The number refers to the size of particles. PM 10 is composed of particles that are 10 microns or smaller in diameter and PM 2.5 is even smaller, being only 2.5 microns. EPA has set National Ambient Air Quality Standards (NAAQS) for PM considered harmful to public health and environment. NAAQS for PM with diameters of 2.5 micrometer or less is 65 $\mu g/m^3$ and for PM with 10 micrometer or less is 50 $\mu g/m^3$. Permissible limit of dust concentration level is the threshold values above which the workers should not be exposed.

Cotton ginnery consists of numerous types of machinery; each is a source of noise and contributes to the overall noise level. Noise surveys of gins conducted in USA indicated that the noise levels inside the gins ranged from 95 to 99 dB(A). Gin noises are usually in the frequency range of 31.5 to 250 cycles/s but many are in the range of 500–2000 cycles/s, which is most damaging to the human ear. (Antony, 1974). Some common sources of noise are lint cleaners, gin stands, vane axial fans, centrifugal fans and seed cotton cleaners, hydraulic pumps and motors (Laird and Baker, 1982). Exposure to such a noise may adversely affect their overall efficiency, safety, and hearing ability. People exposed to intense noise can experience stress related disorders and suffer temporary or permanent threshold shifts in hearing ability (Antony and Laird, 1994).

Dust and noise pollution in cotton ginning has never received attention in the past in India. Central Institute for Research on Cotton Technology (CIRCOT) has very recently conducted an ergonomic survey of Indian ginneries and a study to evaluate dust and noise levels of different machineries and for different workstations in ginneries and suggested appropriate measures to limit them within the permissible limit.

METHODS

Three representative ginneries one each from conventional, semi-automatic and automatic was chosen for the measurement of airborne dust concentration. The dust concentration in mg/m^3 of PM 2.5 μm and PM 10 μm were measured at different workstations in each factory. The data was recorded after every 10 sec for PM 2.5 and PM 10 with the Haz Dust Monitor (EPAM 5000). The data was analysed to find out the Time Weighted Average (TWA) for the particular particle size and the workstation. Haz Dust Monitor (EPAM 5000) is a hand held, portable instrument with battery, charger and carrying case. It measures dust using light scattering technology. It measures airborne dust concentration immediately and continuously and displays it digitally. It measures dust concentration between 0–20 and 0–200 mg/m^3 with ultimate sensitivity upto 0 to 0.01 mg/m^3.

The noise levels were measured by using a sound level meter and expressed in terms of dB(A). A-weighted scale was used to measure the frequencies that human ear can detect. The sound level meter used for measurement of noise was a hand held, portable, battery operated, digital display type instrument with measurement range 20–140 dB with an accuracy of 0.5 dB. The sound level meter used was AZ 8928. Noise levels of widely used makes and models of double roller gin, pre-cleaner, lint cleaner and baling press were measured. Six models of double roller gin, five models of pre-cleaner, four models of lint cleaner and seven models of baling presses were evaluated for noise measurement. Noise was recorded at three different locations *i.e.,* near the machine, 7.5 m away from the machine and in factory premises at 30 m away from the machine. The noise level in eight ginning factories at different workstations while all the machines were in operation was also measured. These factories had 24 double roller gins with auto feeders, pre-cleaner, lint cleaner, baling press and automatic conveying systems for seed cotton, lint and cotton seeds. These factories were automatic composite units confirming to the norms of TMC MM-IV. The measurements were made at 22 different workstations in each factory. Noise levels were measured at the chest height in all the workstations. Three replications were taken for each location, machine and operating condition. The average of three replications was recorded as the noise level for the particular machine and at the workstation. Average noise level at each workstation for all selected factories was also computed.

RESULTS AND DISCUSSION

Dust Levels in Cotton Ginneries

The workers in conventional ginneries were found exposed to TWA of airborne dust upto 2.30 and 3.75 mg/m^3 of PM 2.5 and PM 10 respectively. The maximum level of TWA of airborne dust was observed in press house followed by lint opener. The workers employed in the gin house, press house and near lint opener of conventional ginnery were found exposed to dust levels above permissible limit. In semiautomatic ginneries workers were found exposed to TWA of dust up to 2.49 and 3.82 mg/m^3 of PM 2.5 and PM 10 respectively. The maximum TWA of dust for PM 2.5 and PM 10 was observed at press house and pala house which was found above permissible limit. In semiautomatic ginneries the TWA of PM 2.5 and PM 10 at gin house, pre-cleaner and lint cleaner workstations was found above permissible limit. In automatic ginneries workers were found exposed to TWA of dust levels of PM 2.5 and PM 10 up to 0.20 and 0.58 mg/m^3, respectively. The maximum dust pollution was observed at press house followed by pala house. TWA for PM 2.5 and PM 10 at gin house, pala house, pre-cleaner, lint cleaner and press house was found above permissible limit in automatic ginneries. In automatic baling press the dust level was found lower than in conventional double stage baling press.

The dust pollution was observed more at lint opener, pala and press house. The thoracic dust and respirable dust levels were observed above the permissible limit in all the workstations except at cotton heaps. Workers in automatic ginneries were found exposed to lower levels of dust compared to conventional ginneries. A number of factors are found to affect dust level potential, including cotton variety, area of production, cultural and harvesting practices, and amount of drying and cleaning during ginning.

Table 1: Airborne Dust Concentration of PM 2.5 and PM10 at Different Workstations in Ginneries

S. No.	*Workstation*	*Airborne Dust Concentration, TWA (mg/m^3)*					
		Conventional		*Semiautomatic*		*Automatic*	
		PM2.5	*PM10*	*PM2.5*	*PM10*	*PM2.5*	*PM10*
1.	Gin House	0.23	0.32	0.23	0.44	0.13	0.20
2.	Press House	2.30	3.75	2.49	3.25	0.12	0.58
3.	Pala House	–	–	0.50	3.82	0.20	0.50
4.	Pre-cleaner	–	–	0.32	0.55	0.16	0.19
5.	Lint cleaner	–	–	0.85	0.90	0.14	0.20
6.	Lint opener	1.92	2.14	–	–	–	–
7.	Cotton heap	0.002	0.004	0.02	0.08	0.04	0.06

Fig. 1: Dust Level Measurement in a Ginnery

Table 2: Noise Levels of Different Parts of Double Roller Gin

S. No.	*Parts of DR Gin*	*Noise Level (dB(A))*	
		Near gin	*7.5 m Away from Gin*
1.	Gear Box (Roller and Beater are disconnected)	86.1	75.3
2.	Roller and Gear Box (Beater is disconnected)	87.1	76.8
3.	Gear Box and Beater (Roller is disconnected)	91.8	79.4

Noise Levels of Double Roller Gin, Pre-cleaner and Lint cleaner

Noise measurements were made on different parts of DR gin under no load condition. The noise of 86.1 dB(A) was observed while only gearbox was running and the roller and the beater were disconnected. Noise level of 87.1 dB(A) was observed when roller and gearbox were running. Noise level of 91.8 dB(A) was observed when gearbox and beater were in operation. The data show that the gearbox of the gin was noisier than the other moving parts of the gin.

The noise level near the gin for different models varied from 87.0 to 97.7 dB(A) and 85.9 to 97.7 dB(A) under no load and load condition respectively. The noise levels at 7.5 m away from gin were found between 77.0 to 87.4 dB(A) and 76.7 to 87.5 dB(A) for no load and load condition respectively. Noise level was found more under no load condition than the load condition. The average noise level for all the models was found to be 91.8 dB(A) near gin machine and 82.3 dB(A) at 7.5 m away from the gin at no load condition. The average noise level for all the models was found to be 90.3 dB(A) near gin machine and 81.2 at 7.5 m away from the gin at load condition. Noise levels were observed to be normally low for Jumbo model (roller length 1365 mm) than normal gins (roller length 1016 mm). Jumbo gins are the improved versions of the normal gins. Noise levels near the gin machines were found to be above the permissible limit of 85 dB(A) (Table 3).

Table 3: Noise Levels of Different Makes and Models of DR gin, Pre-cleaner and Lint Cleaner

Sr. No.	*Model*	*Noise Level Under no Load (dB(A))*					
		DR gin		*Pre-cleaner*		*Lint cleaner*	
		Range	*Average*	*Range*	*Average*	*Range*	*Average*
1.	Near m/c	87.0–97.7	91.8	68.4–80.1	72.1	73.3–76.3	74.3
2.	7.5 m away from m/c	77.0–87.4	82.3	58.7–71.7	63.5	66.6–72.0	68.8
3.	Factory premises	63.3–68.0	66.2	54.2–64.1	58.8	55.5–63.4	58.8

Five models of pre cleaners were studied for noise measurement. Noise levels were found to vary from 68.4 to 80.1 dB(A), near the pre-cleaner, 58.7 to 71.7 dB(A), 7.5 m away from the pre-cleaner and 54.2 to 64.1 dB(A), in factory premises for the different models. Four modes of lint cleaners were studied for noise measurement. Noise levels were found to vary from 73.3 to 76.3 dB(A) near the lint cleaner, 66.6 to 72.0 dB(A) 7.5 m away from the lint cleaner and 55.5 to 63.4 dB(A), in factory premises. The average noise level was found to be 72.1 dB(A) near the lint cleaner. The noise generated by the pre-cleaner and lint cleaner were observed below the permissible exposure limit.

Noise Levels of Baling Press

The seven models of bale presses were studied for noise measurement. The selected presses were the conventional double stage presses and the automatic single stage presses with automatic tramping facility. The noise levels were measured during different unit operations of bale pressing. Noise levels were observed between 79.3 and 93.5 dB(A) in the centre of the press house for different makes. Noise levels for different makes were found to vary from 77.5 to 100.5 dB(A) during rams in operation. In operator room the maximum noise of 94.2 dB(A) was observed. The noise level outside the press house was observed in the range of 64.4 to 78.0 dB(A) (Table 4).

Table 4: Noise Levels Different Makes and Models of Baling Presses

S. No	*Location in Press House*	*Noise level (dB(A))*	
		Range	*Average*
1.	Lint suction fan /Belt	72.5–93.3	81.9
2.	Rams in operation	77.5–100.5	87.7
3.	Tramper in operation	78.5–93.4	85.0
4.	Power Pack	81.8–110.3	90.7
5.	Operator/ Control Room	78.1–94.2	86.8
6.	Centre of Press house	79.3–93.5	84.9
7.	Outside the press house	64.4–82.7	72.6

Noise Levels at Different Workstations in Cotton Ginneries

The average noise levels for all the factories, near factory gate, office, Weigh bridge, staff quarter and cotton heaps were found in the range of 62.3 to 66.1 dB(A). The average noise generated by the raw cotton suction fans found to be 88.2 dB(A). The noise level found above the permissible limit of 85 dB(A) for the workstations such as raw cotton suction fan, pre-cleaner, centre of gin house, central platform, end of gin house, seed blower and in press operator room. The maximum average noise level of 96.0 dB(A) was found at the gin house. The noise levels at the centre of the gin house were ranged from 92.5 to 98.5 dB(A) for the different factories. The noise levels at the centre of press house for the different factories were ranged between 72.5 and 93.5 dB(A) with an average of 83.5 dB(A). The noisiest workstations were found to be gin house, central platform, gin house, raw cotton suction fan, pre-cleaners, press house and seed blower. The average noise level of all the workstations for all the factories found to be 78 dB(A) (Table 5).

Table 5: Noise Levels at Different Workstations in Ginneries

S. No.	*Workstation*	*Noise Level (dB(A))*	
		Range	*Average*
1.	Factory Gate	54.5–85.4	64.4
2.	Office	55.2–73.5	64.8
3.	Staff Quarter	53.8–73.5	63.6
4.	Weigh bridge	56.7–70.7	62.3
5.	Kapas heaps	60.6–75.1	66.1
6.	Mouth of suction pipe	58.5–80.8	73.7
7.	Raw cotton suction fan/Cyclone room	81.5–99.5	88.2
8.	Pre-cleaner	80.0–95.7	87.8
9.	Central platform	88.7–92.4	90.7
10.	Centre of gin house	92.5–97.8	96.0
11.	End of gin house	85.5–93.4	89.0
12.	Lint cleaner	77.5–91.3	83.6
13.	Lint suction fan (Gin)	68.1–94.1	81.2
14.	Pala house	72.3–87.0	77.3
15.	Lint Opener	79.6–82.3	81.0
16.	Lint suction fan Press)/Cyclone room	72.5–96.8	83.5
17.	Seed Blower pit/ Seed Bucket elevator	80.5–94.5	87.2
18.	Seed heap	64.6–78.0	69.1
19.	Centre of press house	72.5–93.5	83.5
20.	Press Operator room	78.1–93.5	85.1
21.	Temporary bale storage	66.6–82.7	73.9
22.	Bale Godown	59.3–72.4	65.8

Measures to Protect Workers from Dust and Noise Pollution

To protect workers from disabling respiratory diseases, employers are required to limit the amount of respirable cotton dust in the air to permissible exposure limits. The dust levels would be reduced by adjusting the dust control equipments such as cyclones, ventilation system and by cleaning and repairing the equipment and by use of dust barriers and enclosures Workers should use breathing aids such as dust masks. They must be provided with respirators, if other measures are not sufficient to reduce exposure levels lower than or equal to the permissible exposure limit.

Fig. 2: Noise Level Measurement in a Ginnery

To protect workers from loud noise, modify the noise source to make them quieter, or replace them with quieter machines. Certain components within a machine may be replaced by components that are less noisy. The regular maintenance and replacement of worn out parts at appropriate time may reduce the noise to some extent. The proper size and design of gin house with planned machine layout may reduce the noise level. Noise levels of centrifugal fans can be reduced by locating them in separate fan rooms. Use of dampening or viscoelastic materials can muffle noise. Administrative controls such as reduction in exposure time of employees, stationing employees away from noisy source and use of personnel protective equipment such as earmuffs and earplugs would protect the workers from noise pollution.

To achieve safety, the workers have a responsibility to follow safe and healthy work practices; workers should be safety conscious and must avoid unsafe acts and practices. Factory owners must endeavor to rectify unsafe working conditions. Proper medical care should be provided to all employees. Ginners must provide free annual medical checks, including breathing tests to workers employed in dust generating areas. Safety and health instructions should be given to the workers as often as necessary during the season

CONCLUSIONS

1. In conventional and automatic ginneries TWA of PM 2.5 was found up to 2.30 and 0.20 mg/m^3 respectively and TWA of PM 10 was found up to 3.75 and 0.58 mg/m^3 respectively. Thoracic and respirable dust levels were observed above the permissible limits in all workstations. Dust levels were found above the permissible limits in all workstations. Workers in automatic ginneries were found exposed to lower levels of dust compared to conventional ginneries.

2. The noise levels for double roller gin, pre-cleaner, lint, cleaner and baling presses were found in the range of 79.3 to 97.7 dB(A). Gin house was found to be the noisiest workstation with an average noise level of 96.0 dB(A). The noise level for the workstations such as seed cotton suction fan, pre-cleaner, gin house, central platform, seed blower and press operator room were observed above the permissible exposure limit of 85 dB(A).
3. Exposure of workers to higher levels noise and dust adversely affect overall efficiency and safety. Appropriate technical and personal measures need to be employed to limit dust and noise pollution to an acceptable level.

REFERENCES

[1] Antony, W.S. (1974). "Noise Levels of Cotton Ginning Systems." *Transactions of ASAE.* 21: 574–580, 584.

[2] Anderson, J.D. and Baker, R.V. (1979). "Development of Experimental Cotton Dust Analyzer." *Transactions of the ASAE*, pp. 918–925.

[3] Antony, W.S. and Laird, J.W. (1994). "Noise in Cotton Gins." Cotton Ginners Handbook, Agricultural Research Handbook No. 503. USDA: 279–286.

[4] Eugene, P. (1973). "Experimental Verification of Displacement Equations Used to Predict Trajectories of Cotton Gin Emissions." *Transactions of the ASAE,* pp. 1126–1133.

[5] Laird, Weldon and Baker, R.V. (1982). "Confirmation of Cotton Gin Noise by Multivariate Analysis Techniques." *Transactions of ASAE*, 25: 1050–1056.

Agricultural Injuries in Etawah District of Uttar Pradesh in India

S.K. Patel, M.R. Varma, Adarsh Kumar and P.K. Sundaram

Indian Agricultural Research Institute, New Delhi, India
E-mail: sppiari@gmail.com

ABSTRACT

Increased use of farm machinery along with increased application of other agricultural inputs has enhanced the production and productivity of Indian farms. Simultaneously it has also increased occupational health hazards on the farm. Major occupational health hazards are injuries caused on farm. Very little information is available about the type and nature of agricultural injuries on Indian farms. In the present study, agriculture injuries data were collected for the period 1996–2000 from Etawah district of Uttar Pradesh in India. It has a total population of 1.4 million while 0.75 million were agricultural workers. The main objective of the survey was to collect information related to farm machinery injuries. The blocks having maximum tractor density in the Etawah district were selected for the study. In the selected blocks, villages having extensive use of farm machinery were selected. Total economic burden of fatal and non-fatal injuries was calculated considering YPLL, wage rate, working days per year and treatment cost. The agricultural injuries incidence rate was 0.80 per thousand workers per year in the study area. Out of all agricultural injuries, 9% of the injuries were fatal. The majority of victims (86%) were farmers. The highest percentage of injury victims was in the age group of 30–44 years. The highest percentages of injured victims were literate but non-Matric. The monetary loss due to agricultural injuries in study area was estimated about US $ 1.2 million per year. To minimize agricultural injuries, development of interventions and training courses needs be organized at block levels for proper and safe operation of agricultural equipment.

Keywords: Agricultural Injuries, Agricultural Workers, Farm Machinery, Farm Operation, Monetary Loss.

INTRODUCTION

In India, it is estimated that about 235 million workers (Census, 2001) are engaged in agricultural and allied activities wherein traditionally, agriculture utilised mainly human and animal power sources. In present scenario, electro-mechanical sources of power are being extensively used for various farm operations. The agricultural machinery population in the country has also increased over a period of time. It is estimated to be about 223 million (17th Livestock census) and the number of tractors has crossed 4.2 million (The Hindu Survey of Indian Agriculture 2007) mark. Farm mechanisation along with increased application of other agricultural inputs has enhanced production and productivity on Indian farms. But it has also increased the associated occupational hazards. In global context, agricultural injuries are most hazardous occupation in the world. Agricultural injuries are currently higher than injuries in any other industry (Yoder *et al.*, 1989; Sweeney and Ciesielski, 1990; Helmkamp and Lundstrom, 2002). Agriculture also has the highest work death rate of all the industry (Purschwitz, 1989; Purschwitz and Field, 1990; Hayden *et al.*, 1995; Pratt *et al.*, 1996; Pickett *et al.*, 1999). Severity of agricultural injuries are also associated with very low infrastructure availability, scarce medical facilities, occasional use of machines, no age and gender specific work, no insurance facility, no maintenance of machines, long continuous working hours and no formal training to operate the machines. Poor supervision, unreasonable expectations, financial difficulties and lack of safety devices are also reported to be associated with injury risk (DeMuri and Purschwitz, 2000). An injury on farm not only causes physical inability to a worker but also occasionally loss of human life, which brings sorrow and financial hardship to the victim's family. There are a number of epidemiological studies on agricultural injuries from the developed countries (Nordstorm *et al.*, 1992). However, only a few studies have been reported from developing countries, where tractors and other agricultural machines are mostly imported without any consideration for the human limitations of the agricultural workers in these countries (Tiwari *et al.*, 2002).

Zhou and Roseman (1994) studied the occurrence of agricultural injuries among 1000 farm operators in Alabama and reported that limbs were the body parts most frequently injured in these incidents. Leading external causes were machinery (28.6%), falls (23.2) and animals (12.5%). Pickett *et al.* (1995) studied non-fatal farm injuries in Ontario by mail survey. They found that rate of farm injuries was 58 per 1000 persons per year. High injuries rates were observed in the male of 31–40 years age group (122/1000 persons/year). Langley *et al.* (1997) studied the epidemiology of tractors incidents on New Zealand farms and reported that 18.6% of all fatal injuries occurring on farms were related to tractors and 55% of these deaths were due to overturning of tractors. The death rate due to tractor related injuries was estimated as 0.01/1000 rural residents/year. Rautiainen and Reynolds (2002) studied on mortality and morbidity in agriculture in the United States. They found that the fatality rate in agriculture remained high (about 0.22/1000) and the injuries rates have been 5 to 166/1000 workers per year.

There are very few studies on farm injuries from developing countries, because of non-availability of nation wide compiled information on agricultural injuries and lack of adequate and reliable data on etiology of farm injuries. A few epidemiological studies have been carried out in India that provides some information on agricultural injuries, but their scope and applications are limited to a particular problem.

Mohan and Patel (1992) conducted an epidemiology study in nine contiguous villages in the Sonipat district of Haryana state. In one-year period, a total of 576 agricultural injuries were reported of which 87% were minor injuries, 11% moderate injuries and the remaining severe injuries. Maximum injuries occurred while working with hand tools like spade (24%), followed by sickles (23%), bullock carts and manually operated chaff cutters (19% each) and power operated chaff cutters, tractors and diesel engines (5% each). They estimated that agricultural activities caused about 5,000 to 10,000 deaths, 15,000 to 20,000 amputations and 150,000 to 200,000 serious injuries every year in the states of Haryana, Punjab and Uttar Pradesh alone. A study conducted by Mittal *et al.* (1996) in Punjab state pointed out that out of 36 injury cases caused from agricultural accidents in twelve villages during a one year period, 3 (8.3%) were fatal and 33 (91.7%) were non-fatal. The incidence rate/1000 machines/year was the highest for tractors (23.7) followed by sprayers (15.5), electric motors (7.1), threshers (5.7) and chaff cutters (2.2).

Tiwari *et al.* (2002) conducted a survey on agricultural injuries in Central India. They found that the overall agricultural injuries incidence rate was 1.25/1000 workers/year. About 9.2% of the incidence was fatal and most of these were due to tractors and snakebites (42.9% each). About 77.6% of all incidents were due to farm machinery, 11.8% were due to hand tools and the remaining 10.6% were due to other sources like snakes, wells etc. The total monetary loss due to these injuries was estimated as about US $ 27 million/year. In developing countries, it becomes difficult to quantify the economic burden of these injuries and to chalk out ways and means to minimizing them. To develop safer equipment and to pursue measures for minimise injuries, realistic data on agricultural injuries are essential. Keeping above in view, a study was undertaken with the objectives to assess the magnitude, type of injuries and their causes on Indian farms.

MATERIALS AND METHODS

Etawah district falls in Uttar Pradesh, the biggest state of India located in northern part of the country. It has eight blocks with 694 villages. As the main objective of the survey was to collect information related to farm machinery injuries, the blocks having maximum tractor density were selected from the district. In the selected blocks, villages having extensive use of farm machinery were selected in consultation with Assistant Development Officer (Agriculture) of the block for the survey. In total, 22 villages were selected for the study. Agriculture machinery injuries data were collected for the period of five years from 1996–2000. Contacts were established with village level administration functionaries, *viz.*, *Gram Pradhan* (head of village level elected body), Block Development Committee member, village level medical practitioners and group of the farmers in the selected villages. Village survey was done on a designed proforma in terms of population, number of agricultural machines including hand tools and cropping pattern.

In India, the Factories Act of 1948 (Sethi, 1997) defines an injury causing incident to be reportable if it results in death or any bodily injury by reason of which the injured person is prevented from working for a period of 48 hours or more immediately following the incident. For the present study, the same criteria was adopted. All the twenty-two villages were visited and with the help of key contact persons, detailed information on agricultural injuries were collected on the designed proforma. Farmers of the selected villages were requested to provide the necessary information about the agricultural injuries without any fear or legal implications. They were involved in participatory mode and were informed that the information collected would only be used for making necessary design modifications in the machines or for preparing guidelines for safe operation of agricultural equipment. The preliminary information on agricultural injuries sustained during last five years was collected during the group meeting and discussion with villagers. After this, each injured victim was contacted for collecting detailed information of the injuries. For fatality cases, members of the victim's family were approached for details. The data collected were analyzed.

To assess a quantitative value on the cost of fatal injuries, a procedure described by Lehtola *et al.* (1994) was used for estimating Years of Productive Life Lost (YPLL). This is estimated by subtracting the age at death from the age at which the workers would have stopped productive work. In India, a person superannuate at the age of 60 years in most of the organised sector, but many agricultural workers continue to work on farms or participate in farm related activities beyond the age of 60. Therefore, the YPLL used for the present study was based on a productive life of 65 years. The average number of days of employment in a year for agricultural workers was considered as 290 days based on cropping pattern of the study area. The minimum prescribed wage rate per day is US $ 1.63 (1 US $ = Rs. 46) (GOI, 1991) was considered for farm worker. Total economic burden of fatal injuries was calculated by multiplying number of fatal injuries, YPLL, wage rate and working days per year. The total cost of non-fatal injuries was worked out by multiplying the average cost of medical treatment and cost of recovery time in terms of wages lost was established from the survey.

RESULTS AND DISCUSSION

Injuries Survey

Total agricultural injuries reported in surveyed villages of different blocks during the period of 1996–2000 were 106 (Table 1). Out of 106 injuries, 57 injuries were farm machinery related followed by hand tools (43) and others (6). Out of the total 106 injuries, 9 (8%) were fatal injuries. These data were similar to previous studies conducted by Mittal *et al.* (1996) and Tiwari *et al.* (2002). They reported that about 8 to 10% injuries were fatal in nature and remaining non-fatal. The agricultural hired laborers constituted 14% of the total injury victims. Age-wise and sex-wise classifications of injury victims are given in Figure 1. Male victims constituted 79% of the total victims while percentage of female victims was 21%. This is because male workers operate most of the agricultural machines. Female workers are less exposed to mechanized works and mostly confine to manual transplanting or hand tool operations for intercultural and harvesting. This is the reason for their getting less injured on the farms. These results were similar to Jennings (1986), Hansson *et al.* (1989), Hopkins (1989), Jones and Field (2002) and Tiwari *et al.* (2002). They have reported male victims in the range of 70 to 80% of the total injured.

Table 1: Causes of Injuries (1996–2000) in the Surveyed Villages

Cause	*Number of injury*		*Total (%)*
	Non- fatal (%)	*Fatal (%)*	
Farm machines	51	6	57 (54)
Hand tools	43	–	43 (41)
Others	3	3	6 (5)
Total	97 (92)	9 (8)	106 (100)

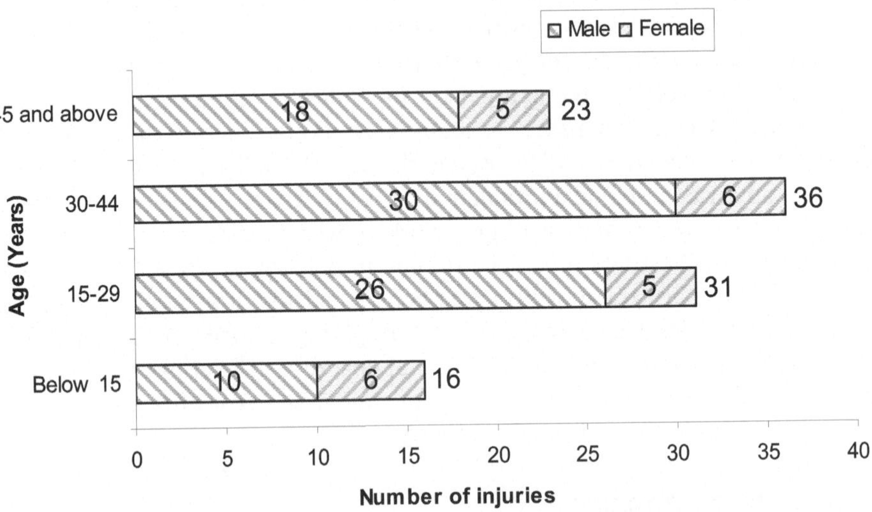

Fig. 1: Age and Sex Distribution of Injured Victims

Highest number of injury victims (36) was in the age group of 30–44 years followed by 15–29 years (31), above 45 years (23) and below 15 years age groups (16). The first two age groups mostly operate tractors and other injury prone agricultural machines. The similar results were also reported by Lloyd (1977), Pickett *et al.* (1995) and Tiwari *et al.* (2002). Figure 2 presents education wise classification of injury victims in selected villages. The highest percentage of injury victims was in the group of literate but below matric (50%) followed by illiterate (26%), matric (14%), Intermediate (7%) and graduates (3%). This trend was same as reported by Tiwari *et al.* (2002). The first two groups mainly constituted the rural population and involved in operation of most of the agricultural machines hence, the involvement of these groups was observed to be more in injury cases.

Injury Incidence Rate

106 agricultural injury occurred in the study area for a farm worker population of 27125, in five year, gave an agricultural injury incidence rate of 0.8 per thousand workers per

year. Reasons for the lower incidence rate in comparison to studies in developed countries may be definition of incidence (at least 48 hours of disability) used in the study and less mechanization levels on farm.

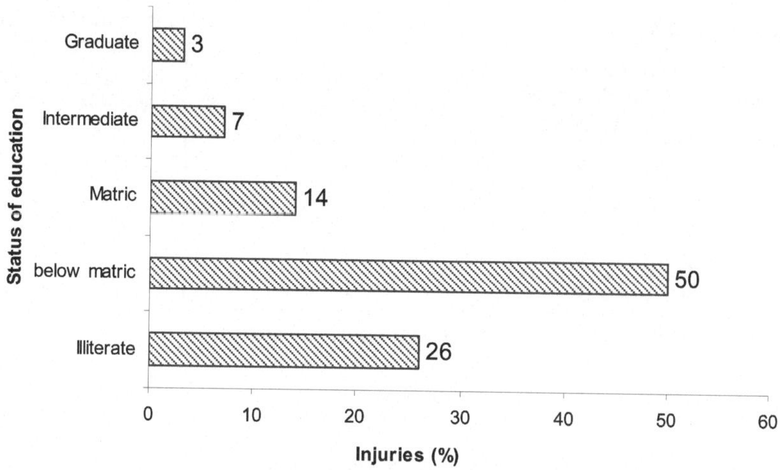

Fig. 2: Education Status of Injured Victims

Monetary Loss

A total of 106 injuries occurred during this period out of which nine were fatal and 97 were non-fatal (Table 1). This data were generalised for the whole district using the observed proportion of injuries in sample work force. The generalized estimates are given in Table 2. The estimated number of agricultural injuries in Etawah district for the year 2000 was 580, out of this 8 per cent (46 in number) turned out to be fatal and 92 per cent (534 in number) as non-fatal. About 71 per cent fatal injuries were caused due to farm machinery. In case of non-fatal injuries, about 56 per cent were caused by farm machinery and 41 per cent by hand tools.

Table 2: Estimated Number of Agricultural Injuries in Etawah District of U.P. for the Year 2000

S. No.	*Cause of Injuries*	*Estimated Number of Injuries*		
		Non-fatal	*Fatal*	*Total*
1.	Farm machinery	282	30	312
2.	Agricultural hand tools	236	–	236
3.	Others injuries	16	16	32
Total		534	46	580

The mean age of fatal injury victims observed in the survey was 34 years. Taking the productive life span of 65 years, the mean value of YPLL worked out to be 31 years. Total costs of treatment of injuries based on village survey for fatal and non-fatal injuries are given in Tables 3 and 4. Taking the above considerations, the estimated monetary loss of agricultural injuries in study area is presented in Tables 4 and 5. Total annual monetary loss due to agricultural injuries in Etawah district was about US $ 0.73 million.

Table 3: Estimate of Monetary Loss of Fatal Agricultural Injuries in the Year 2000

Type of Injury	*Number of Injuries in Surveyed Villages*	*Cost of Treatment in Surveyed Village, US $*	*Estimated Number*	*Total Cost of Injury*, US $*
On the spot death	5	–	26	380996
Treated but could not survive	4	2266	20	304404
Total	9	2266	46	685440

*Estimated number × (YPLL × Wage days × Wage rate + avg. treatment cost)

Table 4: Estimate of Monetary Loss of Non-fatal Agricultural Injuries in the Year 2000

Source of Injury	*Number of Injuries in Surveyed Villages*	*Cost of Treatment @ in Surveyed Village*	*Estimated Number*	*Total Cost of Injury* US $*
Farm machines	51	6950	282	38429
Hand tools	43	625	236	3430
Others	3	360	16	1920
Total	97	7935	534	43779

@ Treatment cost including recovery time cost * Estimated number of injury x avg. treatment cost

CONCLUSIONS

The survey conducted in Etawah district of Uttar Pradesh in India to collect information on agricultural injuries that occurred during a five-year period (1996–2000) revealed that out of 106 injuries, 9% were fatal and 91% non-fatal. The agricultural injuries incidence rate was 0.8/1000 workers/year. Out of total injuries 57 were due to farm machinery, 43 were due to hand tools and six were due to other sources like snakebites, slip and fall in the field. Males constituted 79% of the victims, while 21% of the victims were female. The highest number of victims was in the age group of 30–44 years (36) followed by 15–29 years (31), above 45 years (23) and below 15 years (16). Majority of injuries victims were farmers.

Based on the injury incidence rate, it was estimated that in the year 2000, there would have been about 580 agricultural injuries in the Etawah district, causing 46 deaths and 534 injuries to farm workers. The monetary loss due to agricultural injuries in Etawah district of U.P. for the year 2000 was estimated to be US $ 0.73 million. This is a substantial economic burden, therefore, interventions and injury prevention trainings needs to be given due attention to reduce injuries on Indian farms.

REFERENCES

[1] Census (2001). Government of India.

[2] DeMuri, G.P. and Purschwitz, M.A. (2000). "Fatal Injuries in Children: A Review." *WMJ* 99 (9), pp. 51–5.

[3] GOI (1991). "Report of the National Commission on Rural Labour." Ministry of Labour, Government of India.

[4] Gupta, S.K. (2007). "Inter-twined with Development." The Hindu Survey of Indian Agriculture, pp. 166–170.

[5] Hansson, R., Broberg, E., Johansson, A., Jorner, U., Selander, R. and Karlberg, N.B. (1989). "Study of Accidents in Farming and Forestry in 1987." Stockholm, Sweden: Lantbrukshalsan AB (The Swedish Farmers' Safety and Prevention Health Association), n.p.

[6] Hayden, G.J., Gerberich, S.G. and Maldonado, G. (1995). "Fatal Farm Injuries: A Five Years Study Utilizing a Unique Surveillance Approach to Investigate the Concordance of Reporting between Two Data Sources." *J. Occup. Environ. Med.,* 37(5), pp. 571–7.

[7] Helmkamp, J. and Lundstrom, W. (2002). "Tractor Related Deaths among West Virginia Farmers." *Ann Epidemiol*, 12 (7): p. 510.

[8] Hopkins, R.S. (1989). "Farm Equipment Injuries in a Rural Country, 1980 through 1985: The Emergency Department as a Source of Data for Prevention." *Ann. Emerg. Med.*, 18(7), pp. 758–762.

[9] Jennings, B. (1986). "Arkansas Farm Accidents Study." Little Rock, Arkansas: Cooperative Extension Service, University of Arkansas, 11p.

[10] Jones, P.J. and Field, W.E. (2002). "Farm Safety Issues in Old Order Anabaptist Communities: Unique Aspects and Innovative Intervention Strategies." *J. Agric. Safety and. Health,* 8 (1), pp. 67–81.

[11] Langley, J.D., Clarke, J., Marshall, S.W., Cryer, P.C. and Alsop, J. (1997). "Tractor Fatalities and Accident on New Zealand Farms." *J. Agric. Safety and Health,* 3(4), pp. 237–251.

[12] Lehtola, C.J., Merely, S.J. and Melvin, S.W. (1994). "A Study of Five Years of Tractor Related Fatalities in Iowa." *Applied Engineering in Agriculture*, pp. 627–632.

[13] Lloyd, M. (1977). "Arizona Farm and Ranch Accident Survey, 1976. Arizona: Cooperative Extension Services", 11p.

[14] Mittal, V.K., Bhatia, B.S. and Ahuja, S.S. (1996). "A Study of the Magnitude, Causes, and Profile of Victims of Injuries with Selected Farm Machines in Punjab." *Final report of ICAR ad–hoc research project. Ludhiana, India:* Department of Farm Machinery and Power Engineering, Punjab Agricultural University.

[15] Mohan, D. and Patel, R. (1992). "Design of Safer Agricultural Equipment: Application of Ergonomics and Epidemiology." *International J. Industrial Ergonomics,* 10(4), pp. 301–309.

[16] Nordstorm, D.L., Brand, L. and Layde, P.M. (1992). "Epidemiology of Farm–Related Injuries: Bibliography with Abstracts. Washington, D.C.: U.S." Department of Health and

Human Services, Public Health Service, Centers for Disease Control, National Institute for Occupational Safety and Health.

[17] Pickett, W., Brison, R.J., Niezgoda, H., Chipman, M. L., 1995. Non-fatal injuries in Ontario: a population based survey. *Accid. Anal. Prev,* 27 (4), pp. 425–33.

[18] Pickett, W., Hartling, L., Brison, R.J. and Guernsey, J.R. (1999). "Fatal Work Related Farm Injuries in Canada", 1991–1995. *CMAJ,* 160 (13), pp. 1843–8.

[19] Pratt, S.G., Kisner, S.M. and Helmkamp, J.C. (1996). "Machinery Related Occupational Fatalities in the United States, 1980 to 1989." *J. Occup. Environ. Med.*, 38 (1), pp. 70–6.

[20] Purschwitz, M.A. (1989). "Development of a Data Collection System for Farm Related Accidents Resulting in Injury (Doctoral dissertation)." West Lafayette, Indiana: Purdue University, p. 352.

[21] Purschwitz, M.A. and Field, W.E. (1990). "Scope and Magnitude of Injuries in the Agricultural Workplace." *Am. J. Ind. Med.*, 18 (2), pp. 179–192.

[22] Rautiainen, R.H. and Reynolds, S.J. (2002). "Mortality and Morbidity in Agriculture in the United States." *J. Agrc. Saf. Health,* 8 (3), pp. 259–76.

[23] Sethi, K.L. (1997). "Commentary on Factories Act with M.P. Rules (Act No. LXIII of 1948)." Indore, India: The Lawyers Home.

[24] Sweeney, M.A. and Ciesielski, S. (1990). "Where Work is Hazardous to Your Health: A Survey of Occupational Injuries and Field Sanitation among North Carolina Farm Workers." n.p.

[25] Tiwari, P.S., Gite, L.P., Dubey, A.K. and Kot, L.S. (2002). "Agricultural Injuries in Central India: Nature, Magnitude and Economic Impact." *J. Agric. Saf. Health,* 8(1), pp. 95–111.

[26] Yoder, C., Gartin, S.A., Bean, T.L., Lawrence, L.D. and Odell, K.S. (1989). "Analysis of Farm Worker Injuries in West Virginia in 1988 as Reported by Agricultural Producers." *Proceedings of summer meeting of National Institute of Farm Safety,* Inc., June 18–21; NIFS Paper No. 89–7, p. 15.

[27] Zhou, C. and Roseman, J.M. (1994). "Agricultural Injuries among a Population–Based Sample of Farm Operators in Alabama." *American J. Industrial Medicine,* 25(3), pp. 385–402.

Upper Limb Amputation in Agricultural Sector: Centre Based Survey

Poonam Sachdev, Aruna Ravipati and G.A. Joshi

Composite Regional Centre for Persons with Disabilities (CRC)
(MoSJE, GOI), Punarwas Bhavan,
Khajuri Kalan, Piplani, Bhopal–462 021, India

ABSTRACT

Objective for this paper is to study the risk factors leading to U/L amputation in agricultural sector. Fifty case files of U/L amputation cases who got registered at C.R.C. Bhopal between the years 2002–2007 were selected randomly and studied. The details of the occupation and cause of the disability were studied. 30% of the cases are found to be from agricultural sector. Threshers (66.6%) are found to be the most risky agricultural equipment causing U/L amputations at various levels. It was observed that maximum percentage of victims is middle aged men. Agricultural equipment should be designed with more safety standards to prevent disability among agriculture sector.

Keywords: Upper Limb Amputation, Disability, Agricultural Operations.

INTRODUCTION

Developing countries such as India are made up of a large chunk of rural area. Agriculture is the main industry in rural areas. Human beings have accepted it as a necessity and developed it as an art. Time has changed the ways and means of agriculture making it more productive and also easier. Use of science and technology for improvisation of the sector has provided the farmers with a number of agricultural tools and machineries. Such inventions carry certain risks too, which many a times, escape notice by the agricultural workers. These risks may cause a lifelong disability in the form of amputations and thus, dependency of the workers. Tiwari *et al.* (2002) reported that incidence rate of injuries is around 1.25 for every 1000 workers each year in agriculture in Madhya Pradesh. The study focuses that 77.6% of the injuries were due to farm machinery. Out of 17,480 agricultural accidents in the year 2000, most of them included amputation of limbs burry most etc causing monitory loss estimated to US$ 27 million/years. The purpose of this study is to find out the risk factors resulting in upper limb amputations in workers involved in agricultural sector and enlist their rehabilitation areas and options.

METHODS

Fifty case files of upper limb amputees were randomly selected from the registration counter of Composite Regional Centre for Persons with Disabilities (CRC), Bhopal between the years 2002–2007 (five years). The study was carried in three phases. The first phase consisted of analysis of the data present in the files for availing the number of persons who were from agricultural sector. The second phase consisted of a thorough analysis of the history which gave information on the cause of the upper limb amputation. These included all types of cause manual or mechanical. The data obtained was analysed to derive the possible risk factors leading to the disability with study of type of amputation in the concerned persons. The levels of amputation were taken as criteria for describing the types of amputation *viz.* hand, wrist disarticulation, below elbow amputation (BE amputation), elbow disarticulation, above elbow amputations (AE amputation) and shoulder disarticulation. Data on age and sex was also collected as a part of study. The terminal phase consisted of a detailed study of management in the form of rehabilitation services received from the centre. These services were classified into three categories *viz.*, artificial limbs and training (pre and post-prosthetic fitment) information on government schemes and loans and guidance on vocational aspects. Data thus obtained was analysed by percentage calculations for vocation risk factors, level of amputation, age, sex and rehabilitation services obtained from the centre.

RESULTS AND DISCUSSION

Table 1: Prevalence of Upper Limb Amputations in Agricultural Workers

S. No.	*Vocation*	*Numbers*	*Percentage*
1.	Agricultural workers	15	30
2.	Industrial workers	12	24
3.	Road traffic and railway accident	08	16
4.	Others (Burns, electric shock, disease)	15	30

Table 2: Causes of Amputations

S. No.	*Causes*	*Numbers*	*Percentage*
1.	Thresher	10	66.60
2.	Sugarcane machine	02	13.33
3.	Electricity	02	13.33
4.	Road traffic and railway accident	01	6.66

Table 3: Types of Amputations

S. No.	*Type*	*Numbers*	*Percentage*
1.	Hand	3	20
2.	Wrist disarticulation	2	13.3
3.	Below elbow	6	40
4.	Elbow disarticulation	1	6.66
5.	Above elbow	3	20

Table 4: Age and Sex of Disabled Persons (Total)

S. No.		*Numbers*	*Percentage*
Age group	0–25 Year	5	33.3
	25–50 Year	7	46.6
	50–75 Year	3	20
Gender	Male	9	60
	Female	6	40

Table 5: Rehabilitation Services

S. No.	*Type of Rehabilitation Services Availed from Centre*	*Numbers*	*Per cent*
1.	Prosthesis and training	13	86.6
2.	2nd on Govt. schemes and loans For the disabled.	14	93.3
3.	Vocational guidance	3	17

It is observed from the results that the upper limb amputations are very common in agricultural sector. The high percentage of population engaged in agricultural sector in Madhya Pradesh attributes to this finding. Another factor responsible for this finding may be the amount of risk associated with the agriculture industry. A study by Horsburgh *et al.* (2001) stated that the fatality rate for agricultural industry workers is higher than all other industry workers in any part of the world, because the focus on biomechanics and ergonomics for safety in road, industry and sports has limited applications in agriculture. Threshers were noticed to be the most common cause of amputation in agriculture. Two-thirds of the agricultural workers in this study were disabled due to accidents with the thresher machines during the process of threshing. Mohan and Patel (1992) as quoted by Kumar *et al.* (2002) stated that threshers caused two per cent of total agricultural injuries. The same study quotes a research by Singh and Sinha (1980) who reported 30 thresher injuries out of 50 in a survey from India. This high risk associated with thresher usage is a point of attention in safety standards of agricultural equipment.

The results have shown that the type of upper limb amputation associated in thresher injuries ranges from hand to above elbow level. A high incidence of below elbow and above elbow levels of amputation is noted with percentages of 40% and 20% respectively. This variation reflects that modes of accidents are multi-factorial in case of threshers.

When age and gender of the victims were studied, the results showed that middle aged men were the most affected group. This data may have implications on the financial aspects of the families when the productive age group is affected. 86.66% registered at the centre with upper limb amputation involved with agriculture were benefited by prosthesis fitment and training. 13.4% were reluctant to get prosthesis. 93.3% of the cases took information on government schemes and loans and applicability in agriculture set up. Seventeen per cent of the cases were keen to change their vocation to other light manual work for self employment. This implies that most of the victims seek benefits of government schemes and loans for their sustenance after the onset of disability and generally accept prosthesis which is aimed for cosmetic purpose and perform upper limb functions.

CONCLUSION

The risk of disability especially upper limb amputation is high in agriculture industry. Main cause of amputation is the use of thresher and it holds risk at various functioning aspects. This implies that threshers are highly risky and need to be designed with more safety measures. The study shows that most of the victims seek government help due to their dependency in financial aspects. This emphasizes the improvisation of agricultural equipment for agricultural workers to protect farmers from financial cries. The study provokes thoughts on the insurance for accidents of agricultural workers as mandatory policy making to facilitate economical rehabilitation of affected family.

REFERENCES

[1] Horsburgh, S., Feyer, A.M. and Langley, J.D. (2001). "Fatal Work Related Injuries in Agriculture Production and Services to Agriculture Sectors in New Zealand, 1985–94." *Occup. Environ. Med.,* 58, pp. 489–495.

[2] Mohan, D. and Patel, R. (1992). "Design of Safer Agricultural Equipment Application of Ergonomics and Epidemiology." *Int. J. Ind. Ergonom,* 10, pp. 301–309

[3] Singh, K.P. and Sinha, I. "Accidents in agriculture." *J. Indian Med. Assoc,* 75(1), pp. 4–6.

[4] Tiwari, P.S., Gite, L.P., Dubey, A.K. and Kot, L.S. (2002). "Agricultural Injuries in Central India: Nature Magnitude and Economic Impact." *J. Agric. Safety Health,* 8, pp. 95–111.

Impact of Ergonomic Intervention among Garment Workers

P. Parimalam[1], K. Balakamakshi[1] and A.K. Ganguli[2]

[1]Home Science College and Research Institute
Tamil Nadu Agricultural University, Madurai–625 104, India
[2]Senior Manager, Occupational Health Services, BHEL, Trichy, India

ABSTRACT

Garment manufacturing process involves tasks such as fabric laying, pattern making, cutting, stitching, checking, ironing and packing. The workstation used by the checking section workers indicated that these workers experienced postural discomfort. Hence, new workstation was designed taking into consideration various factors such as type and nature of activity, time spent and productivity of workers. The newly designed workstation was a height adjustable table, with an adjustable tilting surface. This table was evaluated by the workers and the results indicated a reduction in the postural stress and an increase in their productivity.

Keywords: Garment Workers, Musculoskeletal Discomforts, Posture Stress.

INTRODUCTION

The garment sector has been traditionally viewed as a major source of employment generation. Its low skill requirements and large absorption potential have made it an important source of non-agrarian employment. The industry has mushroomed in the developing countries, especially in the urban and suburban areas, because of the ready pool of cheap labour.

Empirical evidences suggest that the workers in the garment units suffer work related musculoskeletal disorders such as Carpal Tunnel Syndrome (CTS), forearm tendinitis, epicondylitis, bicapital tendinitis, low back pain, neck pain, shoulder pain and osteoarthritis of the knees. This is because of the strenuous, fast paced and repetitive nature of their work (Encyclopaedia of Occupational Health and Safety, 1998; Herbert *et al.*, 1999).

Cutting and finishing work in the garment units require prolonged hours of standing or sitting in a forward bending posture. The operation of a sewing machine involves coordination of feet, hand and eye and the machine operators maintain a constant sitting posture on a stool with the body inclined forward. Such stressful postures adopted for eight hours per day would be regarded as health hazards (Bridger, 2003).

Important factors in the field of occupational health which seriously affect workers are workstation design and protective work equipment (Misner *et al.*, 2000). The relationship between poor, ergonomically deficient design in the workplace and diseases of the musculoskeletal system has been demonstrated in several studies (Van Wely, 1979) and illustrative case reports have been published (Perrot, 1961). Prolonged static loads are probably the major factor in modern working life in causing most work-related musculoskeletal disorders (Grandjean and Hunting, 1977; Kelsey *et al.*, 1979; Andersson, 1985; Hettinger, 1985; Monod, 1985). The prevention of such musculoskeletal injuries involves various ergonomic measures aimed at reducing the load on the locomotor system by altering the design of equipment in the workplace (Kant *et al.*, 1990).

A workplace is a location where a person or several persons perform tasks for a relatively long period of time. A workstation is one of a series of workplaces that may be occupied or used by the same person continuously/regularly while performing his or her job. A well-designed ergonomic workstation should assist a person in maintaining a dynamic working posture for a prolonged period of time. Designing of a workstation should therefore be based on Man-Machine-Environment (MME). Workers require appropriately designed tools, equipment and architectural features, which provide adequate protection from occupational risks and limit impedance of job performance.

MATERIALS AND METHODS

Designing of Workstation for Checking Section

The earlier researches conducted among the garment workers indicated the need for redesigning of workstation used by checking section workers. A new workstation for checking section workers was fabricated taking into consideration the following factors, such as nature of work, posture adopted, weight of the load handled, dimensions of the furniture, anthropometric dimensions of workers and productivity of the workers. All these details were gathered by observation and video recording of the work cycle. The newly designed workstation was a height adjustable table with a tilting top (adjustable) with a high stool for the workers to alternate between sitting and standing position during work.

Field Studies of the New Workstation

The new workstation was given for field-testing to the garment workers. Six workers evaluated the study for a period of two months. They were initially trained to adjust the worktable to suit their anthropometric dimensions.

Evaluation of the New Workstation

Rapid Upper Limb Assessment (RULA) method was used to assess the work postures adopted by the workers. The postures adopted by the workers in the conventional table and the new workstation were assessed. The postures were photographed using a digital video camera and they were analysed.

Postural discomfort was assessed using mapping of body part discomfort method. The workers were asked to indicate the body part/body parts that had higher amount of discomfort followed by the next part and so on. This procedure was carried out at regular intervals throughout the day, namely, before starting of work, before mid morning tea, before lunch, before afternoon tea and before end of work. The discomfort levels were recorded with scores – '5' indicating extreme discomfort and '1' indicating no discomfort.

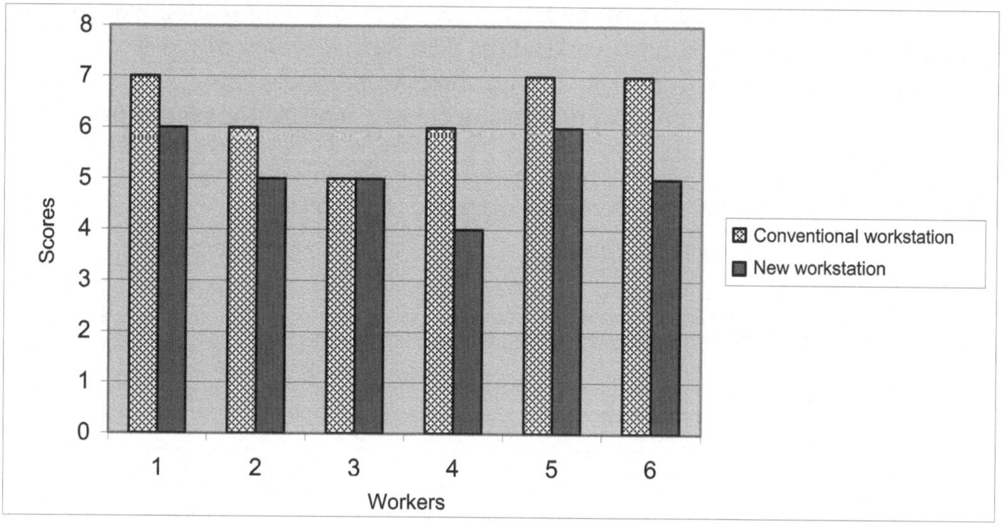

Fig. 1: RULA Scores of Workers Using Conventional and New Workstation

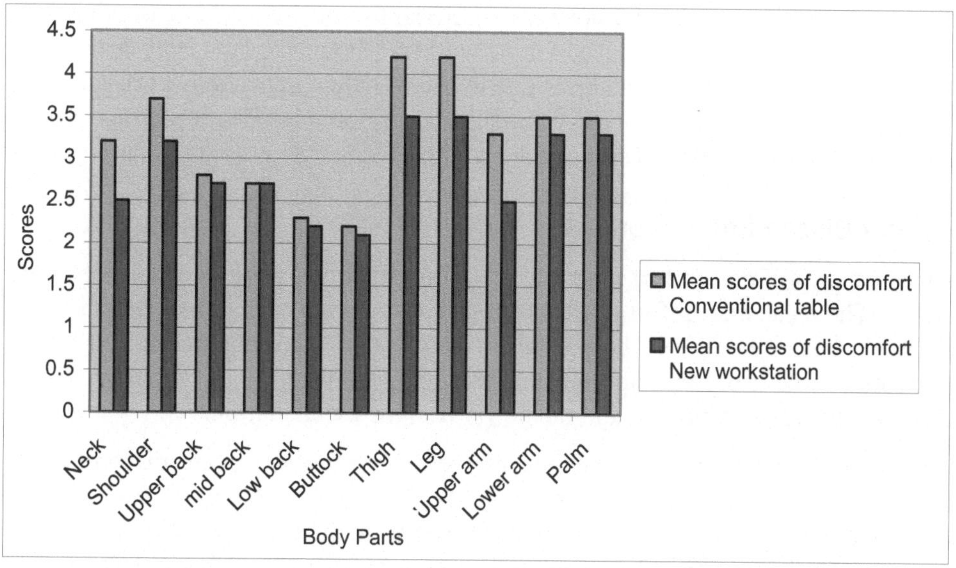

Fig. 2: Discomforts of Workers Using Conventional and New Workstation

RESULTS AND DISCUSSION

The impact of the ergonomic interventions among garment workers is discussed under the following aspects.

Rapid Upper Limb Assessment

Table 1 presents the comparison of the RULA scores of checking section workers using conventional work table and new workstation. The RULA scores of the workers using conventional table ranged from 5 to 7 with three workers having a total score of 7. However the RULA total score of workers using the new workstation ranged from 4 to 6.

Table 1: Comparison of RULA Scores of Checking Section Workers Using Conventional Worktable and New Workstation

Workers	*Conventional Table*			*New Workstation*		
	Arm and Wrist Score	*Neck and Trunk Score*	*Total Score*	*Arm and Wrist Score*	*Neck And Trunk Score*	*Total Score*
I.	5	6	7	5	5	6
II.	5	5	6	5	4	5
III.	5	4	5	5	4	5
IV.	5	5	6	4	4	4
V.	5	6	7	5	5	6
VI.	5	6	7	4	5	5

The scores for arm and wrist did not vary significantly for the workers, but the neck and trunk scores was considerably reduced for the workers who worked in newly designed workstation. The reason for this reduction is due to use of height adjustable workstation with a tilting top. This enabled the workers to maintain comfortable posture during work. As a result of this reduction, the total score of the workers using new workstation was decreased.

Body Part Discomfort of Workers

The mean scores of discomfort were recorded for all the six workers who participated in the study. The mean discomfort level was assessed for every two hours—beginning of the day's work, before tea break, before lunch, before mid afternoon break and by end of day's work. Table 2 presents the mean scores of discomfort of workers (by the end of day's work) using conventional and the newly designed workstation.

While using the conventional table the workers experienced severe discomfort in the thighs and legs, with a mean score of 4.20. Due to the nature of the activity and prolonged standing, the workers experienced higher amount of discomfort in their lower extremities. Neck, shoulder, low back and arm were the other parts of the body that experienced discomfort to a moderate level.

In the newly designed workstation, workers experienced moderate level of discomfort in the lower extremities such as thighs and legs, followed by upper arm, shoulder and upper back, mid back and low back region. The discomfort was reduced to a greater extent in the thighs, legs and neck region.

Table 2: Comparison of Mean Scores of Discomfort of Workers in the Conventional Table and Newly Designed Workstation

Body Parts	*Mean Scores of Discomfort*		*'t' value*
	Conventional Table	*New Workstation*	
Neck	3.20	2.50	8.14**
Shoulder	3.70	3.20	13.80**
Upper back	2.80	2.70	55.00**
mid back	2.70	2.70	–
Low back	2.30	2.20	45.00**
Buttock	2.20	2.10	43.00**
Thigh	4.20	3.50	11.00**
Leg	4.20	3.50	11.00**
Upper arm	3.30	2.50	7.25**
Lower arm	3.50	3.30	34.00**
Palm	3.50	3.30	34.00**

**indicates significance at 1 per cent.

CONCLUSION

Low checking table and inadequate seating arrangement intensified the musculoskeletal problems. Worker education on the proper use of height adjustable furniture and use, care and maintenance of personal protective equipment is necessary. Periodical health surveillance by competent authorities is to be insisted on at the work site. Sit-stand stools are to be designed for each of the units and made mandatory for reducing the postural strain of the workers while at work.

REFERENCES

[1] Andersson, G.B.J. (1985). "Posture and Compressive Spine Loading: Intradiscal Pressures, Trunk Myoelectric Activities, Intra–abdominal Pressures, and Biomechanical Analyses." *Ergonomics*, 28, 91–93.

[2] Bridger, R.S. (2003). *Introduction to Ergonomics*. London: Taylor and Francis.

[3] Encyclopaedia of Occupational Health and Safety (1998). Vol. III, Geneva: International Labour Organisation.

[4] Grandjean, E. and Hunting, W. (1977). "Ergonomics of Posture—Review of Various Problems of Standing and Sitting Posture." *Apllied Ergonomics* , 8 (3), 135–140.

[5] Herbert, R. *et al.* (1997). "Impact of an Ergonomics Program Featuring Adjustable Chairs on Upper Extremity Musculoskeletal Symptoms Among Garment Workers." *Proceedings of Managing Ergonomics in the 1990's.* Available Online http:www.ergoweb.com/resources/reference/manergo/dropkin.ctm

[6] Hettinger, T. (1985). "Occupational Hazards Associated with Diseases of the Skeletal System." *Ergonomics* , 28, 69–75.

[7] Kant, I., Notermans, J.H.V. and Borm, P.J.A. (1990). "Observations of Working Postures in Garages Using the Ovako Working Posture Analysing System (OWAS) and Consequent Workload Reduction Recommendations", *Ergonomics*, 33(2), 209–220.

[8] Kelsey, J.L. *et al.* (1979). "The Impact of Musculoskeletal Disorders on the Population of the United States." *Journal of Bone and Joint surgery*, 61A, 959–964. Cited in Kant, I., Notermans, J.H. and Borm, P.J.A. (1990). "Observations of Working Postures in Garages Using the Ovako Working Posture Analysing System (OWAS) and Consequent Workload Reduction Recommendations." *Ergonomics,* 33(2), 209–220.

[9] Misner, S.T., Hewitt, J.B. and Levin, P.F. (2000). *Women and Occupational Health.* Chicago: University of Illinois.

[10] Monod, H. (1985). "Contractility of Muscle during Prolonged Static and Repetitive Dynamic Activity." *Ergonomics,* 28, 81–89.

[11] Perrot, J.W. (1961). "Anatomical Factors in Occupational Trauma." *Medical Journal of Australia*, 3, 73–82. Cited in Kant, I., Notermans, J.H.V. and Borm, P.J.A. (1990). "Observations of Working Postures in Garages Using the Ovako Working Posture Analysing System (OWAS) and Consequent Workload Reduction Recommendations", *Ergonomics*, 33(2), 209–220.

[12] Van Wely, P. (1979). "Design and Disease", *Applied Ergonomics*, 1, 262–269.

Abstracts of Other Presented Papers

Energy Expenditure—Controlling Factors

Shriram M. Sane[1], Varsha N. Karandikar, Sagar M. Shinde and Prashant P. Rayrikar

Industrial and Production Engineering Department, Vishwakarma Institute of Technology
666, Upper Indira Nagar, Bibawewadi, Pune–411 037, India
E-mail: [1]shriram.sane@vit.edu

ABSTRACT

In the proposed paper an attempt has been made to find out the effect of different parameter on the physical energy consumption, while undertaking predefined physical task. The study was aimed to quantify the energy expenditure. Heart rate measurement data of 40 people were collected by using Heart rate monitor. The test was conducted on treadmill which was set at zero degree inclination from ground and at the speed of 4.8 km/h and 6.4 km/h. This test was of about 11 minutes. Subjects were asked to walk for first 5 minutes @ 4.8 km/h and next 5 minutes @ 6.4 km/h. The purpose of this experimentation was to find out whether the energy consumption for defined task varies according to age or any other anthropometry factor.

The test was conducted on subjects ranging between 21 and 50 years. Collected data indicated that heart rate at relax posture ranged between 65 and 100 beats/min. During the exercise maximum value range was 135 to 190 beats/min. Kilocalories consumed ranged between 60 and 227. Recovery Rate varied from 1 min 20 s to 30 min.

From this study it may be concluded that people performing daily exercise walking, running, yoga, gym, cycling or outdoor sports have minimum recovery time. From the results available till now it seems that no relationship exists between energy consumption and age.

Keywords: Heart Rate, Recovery Time.

Computer Aided Modeling and Simulation for Ergonomic Analysis of Indian Army Vehicle

S. Karmakar[1], M.S. Pal, Deepti Majumdar and D. Majumdar

Defence Institute of Physiology and Allied Sciences
Lucknow Road, Timarpur, Delhi–110 054, India
E-mail: [1]karmakar.sougata@gmail.com

ABSTRACT

The present study investigates the systematic design approach for human factor issues in the development of a military vehicle with the help of Digital Human Modelling (DHM) and simulation software which incorporate virtual human (CAD generated) into the virtual solid model of vehicle workstation to perform ergonomic analysis.

Digital ergonomic studies that was carried out for the army vehicle under study, typically considered spatial accommodation, reach ability, clearance and interference, strength and motion study, joint angles, vision analysis, comfort-discomfort, ingress-egress study and many other standard ergonomic practices.

With the help of DHM software, 3-digital human models: 5th Percentile Gorkha, 50th percentile Army (pooled data) and 95th percentile Sikh-Punjabi which are the representative of smallest, average and largest Indian army personnel respectively, were generated by using corresponding percentile values of body dimensions from anthropometric data base of Indian army (n>11000). These virtual humans were interfaced with virtual model of that vehicle for systematic ergonomic analysis. Such type of design evaluation without real human trial in physical mockups clearly demonstrated the loop holes/ flaws in the vehicle design from the ergonomic viewpoint. Finally design modification for better ergonomic workstation was recommended based on human factor issues, national/ international military and safety standards.

Keywords: Digital Human Modelling, CAD Design, Ergonomics, Human Factors, Vehicle Workstation.

Design of Manually Operated Equipment and Machines for Farm Workers of Central Uttar Pradesh: Anthropometric Data Study

Tarun Kumar Maheshwari[1] **and Devendra Kumar**

FPM, College of Agricultural Engineering and Technology
Etawah–206 001, India
E-mail: [1]maheshwari_tk@yahoo.com

ABSTRACT

Data were collected from central Uttar Pradesh, keeping into consideration the requirements of hand tools and machines. Anthropometric measurement of the identified 79 body dimensions and 16 strength parameters of 350 male and 250 female farm workers have been done. A typical hand tool/equipment consists of the fictional part, handle and a connecting part. The design of the handle depends on the mode of operation and amount of effort required, anthropometric data of working population, handle material etc. It is not enough, as a rule, to design a work place to suit an average person. Since, it is not usually possible to design work place to suit the tallest or shortest workers, one has to be content by meeting the requirement of the majority *i.e.,* 90% of the population (*i.e.,* people having values between 5th and 95th percentile for the particular dimensions.). While operating the thresher, So that the maximum permissible height for feeding chute should be decided considering the 5th percentile value of acromion height and from this data its value is 129.4 cm. For better operation and more human comfort a proper handle height for these equipments should be between 63.7 to 95.8 cm with 63.7 cm being the 5th percentile value for metacarpal height and 95.8 cm being the 5th percentile of elbow height From the recorded value of body dimensions the mean value, standard deviation, 5th and 95th percentile values were computed and recorded for male and female subjects respectively.

Keywords: Agricultural Machinery Design, Anthropometric Data.

Font Type, Size and Readability of Young Adults

J. Banerjee and D. Majumdar

Defence Institute of Physiology and Allied Sciences
Lucknow Road, Timarpur, Delhi–110 054, India
E-mail: jayita_79@rediffmail.com

ABSTRACT

Font type and size characteristics play an important role in understanding the complexities of visual information in man-computer interface. The research reported here comprises an empirical investigation of the phenomenon of typographic (font type and size) influences on reading. In this study, subjects rated the perceptual qualities possessed by different type faces (e.g. Serif and San serif fonts).

Thirty subjects volunteered for this study. The mean age (SD) was 27.6(1.94) years. All participants read English fluently and all of them were postgraduate in sciences. Two types of fonts were used—the serif fonts and the sans serif fonts. Serif font includes Times New Roman (TNR), Georgia and Courier New and sans serif font includes Arial, Verdana and Tahoma. All of these fonts were presented in three different sizes, *i.e.,* 10-point, 12-point and 14-point. So this study used a 6 × 3 (font type × size) within-subjects design to investigate difference in preference for reading a text.

Subjects compared font conditions by reading eighteen passages of approximately the same length written in black on a white background. The text of each passage comprised of a font from one of the eighteen type/size font conditions. Subjects were asked to read the text as quickly and accurately as possible presented in a 17″ TFT monitor. Reading time was recorded by using a digital stopwatch. After reading the passages, they ranked the fonts for general preference on a seven point scale.

Reading time was found to be minimum for Courier New 14-point for most of the participants in comparison to other font type/size combinations. For font size 66.7% of the subjects preferred 14-point and 33.3% of the subjects preferred 12-point. On the other hand 46.7% of the subjects rated Verdana as their preferred font type, followed by Arial (30%) and TNR (10%). Majority of the subjects (46.7%) ranked Verdana 14-point as their most favoured font type with size combination for computer work.

Keywords: Serif Font, San Serif Font, Font Size, Reading, Subjective Preference.

Kinematic Changes of Gait during Load Carriage at Two Walking Speeds

Deepti Majumdar, M.S. Pal and D. Majumdar

Defence Institute of Physiology and Allied Sciences
Lucknow Road, Delhi–110 054, India
E-mail: majum55@yahoo.com

ABSTRACT

Four different forms of loads such as Haversack (HS), Backpack (BP), Rifle (RF) and Light Machine Gun (LMG) are carried separately or in a combination of them by soldiers in Indian army in routine and combat operations. Present study aimed to quantify the effects of carrying such loads on kinematic parameters of gait at slow and fast walking speeds under controlled laboratory conditions.

Ten healthy male Indian soldiers whose mean (SD) age, height and weight were 23.2(2.6) years, 172.6(3.8) cm and 64.3(7.4) kg respectively, walked on a 10 m level ground walkway at two different speeds (3.5 km/h and 6.5 km/h) with and without load. There were eight load conditions: RF (4.2 kg), HS (4.4 kg), LMG (6.8 kg), BP (10.7 kg), HSRF (8.6 kg), HSLMG (11.2 kg), BPRF (14.9 kg), BPLMG (17.5 kg). These loads constitute 6.5%–27.2% of mean body weight of the subjects. Gait data were collected using a 6 Camera based 3D Motion Analysis System. Joint angles at knee, pelvis and trunk at foot strike and toe off and ranges of motion (ROM) of these joints in support and swing phases of gait cycle were recorded. ANOVA with repeated measures and post hoc analysis of the data were made to compare the degree of changes in different loads and speed conditions. There were significant changes in joint angles and ROM in each of the load conditions compared to no load in each of the walking speeds. These angular changes in fast walking were significant compared to slow walking in all the loaded conditions.

These changes are mostly considered as normal compensatory mechanism of gait. However, the degree of angular changes in trunk forward tilt at both the walking speeds with BP indicate a possible source of increasing spinal stress and needs necessary corrective measures.

Keywords: Joint Angles, ROM, Load Carriage, Biomechanical Stress.

Influence of Upper Extremity Support in Computer Keyboard Operation

Swati Pal[1] and P.K. Nag

National Institute of Occupational Health
Ahmedabad–380 016, India
E-mail: [1]pswati27@rediffmail.com

ABSTRACT

The laboratory based simulated study examined the possible impact of forearm and wrist support on the electromyographical activity of forearm and back muscles and associated deviation of the upper extremity postures of the professional computer operators (n = 8). A 3D motion analysis system recorded wrist and elbow angle deviations in different test conditions, *i.e.*, (i) forearms unsupported, (ii) supported and wrists supported on (iii) polystyrene bead-pack and (iv) gel-filled wrist rest. Results indicated that the bilateral elbow extension was substantially influenced by the forearm and wrist support and most of the time the elbow was maintained at anatomical position (upper arm and forearm at 90°). Whereas, the elbow extensions were at 65–68° in unsupported forearms. Wrist deviation was more pronounced in absence of wrist support, and these deviations decreased approaching towards the anatomical wrist position with the use of forearm and wrist supports. The study indicated that the EMG activity of forearm (flexor digitorum superficialis, extensor digitorum) and upper trapezius muscles decreased with the use of forearm support. In contrary, the wrist rest increased upper trapezius load, and gel-filled wrist rest was more effective in reducing forearm muscle activity. In conclusion, supporting of the forearm and wrist at the work surface substantially improved work posture and reduced muscular strains of the keyboard users. Any increase in wrist extension was considered as a risk of tendon strain due to mechanical friction. The study reiterated that the wrist rests were useful to reduce the wrist extension and also the EMG activity of the forearm and shoulder muscles. The polystyrene bead packed wrist rest was more effective to attain the wrist close to the neutral position, and the gel-filled wrist pad appeared better in reducing load of the forearm muscles.

Keywords: Keyboard Operation, Forearm Support, Wrist Rests.

Panel Discussion on "Application of IEA/ILO Ergonomics Checkpoints in Agriculture"

The workshop provided an opportunity to discuss the philosophy and content behind Checkpoints.

CHAIRPERSON

- Prof. David Caple—President, IEA,

Panel Members

- Dr. S.K. Tandon—ADG (Engg.), ICAR, New Delhi
- Dr. C J S Pannu—Research Engineer, AICRP on ESA, PAU, Ludhiana
- Dr. V K Tewari—Professor, IIT Kharagpur
- Dr. Jatinder Kishtwaria - Professor, Home Science, CSKHPKVV, Palampur

Highlights of the Discussion

1. While dealing with checkpoints the realistic illustrations/suggestions are required suitable to the farm environment in India. These include clothing that is worn, hair styles, general farm environment and local language used.
2. It was felt that the structure of the document would be more useful if it was more structured around operational areas rather than ergonomic risk areas. For example, a section on the major sources for fatalities relating to machinery safety should be more prominent. This includes the operation of: threshers, tractors, trolleys, harvesting machines, cotton picking, power tilling, seeding, etc. It could also include other potential fatal risks such as poisonous gases in silos or in wells, and snake bites. An initiative from Uzbekistan relating to gas removal from wells prior to entry, was given as an example. Electrical hazards such as loose wiring on the farm could also be included in this checkpoint.
3. The operational areas in agriculture are based on human powered and animal powered tools and equipments, self powered machines with persons walking behind and self propelled machinery such as tractors, combine harvester. A problem in rural areas relates to the lack of roads that link locations. Consequently in such case, the use of tractors would be limited and man powered or animal powered equipment may be the only feasible options.
4. A separate section of Checkpoints should be developed for family resource management activities/equipments. This relates to the use of information relating to the 24 hour analysis of tasks undertaken by women. These include home duties, as

well as animal husbandry and drudgery prone tasks. Examples should be given of the different application options depending on the natural environment. For example, the use of machinery and trolleys are possible on flat terrain, but not practicable in hill areas. In these areas women carry significant loads on their head due to the lack of suitable options.

5. It would be appropriate to develop two pilot modules for India, one for machine safety relating to advice to prevent fatalities and another for family resource management activities/equipments.
6. Within the Indian rural sector there are many potential target groups. Each would have specific interest in the level of information and Checkpoints provided. Once the target for the Checkpoint is defined such as the researchers or the farmers, then the Checkpoint including the language, illustrations and content should be then developed.
7. Within India, there are a range of legislations that need to be considered. This relates to the Factories Act together with international conventions such as ILO184 (Safety and Health in Agriculture Convention). This has 10 articles that are used by the Indian government as the basis for legislation requirements. ILO/IEA checkpoints should be linked with ILO 184 convention as it will help in wide adoption.
8. There are agriculture universities across India, together with state extension departments. These can be seen as primary disseminators of the Checkpoints to the "agents of change", i.e. extension officials/worker.
9. There was specific discussion on some Checkpoint recommendations. Some salient observations made during this discussion are given below.
    - The Checkpoints should not only show protection from heat, but also shelter from the rain.
    - The use of baskets and bags carried whilst picking fruits and plucking tea leaves, should be part of the Checkpoint content.
    - Farm worker are capable of safely pulling a four-row seeder rather than the eight rows shown in the Checkpoint illustration.
    - A low stool has been found to be safer and more productive by women and men when manually milking cows.
    - Specification of the thresher feeding chute design is required to be clear about hazard management.
    - Mesh guards are lighter and cheaper. Those should be added and illustrated for machinery.
    - Tractors should have air conditioners fitted using excess engine capacity.
    - The option of "diversified tractors" should be considered. This relates to using the tractor for many applications in the local community apart from farm work including fire fighting, and movement of families.
    - Safe design of trolleys pulled by tractors and the option of changing rear wheels for road application needs to explored.

- Gloves should be worn for cotton picking to reduce hand damage.
- Road safety needs to be given more emphasis particularly due to the incidents of fatalities between cars, pedestrians and farm machinery.
- More specific guidance on "emergency first response" should be provided apart from general comments on first aid.

10. Within India, it is identified that the majority of farmers obtain much public information via television. Hence, the selection of illustrations that are easily recognisable within India, would enable the use of television to reinforce the Checkpoint principles.

Author Index

T

V

W

Sponsors

Mahindra & Mahindra Ltd.

Department of Science & Technology, GOI

Indian Council of Medical Research

Council of Scientific and Industrial Research

MP Council of Science and Technology

Tata Steel

Indian National Science Academy

SMI and Noldus

Cosmed Srl

Indus International

Godrej

AIMIL

Esbee Dynamed

Novatech Measurements

Swan International

Dessault Systems

People's Group

About the Editors

Dr. L.P. Gite is the Project Coordinator of All India Coordinated Research Project on Ergonomics and Safety in Agriculture at Central Institute of Agricultural Engineering (CIAE), Bhopal. He has been instrumental in collecting and compiling anthropometric and strength data of agricultural workers from all over the country, and in applying ergonomics interventions in agricultural tools and equipment for improving productivity, reducing drudgery and enhancing safety. His work on ergonomical assessment and refinement of tools and equipment for farm women is a landmark in this area. Dr. Gite is involved in Ergonomics consultancy work with sprayer industry, tractor industry and development organizations. He has to his credit more than 50 publications in various national and international journals and conference proceedings. His major areas of professional interest are Ergonomics, Safety and Occupational health aspects in agriculture, women friendly technologies, farm safety legislations and social security of farmers and agricultural labourers. His present activities are planning, implementation and monitoring of programmes related to Ergonomics and Safety in Agriculture through the AICRP on ESA. He is actively involved in building up of linkages with organizations and individuals working in this area which includes Govt. Departments, Agril Engineering Colleges, other Engineering colleges, Home Science Colleges and Non Govt. Organizations (NGOs).

Dr. C.R. Mehta is a Principal Scientist at CIAE Bhopal having specialization in Ergonomics and Instrumentation. His R&D achievements include optimization of seat design and tractor workplace layout, control of noise and vibration; instrumentation systems such as tractor three point linkage dynamometer, seed drill choke indicator, electronic timer, reaction timer etc. He is also associated in consultancy work of leading tractor manufacturers of country. He has published over 70 papers including 13 in reputed International journals. He has also published 6 books/proceedings, 11 technical bulletins and 11 book chapters.

Dr. Nachiket Kotwaliwale is a Senior Scientist at CIAE, Bhopal. He obtained his Ph. D. in Biosystems Engineering from Oklahoma State University, USA. His major areas of professional interest are design of agro processing machinery, non-destructive quality evaluation of foods, drying, computer aided design, instrumentation and safety in agro processing machines. His present activities are research and development for machines and technology related to post harvest engineering. He has more than 30 publications to his credit in national/international journals and proceedings. He has been conferred with ISAE outstanding book award for 2004–05 and ISAE Distinguished Service Certificate for 2006–07.

Mr. Joydeep Majumder is working as Assistant Research Officer at Morarji Desai National Institute of Yoga, New Delhi. Earlier he was associated with AICRP on Ergonomics and Safety in Agriculture at CIAE, Bhopal as Human Physiologist. His work included application of ergonomics in agricultural tools, machines and workplaces. He has published/presented more than 15 papers in national/international journals and conferences.

About the Editors